Springer Texts in Statistics

Series Editors:
G. Casella
S. Fienberg
I. Olkin

Springer Texts in Statistics

(continued after index)

George Casella

Statistical Design

 Springer

George Casella
University of Florida
Dept. Statistics
103 Griffin-Floyd Hall
Gainesville FL 32611-8545
USA

ISBN 978-1-4419-2614-2   e-ISBN 978-0-387-75965-4
DOI: 10.1007/978-0-387-75965-4

To my mentors, who taught me more than I realized

Preface

Statistical design is one of the fundamentals of our subject, being at the core of the growth of statistics during the previous century. Design played a key role in agricultural statistics and set down principles of good practice, principles that still apply today. Statistical design is about understanding where the variance comes from, and making sure that is where the replication is. Indeed, it is probably correct to say that these principles are even more important today. Fisher (1947) compared a dataset to a sample of gold ore. The finest analysis could only extract the proportion of gold contained in the ore. But a good design could produce a sample with more gold.

There are plenty of "Design of Experiments" books available, many of which do a fine job of describing not only how to design experiments, but also how to analyze them. So why bother with another book? There were two main reasons.

The first reason is the observation that many of our "standard" analyses have become driven by the default setting of one's favorite computer package and, unfortunately, many times these default settings provide an incorrect analysis. More frightening is the fact that sometimes such default analyses have found their way into textbooks.

The second reason is that, although design books have gotten broader in coverage of designs and often have much to say about analysis, the basic theoretical underpinnings are not always covered, and if they are, they are not covered in sufficient detail to understand how to construct the correct analysis and to understand why the computer package default analysis may be incorrect. Without fully understanding what the correct analysis *should be*, it is impossible to design a good experiment.

So ... the goal is to describe the principles that drive good design, which are also the principles that drive good statistics. Moreover, this will be done with detail and attention paid to the theoretical background – only by having more than a passing familiarity with the fundamental theory can one truly understand statistical design.

This book is not an encyclopedia of designs. There is no attempt to cover all designs, and no attempt to teach data analysis at all. Although we will analyze many datasets, we will usually use common anova techniques and not preach about transformations, heterogeneity, missing data, and all the other good stuff that a good data analyst needs to know. We assume that the design has been run with enough success so that rescue techniques are not needed for the analysis. We also assume that the student has been exposed to such data analysis strategies.

We will cover the most popular designs in depth, both with theory and examples, and datasets from consulting sessions and research publications. We emphasize basic principles and careful modeling and, armed with such tools, the student should be able to apply these principles in any situation

This book grew out of a course on Statistical Design taught at Cornell during the 1990s, and at Florida in the 2000s. Most of the examples and datasets are from consulting sessions with graduate students writing theses or professors writing papers, and the subjects span everything from planting alfalfa in the field to harvesting brain stem cells. (Most of my career has been spent at Colleges of Agriculture – Rutgers, Cornell, and split between Arts and Sciences and Agriculture at Florida. This will show in the examples.) I have found that although the data and the lab techniques have changed, the statistical principles remain quite constant.

Now, for the more important details:

o The level of the text is for first or second year graduate students. The students should be familiar with standard statistical methodology (anova, blocking, multiple regression) that one would get from a typical one-year methods sequence (from books such as Ott and Longnecker 2000, Rawlings *et al.* 1998).

o The material in the text is about right for a one-semester course. There is probably a bit more than can be comfortably covered, so some picking and choosing will be necessary. However, marching through the text is a reasonable strategy.

o The chapters cover, for the most part, the standard material of a design book, with mostly real examples, and applications of design in real situations. Although we cover many microarray designs, we do not have a special section, instead treating them as the topics apply. The only unusual chapter is Chapter 4, where the concept of blocking is explored further, and the effect of a random factor is examined. To me, blocks are not about being samples from a larger population (which can be difficult to justify) but rather about the correlation structure that they induce.

o Most chapters have a section *Technical Notes*, which contains the underlying theory in detail. The level in these notes is "anything goes"; we use a lot of matrix algebra, some calculus, and also some statistical concepts such as likelihood and sufficiency. These are in-depth looks that will enhance the understanding of the advanced student, but skipping these sections will not hamper the beginner. To fully appreciate these sections it would be good if the student has had a course from a book like *Statistical Inference* by Casella and Berger (2001).

o The exercises are divided into "Essential" and "Accompaniment". Everyone should do the essentials. The accompaniments tend to be more of a theoretical nature, going into the details of the procedures. I strongly suggest that, if the students are able, they should do these exercises too. I have always found that, for me, true understanding only comes from slogging through the details.

o The datasets are on the book web page found at

www.stat.ufl.edu/~casella

and, for most datasets in the examples there will be an accompanying R program. These will not be sophisticated analyses, but rather will serve as a starting point. I make no guarantees about the R programs. Smaller datasets may only appear in text.

My best advice is that if you really want to understand statistical design, read (or even better, reread) Fisher. His ideas, especially about blocking, have greatly influenced my thinking. In fact, what I perceived as mishandling of the randomized complete block design was one of the driving forces behind the text. Fisher, of course, got it right.

There are many gray areas in design – when to pool, how to replicate, etc. – some of which cannot be fully answered with statistical fact. This is where we enter the realm of opinions, where judgments are made more on anecdotal evidence and experience rather than formal calculations. After doing this stuff for over 20 years opinions form about how to do things – I have taken the liberty of sharing those thoughts.

Finally, thanks to all of my mentors. First there was Leon Gleser, my PhD advisor, who taught me to work hard and try to learn as much as possible. My design mentors were many, starting with Virgil Anderson at Purdue, and Walt Federer at Cornell, who not only made me really understand split-plot designs, but also taught me to say "a model" and *never* "the model". And Carl Lowe, the Plant Breeding Professor at Cornell who taught the course "Field Plot Techniques", and knew more about field plot layouts than anyone on the planet. When he retired he gave me all of his notes and examples, many of which appear as examples and exercises (and in the Appendix). I also thank the students and colleagues who suffered through my learning process, who listened as I thought through all of this, read through notes, and solved problems: Mihai Giurcanu, Jamie Jarabek, David Lansky, Michael Meredith, Deborah Reichert, and Andy Scherrer.

My most special mentor was Myra Samuels. A true sadness in my life is that she died in 1992. She taught me to always ask questions until you really understand. Thanks, Myra.

And to all of my other mentors – you have all helped me more than you know – Bob Bechhoffer, Jim Berger, Larry Brown, Shanti Gupta, Jean-Pierre Habicht, George McCabe, Doug Robson, Bill Strawderman. And all of the others that I have learned from. Thanks.

George Casella
Gainesville, Florida

February 18, 2008

You may be right
I may be crazy
But it just may be a lunatic you're looking for.

Billy Joel
You may be right

I finally fixed it because I had, and still have, persistence.
Richard P. Feynman
Surely You're Joking, Mr. Feynman

Contents

List of Tables

List of Figures

List of Examples

1

Basics

The statistician cannot evade the responsibility for understanding the processes he applies or recommends.

R. A. Fisher
The Design of Experiments

Give me problems, give me work, give me the most abstruse cryptogram, or the most intricate analysis, and I am in my own proper atmosphere.

Sherlock Holmes
The Sign of Four

This is a book about design, and is typically not concerned with analysis. Most designs, unless they are complete disasters, will result in a reasonably straightforward analysis. However, as the purpose of a good design is to result in an efficient analysis, it is important to be familiar with the types of analysis that will be done. Thus, we will spend some time discussing the important parts of analyses, and how the design can impact them. We will also do many analyses and somewhat address what to do when the design does not go as planned.

Throughout the book analyses will typically be presented in an anova framework, complete with anova tables, sums of squares and degrees of freedom. This is done not because the anova is the best way to analyze data, but rather because the anova is the best way to think about data and plan designs. Fisher (1934) first called the anova "a convenient method of arranging the arithmetic", but then explained that it is quite a bit more than that, as rigorously demonstrated by Speed (1987). The ideas of partitioning variation, counting degrees of freedom correctly, and identifying the correct error terms, are fundamental to any data analysis. Focusing on the anova helps us focus on these ideas, and ultimately helps us plan a better design.

This first chapter is a collection of "basics", topics which should seem a bit familiar, but the explanations and interpretations may be somewhat different from what was previously seen. However, since we are assuming some familiarity with these topics, the review will be brief and a little disjointed.

1.1 Introduction

We start with some examples of the basic oneway model

$$(1.1) \qquad Y_{ij} = \mu + \tau_i + \varepsilon_{ij}, \quad i = 1,\ldots,t; \quad j = 1,\ldots,r,$$

where Y_{ij} is the response, μ is the overall mean, τ_i is the effect of treatment i, and ε_{ij} is the error, often taken to be $N(0,\sigma^2)$, *independent and identically distributed* (iid).

Example 1.1. ONEWAY MODEL Suppose that plants are grown in pots, and three different types of fertilizer are applied to the pots. After a period of time the dry weight of the plants is recorded. The data are in Table 1.1.

Table 1.1. Dry weight, in grams, of *Geranium* "Dilys", subject to three fertilizer treatments.

Fertilizer		
A	B	C
1.02	1.00	0.99
0.79	1.21	1.36
1.00	1.22	1.17
0.59	0.96	1.22
0.97	0.79	1.12

For this experiment, a reasonable model is (1.1) with

$$\mu = \text{true overall dry weight}$$
$$\tau_i = \text{true change in dry weight due to fertilizer } i$$
$$y_{ij} = \text{observed yield of plant } j \text{ in treatment } i$$
$$\varepsilon_{ij} = \text{unobserved error}$$

Note that we use a lowercase y_{ij} here, while in (1.1) we used upper case Y_{ij}. This is the distinction between the random variable Y_{ij} which is unobserved, and the observed realized value, the data y_{ij}.   ‖

Model (1.1) is *overparameterized*, in that without further assumptions we cannot separate the effects μ and τ_i. (Formally, such a model is *nonidentifiable*.) It is typical to impose a restriction on the parameters to make the model identifiable, with the most common being the restriction $\sum_i \tau_i = 0$.

Alternatively, we can use the *cell means model*

$$(1.2) \qquad Y_{ij} = \mu_i + \varepsilon_{ij}, \quad i = 1, \ldots, a; \quad j = 1, \ldots, r$$

in which the μ_i are identifiable and the theory is somewhat easier. However, model (1.1) tends to be more popular. Here we will typically use the over-parameterized model (1.1) but will not always implicitly assume $\sum_i \tau_i = 0$. Rather, we will sometimes keep track of the parameters so it is clear exactly what parameters can be estimated.

Note: Define the "dot" notation by $y_{i\cdot} = \sum_j y_{ij}$, so the "·" signifies summing over that index. The "bar" means averaging, so that $\bar{y}_{i\cdot} = (1/r) \sum_j y_{ij}$. If there is no chance for confusion, for the sake of simplicity we will write $\bar{y}_i$ instead of $\bar{y}_{i\cdot}$.

Then, for example, from (1.1)

$$\mathrm{E}\, \bar{Y}_{i\cdot} = \frac{1}{r} \mathrm{E} \left(\sum_j \mu + \tau_i + \varepsilon_{ij} \right) = \mu + \tau_i,$$

$$(1.3)$$

$$\mathrm{E}\, \bar{Y} = \frac{1}{rt} \mathrm{E} \left(\sum_{ij} \mu + \tau_i + \varepsilon_{ij} \right) = \mu + \bar{\tau},$$

showing that the parameters $\mu + \tau_i$ and $\mu + \bar{\tau}$ are estimable and that they have *unbiased* estimators. From here, a number of interpretations are possible:

(1) First note that $\mathrm{E}(\bar{Y}_{i\cdot} - \bar{Y}_{i'\cdot}) = \tau_i - \tau_{i'}$, so treatment differences are always estimable.
(2) We can always assume, without loss of generality, that $\bar{\tau} = 0$, so then μ and τ_i are estimable.
(3) Equivalent to (2), define $\mu^* = \mu + \bar{\tau}$ and $\tau_i^* = \tau_i - \bar{\tau}$ and use the model $Y_{ij} = \mu^* + \tau_i^* + \varepsilon_{ij}$ with $\sum_i \tau_i^* = 0$.

Thus, these are all reparameterizations, and are equivalent. What is important is that the experimenter is aware of the parameterization, knows what is being estimated or tested, and can make a meaningful conclusion.

Perhaps the most important concept in statistical design is the *experimental unit*.

Definition 1.2. The *experimental unit* is the unit (subject, plant, pot, animal) that is randomly assigned to a treatment.

The experimental unit, as the name implies, is the basic unit of the experiment, and *defines the unit to be replicated to increase degrees of freedom.*

Note: In the definition of experimental unit the phrase "randomly assigned" is of crucial importance.

Example 1.3. EXPERIMENTAL UNIT   Using model (1.1),

(1) In Example 1.1, the treatment (fertilizer) is applied to the pots, which are the experimental units. The plants are not the experimental units.
(2) An experimenter investigating the effect of different food for a species of fish (τ_i) places the food in tanks containing the fish. The weight increase of the fish is the response (y_{ij}). The experimental unit is the tank, as the treatment is applied to the tank, not to the fish. (If the experimenter had taken the fish in hand, and placed the food in the fish's mouth, then the fish would have been the experimental unit – as long as each fish got an independent scoop of food!)
(3) In a microarray experiment (see Miscellanea 1.9.1), RNA from two groups of people (with and without a certain disease) are applied to the microarray ("chip"), with the response being gene expression. The experimental unit is the person. (See also Section 1.6.)

Replicate the experimental unit to increase df

In (3), no treatment is actually applied, as the "treatment" is the group membership, and the members of the group become the experimental unit. In (1) and (2), a treatment is actually applied to a unit. We do not dwell too much on this distinction, as it becomes less important as long as we correctly identify the experimental unit.

See Miscellanea 2.9.2 for more on this.

For example, in (1), if the experimenter mixes one batch of fertilizer, and applies it to five pots, there is only one experimental unit, not five. This is because the effect of using one batch of fertilizer five times will induce a correlation in the five responses. There would only be five experimental units if the fertilizer for each pot were mixed independently. In (2), if the food is placed directly in the fish's mouth, the food must be prepared independently for each fish.   ‖

Definition 1.4. A *sampling unit* is the object that is measured in an experiment. It may be different from the experimental unit.

Definition 1.5. *Replication* is the repetition of the experimental situation by replicating the experimental unit.

In any design, we must be aware of the consequences of the planned replication. The importance of identifying, and replicating the experimental unit becomes apparent if we look at an anova table.

Example 1.6. REPLICATION Referring to Example 1.3

(a) In (1), replication is obtained by using more pots. Increasing the plants per pot has no effect.
(b) In (2), as an illustration, suppose there are 3 different diets, 4 tanks per diet, and 6 fish per tank. The anova table is

Source	df	Mean Square	F-ratio
Diets	2	MS(Diet)	MS(Diet)/MS(Tank)
Tanks (in Diets)	9	MS(Tank)	
Fish (in Tanks) (subsampling)	60	MS(Fish)	

so the F-test on diets has 9 degrees of freedom from the replication of the tanks. The fact that there were 6 fish per tank does not bear at all on the estimation and testing of diets, and the response (weight gain) is summed over the experimental unit. Replicating the fish is *subsampling* or *pseudoreplication*, and does not affect the main test.
This is an example of a *nested* design, where Tanks are nested in Diets and Fish are nested in Tanks. In such designs the testing is straightforward – the nested factor provides the error mean square for the factor in which it is nested. (See Section 1.5.) Of course, we can test the significance of tanks using MS(Tank)/MS(Fish), but this is wasted effort. There is typically no interest in assessing the significance of tanks; they are merely there to hold the fish! (but see Exercise 1.5).
(c) In (3), replication is obtained by having RNA from more people. Just using more chips, or splitting one person's RNA across two chips, is again wasted effort. See Section 1.6.

‖

Example 1.6 illustrates a key principle of experimental design – knowing the correct denominator in the F-test. If the experimenter comes to us at the design stage, and is interested in getting the "best" test on diets, that would typically mean maximizing the df for the test on **It's all about the denominator!**
diets. From the ANOVA above we see that increasing the number of fish per tank has no effect on the test statistic for diets. To increase df in that test, we must increase the number of tanks. Thus, if resources can be allocated between fish and tanks, it is best to maximize the number of tanks in the experiment. This is just another way of saying that to increase the df (and hence the power of a test) we need to increase the replications of the experimental unit – which, for the test on diets, is the tank.

Example 1.7. FISH ONE LAST TIME The fish weight gain data for Example 1.6 is in the dataset `FishTank`. A default anova statement, in some computer packages, could give the anova

	df	Sum Sq	Mean Sq	F-value
Diet	2	21247.7	10623.9	$\frac{10623.9}{51.4} = 206.606$
Tank	9	107.8	12.0	$\frac{12.0}{51.4} = 0.233$
Within	60	3085.3	51.4	
(Fish in Tanks)				

which is incorrect. The correct anova will test Diet against Tank, yielding

	df	Sum Sq	Mean Sq	F-value
Diet	2	21247.7	10623.9	$\frac{10623.9}{12.0} = 886.85$
Tank	9	107.8	12.0	0.233
Within	60	3085.3	51.4	
(Fish in Tanks)				

The test of Tank against "Residuals" (or, more precisely, fish within tanks) is correct, as the fish are giving replications of the measurement corresponding to the tank.

Unless otherwise specified, computer packages will use the last line of the anova table as the denominator for all tests. This example is typical, in that if you do not specify the error term, the default analysis may get it wrong. ‖

As mentioned, the fish tank design illustrates *nested* factors. The tanks are nested within diets, and the fish are nested within tanks. When a factor is clearly nested in another, then the F-ratio is always the ratio of the mean square of the top factor divided by the nested factor. Thus, in the fish tank experiment, all of the tests are clear (Exercise 1.8).

If the same four tanks had been used in each level of diet, then tanks would have been *crossed* with diets, and the anova would have been

	df	Sum Sq	Mean Sq	F-value
Diet	2	21247.7	10623.9	206.606
Tank	3	91.5	30.5	0.233
Diet × Tank	6	16.3	2.7	
Within	60	3085.3	51.4	
(Fish in Tanks)				

Note that here, with tanks crossed with diets, the anova would become a randomized complete block design (RCB) with tanks as blocks. Diet is then tested against the Diet × Tank interaction. And again, the within error due to fish is not of much use. (See Chapter 3.)

1.2 Variance and Covariance

We have already seen the calculation of an expected value in (1.3), and now we review some basics of calculating means, variances, and covariances. As the statistics in this book are mainly based on linear models with normal errors, an example being model (1.1), we restrict our treatment to those cases.

In model (1.1), the random variable Y_{ij} is related to the random variable ε_{ij}, which has the property that

$$E(\varepsilon_{ij}) = 0, \qquad \text{Var}(\varepsilon_{ij}) = \sigma^2.$$

As $Y_{ij} = \mu + \tau_i + \varepsilon_{ij}$, it then follows that

$$E(Y_{ij}) = E(\mu + \tau_i + \varepsilon_{ij}) = \mu + \tau_i + E(\varepsilon_{ij}) = \mu + \tau_i,$$

which shows that Y_{ij} is an *unbiased* estimator of $\mu + \tau_i$, as it equals $\mu + \tau_i$ in expectation. Also,

$$\begin{aligned}
\text{Var}(Y_{ij}) &= E\left[Y_{ij} - E(Y_{ij})\right]^2 \\
&= E\left[(\mu + \tau_i + \varepsilon_{ij}) - (\mu + \tau_i)\right]^2 \\
&= E(\varepsilon_{ij})^2 = \text{Var}(\varepsilon_{ij}) = \sigma^2,
\end{aligned}$$

where, in both calculations, we have used the fact that μ and τ_i are constants.

When data are collected according to model (1.1), we usually estimate μ and τ_i with their least squares estimators (Exercise 2.5). These are the means $\bar{Y}_{i\cdot}$ and have expected value

$$E\left(\bar{Y}_{i\cdot}\right) = \frac{1}{r} \sum_{j=1}^{r} EY_{ij} = \mu + \tau_i$$

and

$$E(\bar{\bar{Y}}) = \frac{1}{tr} \sum_{i=1}^{t} \sum_{j=1}^{r} EY_{ij} = \mu + \bar{\tau},$$

showing that an unbiased estimator of the treatment effect is $\bar{Y}_{i\cdot} - \bar{\bar{Y}}$, that is,

(1.4) $$E\left(\bar{Y}_{i\cdot} - \bar{\bar{Y}}\right) = \tau_i - \bar{\tau}.$$

It is typical to assume that $\bar{\tau} = 0$, modeling τ_i as deviations from the overall mean level μ. Doing that, the variance of our estimator is

(1.5) $$\text{Var}\left(\bar{Y}_{i\cdot} - \bar{\bar{Y}}\right) = \text{Var}(\bar{Y}_{i\cdot}) - 2\,\text{Cov}(\bar{Y}_{i\cdot}, \bar{\bar{Y}}) + \text{Var}(\bar{\bar{Y}}) = \frac{\sigma^2}{r}\left(1 - \frac{1}{t}\right).$$

Recall that the covariance of two random variables Y and X is

$$\text{Cov}(Y, X) = E[(Y - EY)(X - EX)]$$

(see Section 1.8). Evaluation of this final expectation is a bit lengthy and is left to Exercise 1.7.

So, from (1.4) and (1.5) we have an estimate of τ_i and the variance of the estimate. What remains is to be able to estimate the variance of $\bar{Y}_{i\cdot} - \bar{\bar{Y}}$, which means that we have to estimate σ^2. In the oneway anova this is usually done with an unbiased estimator that comes from the *within* variation. Note that the variables inside a treatment all have the same mean, so if we only work with those variables, differences will be free of τ_i.

Within cell i we have that $EY_{ij} = E\bar{Y}_{i\cdot} = \mu + \tau_i$. Thus, we can actually apply (1.5) within a cell to establish that

$$E(Y_{ij} - \bar{Y}_{i\cdot})^2 = \text{Var}(Y_{ij} - \bar{Y}_{i\cdot}) = \sigma^2\left(1 - \frac{1}{r}\right)$$

and hence

(1.6)
$$\sum_{j=1}^{r} E(Y_{ij} - \bar{Y}_{i\cdot})^2 = r \times \sigma^2\left(1 - \frac{1}{r}\right) = (r-1)\sigma^2.$$

If we do a similar calculation for each cell and sum, and recognize that

$$\sum_{i=1}^{t}\sum_{j=1}^{r}(Y_{ij} - \bar{Y}_{i\cdot})^2 = \text{SS(WithinTrts)},$$

and we have that

$$E(\text{SS(WithinTrts)}) = t(r-1)\sigma^2$$

(1.7) and

$$E\left(\frac{\text{SS(WithinTrts)}}{t(r-1)}\right) = E(\text{MS(WithinTrts)}) = \sigma^2,$$

showing that MS(WithinTrts) is an unbiased estimator of σ^2.

Thus, we can estimate τ_i with the unbiased estimator $\bar{Y}_{i\cdot} - \bar{\bar{Y}}$ with estimated variance $((r-1)/rt)\text{MS(WithinTrts)}$, and use this to construct tests and confidence intervals (see, for example, Section 2.4).

One important message here is that to estimate the variance of our treatment estimate, we use the within SS. What we will see is that in all designs, the variance estimate comes from the mean square in the denominator of the anova F-test, which is found through calculation of Expected Mean Squares (EMS). As we will see in the coming chapters, calculation of EMS is very important.

1.3 Partitioning Variation

An *analysis of variance (anova)* done on data collected according to model (1.1) is a partitioning of the variation in the data into pieces that can be attributed to different factors. Next, in our introductory mode, let us calculate the sums of squares for the data layout of Table 1.1. The total sum of squares is partitioned as

$$SS(\text{Total}) = SS(\text{Trt}) + SS(\text{Within Trts}),$$

(1.8)

$$\sum_{i=1}^{t}\sum_{j=1}^{r}(y_{ij} - \overline{\overline{y}})^2 = \sum_{i=1}^{t} r(\overline{y}_{i.} - \overline{\overline{y}})^2 + \sum_{i=1}^{t}\sum_{j=1}^{r}(y_{ij} - \overline{y}_{i.})^2.$$

We see that the total sum of squares measures the variation in the data where the treatment is not accounted for, and is partitioned into the variation that the treatment can account for (the variation in treatment means) and the variation that the treatment cannot account for (the variation within the treatment levels). This latter *within* variation is a true experimental error in that we have replications of the experimental unit under the same treatment conditions. If there were no experimental error, these replications should give the same value.

Equation (1.8) is related to an identity between unconditional and conditional variance. If X and Y are two random variables, then

(1.9) $$\text{Var}(Y) = \text{Var}[E(Y|X)] + E[\text{Var}(Y|X)].$$

A most important identity

If Y is the response variable and X is the treatment level, then the left side of (1.9) is the variability in Y without accounting for X, analogous to the total sum of squares. The quantity $E(Y|X)$ is the expected value of Y conditional on the treatment level, analogous to the treatment mean,

and the treatment sum of squares measures this variability. The average variability of Y inside each treatment level, the within variability, is the term $E[\text{Var}(Y|X)]$. (See Exercise 1.9.)

We next take a look at the partition of variance in a Randomized Complete Block Design (RCB), to illustrate the contrast with the Completely Randomized Design (CRD). We typically do not calculate the "within treatment" sum of squares in an RCB, but it is useful to see it and to explicitly see where we get SS(Blocks) and SS(T × B).

Example 1.8. STRAWBERRY BLOCKS   A field experiment is conducted to study the adaptability of three varieties of strawberries to Venezuelan soil, which are blocked on four different plots of land. The data are given in Table 1.2.   ‖

We now calculate the sums of squares for the data layout of Table 1.2. The total sum of squares can first be partitioned as

Table 1.2. Yields in kilograms from four blocks of land over a 2-week period

		Blocks		
		1	2	3	4
	A	10.1	10.8	9.8	10.5
Variety of Strawberry	B	6.3	6.9	5.3	6.2
	C	8.4	9.4	9.0	9.2

$$SS(\text{Total}) = SS(\text{Trt}) + SS(\text{``Within Trts''}),$$

(1.10)

$$\sum_{i=1}^{t}\sum_{j=1}^{b}(y_{ij} - \overline{\overline{y}})^2 = \sum_{i=1}^{t} b(\overline{y}_{i.} - \overline{\overline{y}})^2 + \sum_{i=1}^{t}\sum_{j=1}^{b}(y_{ij} - \overline{y}_{i.})^2,$$

and

$$SS(\text{``Within Trts''}) = SS(\text{Blocks}) + SS(T \times B),$$

(1.11)

$$\sum_{i=1}^{t}\sum_{j=1}^{b}(y_{ij} - \overline{y}_{i.})^2 = \sum_{j=1}^{b} t(\overline{y}_{.j} - \overline{\overline{y}})^2 + \sum_{i=1}^{t}\sum_{j=1}^{b}(y_{ij} - \overline{y}_{i.} - \overline{y}_{.j} + \overline{\overline{y}})^2.$$

Comparing this partition to that of the oneway anova (1.8) should clarify the relationships of the sums of squares.

In the RCB the interaction always tests treatments	Equation(1.11) is sometimes interpreted as showing a possible advantage to blocking, that is, the sum of squares used for estimating within error is reduced, possibly leading to a smaller error estimate and hence more significant results.

The reduction is not certain, however, because it is the mean squares that are used and the error degrees of freedom have been reduced from $t(b-1)$ in a oneway anova to $(t-1)(b-1)$ in a RCB. So, even though from (1.11) it is always true that

$$SS(\text{Within Trts}) > SS(T \times B),$$

it does not necessarily follow that

$$\frac{SS(\text{Within Trts})}{t(b-1)} > \frac{SS(T \times B)}{(t-1)(b-1)}.$$

However, realize that (1.11) is just a "what if" scenario, as the *design* of the two experiments are different, and the design is typically a function of the physical setup or limitations. So (1.11) just relates numbers and cannot be used to decide how to analyze a particular anova. The oneway anova and the RCB anova are quite different in a most fundamental way. The oneway anova is a special case of a *completely randomized design*; the data are collected in a random order throughout the experiment. By its very nature, a block design restricts randomization to within the blocks and, hence, cannot be a completely randomized design. Any attempt to analyze it as such can only create bias in the analysis.

1.4 Contrasts

Typically, the goal in a statistical experiment is to understand the effect of a treatment, or to compare and contrast treatment effects. With good design, we strive to optimize the variance that we use for our comparisons, but we also need to understand how to make these comparisons of interest. We use *contrasts* and, if possible, *orthogonal contrasts* to compare treatment effects. We review these concepts in this section.

Definition 1.9.   For parameters $\theta_1, \ldots, \theta_t$ and constants $a_1, \ldots, a_t$, the quantity

$$\sum_{i=1}^{t} a_i \theta_i$$

is a *linear combination*. If $\sum_{i=1}^{t} a_i = 0$, it is a *contrast*.

Contrasts allow us to compare parameters corresponding to treatments. The simplest contrast is $a_1 = 1, a_2 = -1$, and all other $a_i = 0$, which measures the difference between treatments 1 and 2.

Example 1.10. SIMPLE CONTRASTS   Consider the following contrasts among four means:

μ_1	μ_2	μ_3	μ_4
1	−1	0	0
0	0	1	−1
1	1	−1	−1

These three contrasts allow us to compare the four means by (1) comparing the first and second means, (2) comparing the third and fourth means, and (3) comparing the average of the first two to the average of the last two.

Contrasts are not unique; there are an infinite number of them. In any particular experiment they should be chosen to reflect the questions that are of interest.

Another simple set is

$$
\begin{array}{cccc}
\mu_1 & \mu_2 & \mu_3 & \mu_4 \\
\hline
1 & -1/3 & -1/3 & -1/3 \\
0 & 1 & -1/2 & -1/2 \\
0 & 0 & 1 & -1
\end{array}
$$

Here, the first contrast compares μ_1 to the average of the others. This can be a comparison of interest if μ_1 is a control and the others are treatments, so $(1, -1/3, -1/3, -1/3)$ contrasts the control with the average treatment effect. This set of contrasts is called *Helmert contrasts*. ‖

The two sets of contrasts in Example 1.10 are sets of *orthogonal* contrasts.

Definition 1.11. For parameters $\theta_1, \ldots, \theta_t$, two contrasts

$$
\sum_{i=1}^{t} a_i \theta_i \text{ and } \sum_{i=1}^{t} b_i \theta_i
$$

are *orthogonal contrasts* if $\sum_{i=1}^{t} a_i b_i = 0$.

Note: Both Definitions 1.9 and 1.11 are about parameters. We now estimate these contrasts by using the data $\bar{y}_i$, and then there can be confusion between *orthogonal* and *uncorrelated*, a difference that is important to understand.

Although this distinction is often glossed over, it can have very important implications for inference.

Definition 1.12. Two contrasts $\sum_{i=1}^{t} a_i \bar{y}_i$ and $\sum_{i=1}^{t} b_i \bar{y}_i$ are *orthogonal* if $\sum_i a_i b_i = 0$. The same two contrasts are *uncorrelated* $\sum_i a_i b_i / r_i = 0$, where $\text{Var}(Y_i) = \sigma^2$ and $\bar{Y}_i$ is based on r_i observations.

If an anova has the same number of observations per cell ($r_i = r$), then, under the usual assumptions, orthogonality and uncorrelated are exactly the same. However, if this is not the case, then we must be more careful. Some details are given in Technical Note 1.8.2

We start with an example.

Example 1.13. REHABILITATION TIME In a study of the relationship between physical fitness prior to knee surgery and rehabilitation time, data were collected on 24 men, aged 18-30 years who had similar corrective surgery, all within the same year. The data are in dataset RehabTime and look like

	Physical Condition			
	Poor Condition	Below Average	Above Average	Excellent Condition
	42	29	28	26
	$\vdots$	$\vdots$	$\vdots$	$\vdots$
	42	31	33	22
r	6	6	6	6

An anova table for these data is

	df	Sum Sq	Mean Sq	F-value	p-value
Condition	3	861.0	287.00	25.28	$< .0001$
Within	20	227.0	11.35		

To further partition the variation we can break down the treatment sums of squares into components that are attributable to contrasts. If the contrasts are orthogonal, we can completely partition the treatment sum of squares, with no overlap, into additive components that represent the variation due to the contrasts. This is the advantage of having orthogonal contrasts.

	df	Sum Sq	Mean Sq	F-value	p-value
Condition	3	861.0	287.0	25.28	$< .0001$
Poor vs. Others	1	512	512	45.11	$< .0001$
Below vs. Above/Exc.	1	132.25	132.25	11.65	.0028
Excellent vs. Above	1	216.75	216.75	19.09	.0003
Within	20	227.0	11.35		

These are the Helmert contrasts as shown in Example 1.10. Note that the sums of squares of the three contrasts add to the treatment sum of squares, so the variation in treatment is partitioned into three nonoverlapping pieces.
$\|$

The formula for the contrast sum of squares may not, at first sight, be intuitive, but it is the way to partition the total sum of squares (see Technical Note 1.8.2).

Definition 1.14. For a contrast $\sum_i a_i \theta_i$ that is estimated with $\sum_i a_i \bar{y}_i$, where $\bar{y}_i$ is based on r_i observations, the *sum of squares due to the contrast* $\sum_i a_i \bar{y}_i$ is

$$\frac{(\sum_i a_i \bar{y}_i)^2}{\sum_i a_i^2 / r_i}.$$

Note that the formula involves r_i. This has the unfortunate consequence that orthogonal contrasts, such as Helmert contrasts, will not partition the treatment sum of squares orthogonally unless $r_i = r$.

Example 1.15. REHABILITATION TIME CONTINUED   Suppose that the data are unbalanced; that is, the four groups have unequal number of subjects. These data are in dataset `RehabTime2` and have cell sizes

	Physical Condition			
	Poor Condition	Below Average	Above Average	Excellent Condition
r	5	8	7	4

There are still 23 total degrees of freedom, split as 3 for treatments and 20 for within error, but the Helmert contrasts no longer partition the treatment sum of squares into nonoverlapping pieces. Here is a set of contrasts that are uncorrelated and partition the treatment sum of squares, but they are *not* orthogonal.

μ_1	μ_2	μ_3	μ_4
1	$-8/19$	$-7/19$	$-4/19$
0	1	$-4/11$	$-7/11$
0	0	1	-1

Although these look vaguely similar to the Helmert contrasts, they no longer have a nice interpretation. We have kept the same third contrast, but cannot preserve the other two.

Note: In general, we can always start with *one* contrast of interest and partition the treatment sum of squares into two nonoverlapping pieces, one due the contrast of interest and one due to everything else.

This is usually the more meaningful course, rather than to try to make sense of the uncorrelated contrasts when there is unequal r_i.

Alternatively, we can just look at the sums of squares and F-ratios for the orthogonal set of contrasts and realize that we are not partitioning the treatment sum of squares orthogonally, and interpret these more meaningful contrasts in that light. (See Exercise 1.13.)   ∥

There are many different sets of orthogonal contrasts, each having different interpretations and, depending on the experiment, some are more appropriate than others. Here is an example with polynomial contrasts.

Example 1.16. FISH MICROARRAY EXPERIMENT   The goal of this experiment was to find the optimum combination of two treatments, Tissue Mass and presence or absence of hCG (an endogenous stimulant of hormone synthesis) for gene expression in gonadal tissue in fish. The treatment design is the twoway crossed design

Tissue Mass (mg)

	50	100	150	200
hCG Yes	x	x	x	x
No	x	x	x	x

The quantitative levels of *Tissue Mass* suggest using polynomial contrasts. For a factor with 3 levels we can take out linear and quadratic trends with the contrasts[1] given by

$$\text{linear} \quad : -1 \quad 0 \quad 1$$
$$\text{quadratic} : \quad 1 \; -2 \; 1$$

A full set of orthogonal contrasts for this experiment is given by

Linear Tissue Mass (mg)

	−3	−1	1	3
hCG 1	−3	−1	1	3
−1	3	1	−1	−3

Quadratic Tissue Mass (mg)

	1	−1	−1	1
hCG 1	1	−1	−1	1
−1	−1	1	1	−1

Cubic Tissue Mass (mg)

	−1	3	−3	1
hCG 1	−1	3	−3	1
−1	1	−3	3	−1

The dataset **FishTissueMass** has the data for this experiment. The following anova table can be constructed:

Source	df
Tissue Mass	3
Linear	1
Quadratic	1
Cubic	1
hCG	1
Tissue Mass × hCG	3
Linear × hCG	1
Quadratic × hCG	1
Cubic × hCG	1
Within	4
Total	11

The details of the analysis are left to Exercise 1.15.   ‖

---

[1] Many texts, such as Dean and Voss (1999) have tables of polynomial contrasts. Contrasts can also be generated in R with statements such as **contr.poly** or **contr.helmert**. Note that R automatically scales the contrasts so that $\sum_i a_i^2 = 1$.

1.5 Randomization, Layouts, and Designs

In this section we discuss some fundamentals in the collection of experimental data, and clarify a number of terms relating to the actual data that we see. We must be aware of what information we will need to be able to fully analyze the data. For example:

(1) In an agricultural field, we must take account of how treatments are assigned to plots.
(2) If our experimental unit is a subject, are treatments applied in a totally random order? Are groups of subjects treated in a similar way?
(3) In a laboratory experiment, are all treatments run in one day? If not, do conditions change enough between days that we need to account for this factor as a block?
(4) In a microarray experiment, are the microarrays acting as blocks? That is, is there more than one treatment on a microarray?

Randomization

Perhaps the most fundamental principle of design is randomization, that is, obtaining the observations (or, more precisely, the experimental units) in a random manner that is as free from bias as possible. Consider the following example.

Example 1.17. PROBLEMATIC INFERENCE   A researcher in Forestry received a grant to investigate five newly developed varieties of pine trees. She carefully and painstakingly planned an elaborate greenhouse experiment to determine which variety has the greatest annual growth. After 4 years, she analyzed the results of her experiment and recommended variety B as the best (for maximum annual growth). Her results were so overwhelming that 10 lumber companies planted the recommended variety on half of their replacement acreage. After 4 years, 8 of the companies complained that variety B pine trees were only 75% as tall as "an old standby variety". What happened?
∥

There are many possible explanations. Some are

(1) This all happened by chance.
(2) Somehow the trees were not randomly assigned in the greenhouse, and variety B received more light, or better soil, or optimal temperature.
(3) The greenhouse experiment was properly done, but the soil used was not representative of that in the replacement acreage.

Although randomization cannot do much about (1) or (3), which reflects a Block×Treatment interaction, proper randomization should guard against (2).

The basic idea of randomization is that, given a design, the actual assignment of experimental units to treatments should be chosen at random, with equal probability, from all possible assignments. Such a strategy would result in a *simple random sample*. Although, in practice, we do not list out all possible arrangements, the point is that the assignment of experimental units to treatments can never be based on any subjective information, and must be left to a random mechanism.

Example 1.18. A RANDOM ASSIGNMENT   Recall Example 1.1. There are 15 potted plants, which are the experimental units. These must be randomized to the three fertilizer treatments. Formally, there are $\frac{15!}{5!5!5!} = 756756$ ways of assigning the pots to the fertilizers. To obtain a random sample, we could list all of these assignments, then draw a random number between 1 and 756756, and choose that arrangement. In practice, we would choose a pot at random (using a random number generator[2]) and then assign it to one of the three fertilizers at random.

If the 15 potted plants are on a greenhouse bench, then the actual placement of the plants on the bench should entail another random assignment. (If there is a known gradient, for example a light or temperature gradient, this also could be accounted for by blocking.)   ‖

The random assignment of experimental units to treatments should result in the following desirable outcomes.

(1) **Elimination of systematic bias.** Bias comes in many forms, and some of it is unknown. In addition to gradients of light or temperature, we could have things like dye bias in microarray experiments, interviewer bias in surveys, and other unaccountable occurrences. Randomization is one way to break any systematic effect.

(2) **Obtaining a representative sample.** Since our ultimate goal in any experiment is to make a valid inference to a population, our data must be representative of that population. Randomization is needed to obtain a representative sample.

(3) **Accounting for extraneous (unknown) confounding variables.** Confounding variables exist in all experiments. In Example 5.1 we will look at the effect of diets on blood pressure, measured on 12 subjects. Although the subjects are to be of similar health status, confounders such as lifestyle, race, genetic disposition, or many other factors, could influence the results. Although it many be possible to control for some factors, such

---

[2] It was common for statistics texts to include tables of random numbers, but these have been replaced by the ubiquity of computer-generated random (or, more precisely, pseudo-random) numbers. Most established statistical software has such generators built in, and they can produce streams of deterministic numbers that are impossible to distinguish from random. For more details, see Robert and Casella (2004, Section 2.1).

as race, other factors are uncontrollable, even unknown, such as genetic disposition. Randomization will help to distribute this unknown variation throughout the experiment, breaking the confounding influence.

Fisher (1971, Section II.9) wisely noted that "...the uncontrolled causes which may influence the result are always strictly innumerable". This reflects the fact that, in any experiment, there will be confounding factors that we will not be aware of, and randomization is our only method of obtaining a valid inference in the face of these confounders.

Treatment Design

The title of this book, "Statistical Design", was chosen purposefully. Note that the title of this book is *not* "Experimental Design". The reason for this is that there are two pieces to a design, which we separate into *Treatment Design* and *Experiment Design*. A *Statistical Design* contains both of these pieces.

Definition 1.19. A *treatment design* is the manner in which the levels of treatments are arranged in an experiment. Typically, treatments are either *crossed* or *nested*, and this relationship can be either *complete* or *incomplete*.

Example 1.20. CROSSED OR NESTED   Treatments are *crossed*, or in a *factorial arrangement*, if each level of one treatment appears with each level of the others. If Treatment A has 6 levels and Treatment B has 3 levels, here is a crossed treatment design:

$$
\begin{array}{c|cccccc}
 & \multicolumn{6}{c}{A} \\
 & 1 & 2 & 3 & 4 & 5 & 6 \\
\hline
1 & x & x & x & x & x & x \\
B\ 2 & x & x & x & x & x & x \\
3 & x & x & x & x & x & x \\
\end{array}
$$

which might also be portrayed as

			A			
1	2	3	4	5	6
B	B	B	B	B	B
1 2 3	1 2 3	1 2 3	1 2 3	1 2 3	1 2 3
x x x	x x x	x x x	x x x	x x x	x x x

If both A and B are crossed with C, at two levels, this could look like

				C							
		1						2			
		A						A			
1	2	3	4	5	6	1	2	3	4	5	6
B	B	B	B	B	B	B	B	B	B	B	B
1 2 3	1 2 3	1 2 3	1 2 3	1 2 3	1 2 3	1 2 3	1 2 3	1 2 3	1 2 3	1 2 3	1 2 3
x x x	x x x	x x x	x x x	x x x	x x x	x x x	x x x	x x x	x x x	x x x	x x x

Alternatively, we could represent this as a three-dimensional rectangle with sides of 2, 3, and 6, where each of the $2 \times 3 \times 6 = 36$ cells is a treatment combination.

Treatment B is *nested* in Treatment A if different levels of B appear with the levels of A; for example,

	A		
1	2	3	4
B	B	B	B
1 2 3	4 5 6	7 8 9	10 11 12
x x x	x x x	x x x	x x x

Here B, with 12 levels, is nested in the 4 levels of A. This nesting could be unbalanced if, for example, level 3 of A only had 2 levels of B, and the other levels of A all had 3 levels of B.   ‖

With B nested in A, we are only able to compare the levels of B in a particular level of A, which limits our inferences. Thus, it is a good general principle that factors of interest should be crossed rather than nested, to allow for better comparisons of treatments and interactions.

Nested factors have no interaction

In practice, nested factors are usually random – for example, subjects could be nested in a treatment; and with a random factor the interest is rarely about comparisons of means.

Example 1.21. CROSSING SUBJECTS AND TREATMENTS   Crossing a random factor and a fixed factor, as in the case of subjects and treatments, results in correlation in the observations. This is because when we draw the subject at random and apply the treatment, we have used up that experimental unit. The only way we can "cross" subjects and treatments is to apply all of the treatments to the same subject (if possible). If we did this with, for example, 3 treatments and 5 subjects, we could display the data as

	Treatment	
1	2	3
Subject	Subject	Subject
1 2 3 4 5	1 2 3 4 5	1 2 3 4 5
x x x x x	x x x x x	x x x x x

which "looks like" a oneway anova on treatments. However, we know that we do not have 15 independent experimental units. However, if we display the data as

	Subject				
	1	2	3	4	5
1	x	x	x	x	x
Treatment 2	x	x	x	x	x
3	x	x	x	x	x

it should be clear that this is a RCB design. Thus, the treatment design of an RCB is a twoway crossed design. This distinction is explored further in Exercise 3.7.   ‖

Note: The treatment design shows us how to count degrees of freedom and calculate sums of squares in the anova table. But we need more information before F-ratios can be formed. That information comes from the experiment design.

Layouts _____

The treatment design is reflected in the *data layout*.

Example 1.22. DATA LAYOUT   The following data were collected on three hybrids of corn subjected to four different fertilizer treatments:

		Hybrid		
		M4	G10	M15
Fertilizer	a	70.8	57.1	54.2
	b	73.9	68.1	75.4
Type	c	56.8	56.8	67.5
	d	92.9	84.3	90.4

The *treatment design* is a twoway crossed layout or *factorial* arrangement – every level of one treatment appears with every level of the other. This is reflected in the *data layout* given above, which is the data file that is given to the computer. Note that the treatment design does not take account of any error structure or distributional assumptions.   ‖

Example 1.23. FIELD LAYOUT   Although the treatment design of Example 1.22 is apparent from the data layout, the data could have come from many different experiment designs. If the *field layout* of the data are

M15	M4	M4	M15	G10	M15	G10	M4	G10	M15	G10	M4
a	c	d	c	d	b	c	b	a	d	b	a
54.2	56.8	92.9	67.5	84.3	75.4	56.8	73.9	57.1	90.4	68.1	70.8

this would correspond to a design in which each of the treatments was completely randomized to the 12 plots; randomization was complete throughout the data layout and the field layout. Thus, the experiment design is a *completely randomized design*.

Contrast this field layout to an alternative such as

G10	M4	M15	M15	G10	M4	G10	M15	M4	M15	G10	M4
c	c	c	a	a	a	d	d	d	b	b	b
56.87	56.8	67.5	54.2	57.1	70.8	84.3	90.4	92.9	75.4	68.1	73.9

In this field layout the treatments are not randomized throughout, but rather the randomization is carried out in blocks of the fertilizer treatment. So, although the data layout is the same, and the treatments are still in a twoway crossed arrangement, the field layout is totally different. So we can have the same treatment design (data layout) corresponding to different experiment designs.

As the fertilizer is applied to three adjacent plots simultaneously, this *block* places a restriction on the randomization and induces a correlation structure. This randomization restriction is actually

> Implications of fixed vs. random

the defining difference between *fixed factors* and *random factors*. See the discussion in Section 3.2.   ‖

Experiment Design

Note that the term "field layout" is somewhat historical, as much of statistical design originated with agricultural field experiments. However, this term really refers to all restrictions imposed on the collection of the data.

Example 1.24. FIELD LAYOUT CONTINUED   For both field layouts from Example 1.23, an anova table looks like

	df	Sum Sq
Fertilizer	3	1656.44
Hybrid	2	107.22
Fertilizer × Hybrid	6	195.19

However, unless we know how the treatments were assigned and how the randomization of treatments was done, we cannot correctly analyze this data.
‖

Definition 1.25. The *experiment design* is the manner in which the randomization of experimental units to treatments is carried out and how the data are actually collected. The error structure of the experiment is a consequence of the experiment design.

In a twoway anova, there are many "choices" of how to form F-ratios; however, only one "choice" is the correct one for any given experiment design. One of the main goals of this book is to make this idea clear, so we will always know the proper F-ratio and, hence, the proper variance estimate for treatment comparisons.

Example 1.26. TWOWAY F-RATIOS   For the twoway crossed treatment design, we can have the following choices for F-ratios: The three different choices of F-ratios correspond to three different experiment designs which would result in three different field layouts. Referring to the columns of Table 1.3

> **Column** (a): These tests could result from a completely randomized design, in which the treatment combinations are randomly allocated throughout the design.
>
> **Column** (b): These tests could result from a randomized block design, in which the levels of A are randomized within the levels of B, so B restricts the randomization of A.
>
> **Column** (c): These tests could result from a design in which the randomization of both factors are restricted. Note that B is randomized in the columns and A is randomized in the rows (see Section 5.6.1).

Look carefully at the layouts in Table 1.4. These (and almost all layouts) will often be dictated by physical constraints. For example, in (c), the A treatment is constant in rows and the B treatment is constant in columns (see Exercise 1.17).   ∥

Table 1.3. Possible anovas corresponding to *(a)* complete randomization, *(b)* restriction of randomization of one factor, and *(c)* restriction of randomization of both factors.

			Choices		
				F-ratio	
Source	Df	Mean Square	(a)	(b)	(c)
A	$a-1$	MS(A)	$\frac{\text{MS(A)}}{\text{MS(Within)}}$	$\frac{\text{MS(A)}}{\text{MS(Within)}}$	$\frac{\text{MS(A)}}{\text{MS(AxB)}}$
B	$b-1$	MS(B)	$\frac{\text{MS(B)}}{\text{MS(Within)}}$	$\frac{\text{MS(B)}}{\text{MS(AxB)}}$	$\frac{\text{MS(B)}}{\text{MS(AxB)}}$
A × B	$(a-1)(b-1)$	MS(A × B)	$\frac{\text{MS(AxB)}}{\text{MS(Within)}}$	$\frac{\text{MS(AxB)}}{\text{MS(Within)}}$	$\frac{\text{MS(AxB)}}{\text{MS(Within)}}$
Within	$ab(r-1)$	MS(Within)			

Table 1.4. Corresponding to Table 1.3, possible field layouts for *(a)* complete randomization, *(b)* restriction of randomization of one factor, and *(c)* restriction of randomization of both factors.

(a)	(b)	(c)
A1B1 A2B1 A1B3 A1B2 A3B3 A3B2 A3B1 A2B2 A2B3	A2B1 A3B2 A1B3 A1B1 A2B2 A3B3 A3B1 A1B2 A2B3	A1B1 A1B2 A1B3 A3B1 A3B2 A3B3 A2B1 A2B2 A2B3

The three layouts in Table 1.4 are well-known designs:

(a) Completely Randomized Design (CRD)
(b) Randomized Complete Block Design (RCB)
(c) Strip Plot Design

1.6 Replication: True and Technical

In Example 1.3(3) we saw an example where RNA from two groups of people (with and without a certain disease) are applied to a microarray. Here we give a number of detailed examples of experiments to discuss the difference between *true replication*, where the experimental unit is replicated, and *technical replication*, where the experimental unit is subsampled. We will not be very rigorous here, but will leave that to later sections (for example, Section 3.5).

Example 1.27. IRRADIATION MICROARRAYS   In an experiment described by Tusher *et al.* (2001), RNA was harvested from wild-type human lymphoblastoid cell lines, designated 1 and 2. They were then grown in an unirradiated state (U) or in an irradiated state (I), where cell lines and irradiated state are crossed treatments. RNA samples were labeled and divided into two identical aliquots for independent hybridizations, denoted A and B. Thus, data for 6,800 genes on the microarray were generated from 8 hybridizations (U1A, U1B, U2A, U2B, I1A, I1B, I2A, and I2B) according to the treatment design.

The analysis done by Tusher *et al.* (2001) interpreted the design for one gene as

Treatment			Anova	
U	I		Source	df
x	x		Treatments(U/I)	1
x	x		Within	6
x	x		Total	7
x	x			

which treats the design as a two-sample t-test (a oneway CRD).

The six degrees of freedom for error are a fiction here. There were only two cell lines used, which are the experimental units, and here they act as blocks. This reasoning leads to the analysis

<table>
<tr><td colspan="3" align="center">Treatment</td><td colspan="2" align="center">Anova</td></tr>
<tr><td></td><td>U</td><td>I</td><td>Source</td><td>df</td></tr>
<tr><td></td><td>x</td><td>x</td><td>Blocks(Lines)</td><td>1</td></tr>
<tr><td>Line 1</td><td>x</td><td>x</td><td>Treatments(U/I)</td><td>1</td></tr>
<tr><td></td><td>x</td><td>x</td><td>B × T</td><td>1</td></tr>
<tr><td>Line 2</td><td>x</td><td>x</td><td>Subsampling</td><td>4</td></tr>
<tr><td></td><td></td><td></td><td>Total</td><td>7</td></tr>
</table>

The two lines are blocks, and to each one the two treatments are applied. The aliquots are not a replication of the experimental unit, but rather a subsample of the RNA lines on the same experimental units. Although subsampling can increase the precision of the experiment, it does not increase the degrees of freedom for the treatment test.

In the first analysis, as a oneway CRD, the observations within a treatment group are correlated, as two of them come from the same line. This is a surefire red flag that the design may not be appropriate, and a signal that a blocking factor may be overlooked.   ‖

In the above experiment the replications actually come from the same RNA line - it is merely split in two. Thus, at the lowest level of the experiment, we can expect correlation in the cells.

In the following experiment there appears to be true replication at the cell level - the observations in the cell are conditionally independent. However, it is still subsampling, so we will see that the replication still does not produce usable replication. (A more detailed treatment of these models is given in Section 3.5.)

Example 1.28. PSEUDOREPLICATION

An experiment was done to assess the effect of shipping and storage on the acceptability of avocados. Three shipping methods (increasingly expensive) and two storage methods (also increasingly expensive) were used. Three different shipments (which act as blocks) were used in the experiment. Within each shipment, for each Shipping × Storage combination, four crates of avocados were measured for the percent acceptable. A schematic of the data is

An RCB with treatments in a factorial arrangement

Shipment

	1		2		3	
	Storage		Storage		Storage	
	1	2	1	2	1	2

Shipping 1

| x x x x | x x x x | x x x x |
| x x x x | x x x x | x x x x |

Method 2

| x x x x | x x x x | x x x x |
| x x x x | x x x x | x x x x |

3

| x x x x | x x x x | x x x x |
| x x x x | x x x x | x x x x |

This is a threeway crossed design, and it was decided that it was reasonable to assume that there were no interactions. The design was analyzed according to the following anova table:

Anova

Source	df
Blocks (Shipments)	2
Shipping Method	2
Storage	1
Shipping × Storage	2
Residual	64
Total	71

and all tests were done against the "residual".

This analysis again treats subsamples, or technical replicates, as true replications. Note that the treatment (Shipping Method × Storage) is applied to the *group* of four crates, which constitute the experimental unit. Thus, the four individual members are subsamples, and all that matters in the analysis is their average.

Moreover, even if there is a Block × Treatment interaction, these interaction terms still provide the correct errors for the treatment tests. The residual term above can be decomposed into the appropriate pieces as follows:

Source	df	
Blocks(Shipments)	2	
Shipping	2	
Storage	1	
Shipping × Storage	2	
Residual	64	
B × Ship		4
B × Stor		2
B × Ship × Stor		4
Within		54
Total	71	

Where the treatment tests are done against the appropriate interaction with blocks, for example, to test H_0: no shipping effect, we use $F = \frac{\text{MS(Shipping)}}{\text{MS(BxShip)}}$. If there is truly no block by treatment interaction, we can pool the three interaction terms to get one error term with 10 degrees of freedom that can be used for all treatment tests. We want to emphasize a few points about this:

(1) Pooling the three interaction terms to get one error term with 10 degrees of freedom results in conservative tests;, that is, the nominal α level tends to be an overestimate and the tests are less powerful. This means that rejection of H_0 carries a lot of weight. This is because the denominator of the test tends to be an overestimate.

(2) Pooling the interaction terms with the within error, to get an error term with 64 degrees of freedom yields anticonservative tests – tests where the nominal α level is an underestimate – and rejection of H_0 does not carry much weight. This is because the denominator of the test tends to be an underestimate.

(3) No matter what, the 54 degrees of freedom for within error do not contribute to, and cannot be used for, the tests on treatments without some unpleasant assumptions.

The replication in the cells is somewhat of a wasted effort. It is possible that virtually the same precision can be obtained with fewer crates per cell. On the other hand, a superior design would be attained if the number of blocks were increased, without increasing the total number of crates.   ‖

A last point to make is about the other kind of pooling – *pooling* of experimental units, that is, combining more than one experimental unit and obtaining one observation. This practice has become more popular because of microarray experiments, where RNA from different subjects is mixed together and the combined sample is then hybridized (see Miscellanea 1.9.1). This is sometimes unavoidable in some experiments because not enough RNA can be obtained from one subject, but sometimes pooling is employed as a cost-cutting measure, as the microarray chips can be expensive.

The effect of pooling is to change the experimental unit from the subject to the pool of subjects. This has two consequences:

(1) The experimental error (the between subject variation) is reduced

(2) The degrees of freedom are based on the number of pools, not the number of subjects

So there is a positive effect, but it can be somewhat minimal. Suppose that we have a model where the between-subject variance is σ^2 and the observation is made on a pool of p subjects. A model is then

$$(1.12) \quad Y_{ij} = \mu + \tau_i + \varepsilon_{ij}, \quad i = 1, \ldots, t, \quad j = 1, \ldots, r, \quad \varepsilon_{ij} \sim N\left(0, \frac{\sigma^2}{p}\right).$$

The anova for this model has $t(r-1)$ degrees of freedom for error; the fact that p subjects were pooled does not change the anova: There is *no* effect on the degrees of freedom for error. The positive effect of pooling, which is somewhat hidden, is in the variance of a treatment mean, $\text{Var}(\bar{Y}_i.) = \sigma/rp$. So we expect that the variance in the pooled experiment would be smaller than the variance if there were no pooling. However, we still estimate the variance with MS(Within) with $t(r-1)$ degrees of freedom.

Note also the effect of technical replication. The variance of a subject, σ^2, is composed of two pieces; one for the between subject variation and one for the within subject variation; that is,

$$\sigma^2 = \sigma_B^2 + \sigma_W^2,$$

where σ_B^2 and σ_W^2 represent between- and within-subject variation, respectively. Now, the effect of true replication, pooling and technical replication can be summarized as follows. If the experimental unit is composed of a pool of p subjects, and this is replicated r times (true replication), and each experimental unit is subsampled s times (aliquot or technical replication), then the variance of a treatment mean is

(1.13)
$$\text{Var}(\bar{Y}_i..) = \frac{1}{rp}\left(\sigma_B^2 + \frac{\sigma_W^2}{s}\right).$$

> The effect
> of replication
> on variance

So the effect of the technical replication is to cut down σ_W^2, which we expect to be much smaller than σ_B^2, as there is almost always less variability within a unit than between units. We also see that pooling and true replication have a similar effect, but remember that we do not gain error degrees of freedom from pooling.

The moral of the story is that, whenever possible, increase the *true* replication at the expense of all other types of replication. (See Exercise 1.21.)

1.7 Exercises

Essential

1.1 Describe a simple experiment and explain what is the experimental unit.

1.2 Consider an experiment that was designed and performed as an RCB but analyzed as a CRD. If the blocks are effective, would the standard errors of estimates from the CRD analysis be too large or too small?

1.3 Give an example of a twoway layout that is an RCB. Give an example of a twoway layout that is not an RCB. For each example use two replications per cell. Provide a plan or schematic of each example and write the anova table.

1.4 Referring to Examples 1.3 and 1.6:
   (a) Write out the anova table for part (1) assuming that there are 6 fertilizers and 4 pots per fertilizer.

(b) Analyze the dataset `FishTank` to produce the anova table in part (2). Make sure the test on diets has the correct denominator.

(c) Write out the anova table for part (3) assuming that there are 10 people in each group.

1.5 Referring to (b) of Example 1.6, some may disagree with the statement "There is no interest in assessing the significance of tanks; they are merely there to hold the fish", arguing that we should want to make sure that there are truly no differences in the tanks. Here are two sides to this argument:

(1) We should be aware of any large tank effect, as that could mask the effect of the diets

(2) The size of the tank effect is immaterial. If it is so large as to mask the diet effect, then the size of the diet effect is of no practical significance.

Now suppose that you have available 144 fish and 24 tanks to test the three diets:

(a) Design an experiment, setting up an anova table as in Exercise 1.6, that addresses the concerns in (1).

(b) Do the same, but now address the concerns in (2).

(c) Take a stand. Do you agree with the sentiment in (1) or (2)? Which design would you suggest to the experimenter? (You might want to look at the data in `FishTank` to help your argument.

Note: In reality, the limiting factor is money. The experimenter will have some amount of funds to spend and you must advise him or her how to allocate the money between fish and tanks.

1.6 (a) Establish the identity for the partitioning of the sums of squares in (1.8). (Add $\pm \bar{y}_{i\cdot}$ inside the square on the left side, expand, and verify that the cross term is zero.)

(b) Write out the complete anova table, demonstrating the partitioning of the sums of squares, for the dataset `SmogOzone`, the record of an experiment done to measure the effects of ozone, a component of California smog. The data look like

Controls (0 ppm)	Ozone (.3 ppm)	Ozone (.6 ppm)
41.0	26.6	−9.0
$\vdots$	$\vdots$	$\vdots$
21.4	6.1	−9.0

where a group of 45 rats were kept in one of three ozone environments for seven days, with their weight gains (in grams) recorded.

(c) Partition the treatment sums of squares into two orthogonal components in two different ways:

(i) Compare control vs. treatments, and also compare treatments.

(ii) Test the shape of the Ozone curve: Linear? Quadratic?

1.7 Referring to Section 1.2:

(a) Verify (1.4).

(b) Verify (1.5) by showing

(i) $\mathrm{Var}\left(\bar{Y}_{i\cdot} - \bar{\bar{Y}}\right) = \mathrm{Var}(\bar{Y}_{i\cdot}) - 2\,\mathrm{Cov}(\bar{Y}_{i\cdot}, \bar{\bar{Y}}) + \mathrm{Var}(\bar{\bar{Y}})$,

(ii) $\mathrm{Var}(\bar{Y}_{i\cdot}) = \sigma^2/r$, $\mathrm{Var}(\bar{\bar{Y}}) = \sigma^2/rt$,

(iii) $\mathrm{Cov}(\bar{Y}_{i\cdot}, \bar{\bar{Y}}) = \frac{1}{t}\mathrm{E}\left[(\bar{Y}_{i\cdot} - \mu - \tau_i)\sum_{i'=1}^{t}(\bar{Y}_{i'\cdot} - \mu - \tau_{i'})\right] = \sigma^2/rt$, where the first part of the calculation uses the fact that $\sum_i \tau_i = 0$ and the second part uses the fact that $\mathrm{Cov}(\bar{Y}_{i\cdot}, \bar{Y}_{i'\cdot}) = 0$ if $i \neq i'$.

(c) Verify (1.6) and that MS(Within) is an unbiased estimator of σ^2.

1.8 (a) The anova table shown in Example 1.6 is a nested design, but the partitioning of the sums of squares still holds. If y_{ijk} is the weight gain of Fish k in Tank j eating Diet i, show that

$$\text{SS(Total)} = \text{SS(Diets)} + \text{SS(Tanks within Diets)}$$
$$+ \text{SS(Fish within Tanks within Diets)},$$

that is,

$$\sum_{i=1}^{t}\sum_{j=1}^{b}\sum_{k=1}^{r}(y_{ijk} - \bar{\bar{y}})^2 = br\sum_{i=1}^{t}(\bar{y}_{i\cdot\cdot} - \bar{\bar{y}})^2 + r\sum_{i=1}^{t}\sum_{j=1}^{b}(\bar{y}_{ij\cdot} - \bar{y}_{i\cdot\cdot})^2$$

$$+ \sum_{i=1}^{t}\sum_{j=1}^{b}\sum_{k=1}^{r}(y_{ijk} - \bar{y}_{ij\cdot})^2.$$

(b) If Tanks were crossed with Diets, the partitioning of the sum of squares still holds. It is perhaps easiest to see by showing that

$$\text{SS(Tanks within Diets)} = \text{SS(Tanks)} + \text{SS(Tank} \times \text{Diet)},$$

that is,

$$r\sum_{i=1}^{t}\sum_{j=1}^{b}(\bar{y}_{ij\cdot} - \bar{y}_{i\cdot\cdot})^2 = rt\sum_{j=1}^{b}(\bar{y}_{\cdot j\cdot} - \bar{\bar{y}})^2 + r\sum_{i=1}^{t}\sum_{j=1}^{b}(\bar{y}_{ij\cdot} - \bar{y}_{i\cdot\cdot} - \bar{y}_{\cdot j\cdot} + \bar{\bar{y}})^2.$$

1.9 Referring to (1.9):

(a) Show that

$$\text{Var}(Y) = \text{E}[Y - \text{E}(Y)]^2 = \text{E}[Y \pm \text{E}(Y|X) - \text{E}(Y)]^2$$
$$= \text{E}[Y - \text{E}(Y|X)]^2 + \text{E}[\text{E}(Y|X) - \text{E}(Y)]^2,$$

by verifying that the cross term is zero.

(b) Interpret the final two terms in (a) to arrive at (1.9)

(c) Suppose that, given $X = x$, the random variable Y takes on values $y_{x1}, \ldots, y_{xr}$, each with probability $1/r$, and $P(X = x) = 1/t$, $x = 1, \ldots, t$. By calculating $\text{Var}(Y)$, $\text{Var}[\text{E}(Y|X)]$, and $\text{E}[\text{Var}(Y|X)]$, show that the partitioning of the sums of squares in (1.8) is a special case of (1.9).

1.10 Partitioning sums of squares.

(a) Illustrate the partitioning of the sums of squares in the RCB anova by calculating the complete anova table for the data of Example 3.6, which can be found in dataset **Anticoagulant**.

(b) Analytically verify the partitioning of the RCB anova sums of squares by verifying (1.10) and (1.11). (For (1.11), add $\pm(\bar{y}_{\cdot j} - \bar{\bar{y}})$ inside the square on the left-hand side, expand, and show that the crossterm is zero.)

1.11 A researcher brings you the following set of contrasts of interest. Assume equal replication of all treatments:

Contrasts	Treatments					
	A	B	C	D	E	F
1	3	0	3	−2	−2	−2
2	1	0	−1	0	−1	1
3	0	4	0	−4	0	0
4	0	0	0	2	−1	−1

(a) Are contrasts 1 and 2 orthogonal?

(b) Are contrasts 1 and 3 orthogonal?

(c) Give a fifth contrast that is orthogonal to contrast 4. It cannot be one of the first three, but may be orthogonal to any of the first three.

1.12 A plant scientist investigated plant uptake of heavy metals (for example, nickel) when four rates of sludge applications were used. For a single variety of sweet corn, a total of 20 plants were established. The plants were individually potted and 5 were randomly chosen to receive each rate of sludge. After a designated period of time following the sludge application, the leaves were taken from each plant and analyzed for presence of the heavy metal.

(a) Identify the experimental unit and specify a model equation. Identify all terms.

(b) Set up the anova table (sources and df).

(c) Identify a contrast that the experimenter might be interested in testing. Explain how to estimate the contrast and its variance.

(d) In addition to differences in uptake due to different rates of sludge, there was interest in variation among plants treated alike as well as variation among leaves of the same plant. How are these concerns answered by the analysis?

(e) Before the data were collected, the experimenter's assistant misplaced four of the plants from the high rate of sludge. Redo parts (b) and (c) using this information.

1.13 Contrasts.

(a) Verify the partitioning of the treatment sum of squares into orthogonal contrasts given in Example 1.13 (dataset RehabTime)

(b) Show that for Example 1.15, the original Helmert contrasts are not uncorrelated, but the variation given in that example are. Use these contrasts to partition the treatment sum of squares for the dataset RehabTime2.

(c) We can always preserve one contrast and find an uncorrelated set. If we want to preserve the first Helmert contrast, an uncorrelated set is

μ_1	μ_2	μ_3	μ_4
1	$-1/3$	$-1/3$	$-1/3$
$\frac{5}{203}$	$-\frac{208}{203}$	1	0
1	$-\frac{6872}{1795}$	$-\frac{6223}{1795}$	$\frac{2260}{359}$

Verify that these contrasts are uncorrelated and partition the treatment sum of squares using these contrasts.

(d) Compare the partitions in (b) and (c) to that obtained for dataset RehabTime2 using the original Helmert contrasts. Decide on how you would explain the results to an experimenter, and decide which set of contrasts you believe would give the more meaningful conclusion.

1.14 More on contrasts.

    (a) Verify the covariance in (1.15)

    (b) Show that, in the usual oneway anova, two contrasts are uncorrelated if $\sum_i a_i b_i \sigma^2 / r_i = 0$.

1.15 The data for the experiment described in Exercise 1.16 can be found in dataset **FishTissueMass**. Using these data, complete the anova table that was started in the exercise. Note that the data are unbalanced, so orthogonal contrasts are not uncorrelated.

    (a) Complete the anova table using the orthogonal contrasts, which will not partition the sums of squares.

    (b) Complete the anova table using the first orthogonal contrast, then using uncorrelated contrasts, so the sums of squares will be partitioned.

    (c) Discuss the differences in interpretations of the analyses of parts (a) and (b)

1.16 Here we will look at different ways of obtaining a random assignment of treatments.

    (a) Referring to Example 1.1, also discussed in Section 1.5, detail a plan of assigning 15 pots to 3 fertilizer treatments, arranging the pots at random on a greenhouse bench. Use a random number generator to carry out one such assignment.

    (b) An experiment is conducted to test the effect of ozone on plants. The researcher assigned two environmental chambers to each of four levels of ozone (a total of eight chambers). Twelve plants were placed in each chamber. Explain how to randomly assign the chambers to the ozone treatments and the plants to the chambers.

    (c) An agronomist wanted to compare the effect of five different sources of nitrogen on the dry matter yield of barley used as a forage crop. Because he wanted the results to apply over a range of conditions, he decided to conduct the experiment on four types of soil. He located six plots on each of the four soil types, then randomly assigned the treatments to the plots within types. Explain how to carry out such a randomization.

1.17 Consider the following two experiments:

    (1) Treatment A, three varieties of alfalfa, is crossed with treatment B, three types of fertilizer. The response variable is dry weight.

    (2) Blood pressure of human subjects is measured. Classification A, consisting of three age classes, is crossed with classification B, consisting of three weight classes.

These experiments, and the resulting randomization, can be carried out in many ways. For each of these experiments, describe three ways to perform them, where each way would correspond to one of the field layouts in Table 1.4.

1.18 A biologist was interested in the effect of different colors of light on the growth of bacteria. She had 40 lights with filters to allocate to various treatments. An equal amount of bacteria was placed on each of 40 Petri dishes and the dishes were randomly allocated using the following treatments:

A: "White" or unfiltered light for 30 hours,

B: Blue light for 30 hours,

C: Green light for 30 hours,

D: Blue light for 15 hours and Green light for 15 hours.

The contrasts of interest are

A	B	C	D
White	Blue	Green	Blue/Green
0	1	1	−2
0	1	−1	0
3	−1	−1	−1

Although the original allocation put 10 dishes in each treatment, when changing the filters in treatment D two dishes were upset. The data are

	A	B	C	D
	White	Blue	Green	Blue/Green
r	10.0	10.0	10.0	8.0
$\bar{y}$	9.4	6.8	5.8	8.2

with MS(Within) = 0.725.

(a) Explain, in words, the meaning of each contrast.

(b) Show that these contrasts are orthogonal but that the estimated contrasts are not uncorrelated under the usual anova model.

(c) Show that the following set of contrasts are not orthogonal, but are uncorrelated under the usual anova model.

A	B	C	D
White	Blue	Green	Blue/Green
0	1	1	−2
0	1	−1	0
28	−10	−10	−8

(d) The contrasts in part (c) are constructed to be "close" to the original ones. Estimate both sets of contrasts and their standard errors.

(e) Discuss the differences in the estimates in part (d), and decide which set of contrasts you would advise the experimenter to use.

1.19 Refer to the experiment in Example 1.28. The data can be found in dataset Avocado.

(a) Analyze the data according to the first anova, using the "residual" term for the error.

(b) Analyze the data according to the second anova, using the correct error terms for each treatment test. Show that the sum of squares for "residual" decomposes as indicated in Example 1.28. Compare the conclusions from analyses (a) and (b).

(c) Analyze the data according to the second anova, but now pool the block interactions to get one error term. Compare the conclusions from analyses (a), (b) and (c).

(d) Write out an anova table (Source and df) for an experiment with the same number of crates, but with six shipments. Do the same for an experiment with three shipments and half the number of crates. Comment on these designs.

1.20 For the data of Example 1.22, assume that the experiment was done according to the first field layout, that is, a completely randomized design (the data are in dataset DataLayout).

(a) Reproduce the anova table in the example and carry out the F-tests.

(b) The fertilizer treatments were composed of nitrogen (N) and potassium (K) in the following factorial arrangement:

$$
\begin{array}{cc}
 & \text{Levels} \\
 & \text{of K} \\
 & \begin{array}{c|cc} & 0 & 60 \end{array}
\end{array}
$$

		0	60
Levels	60	x	x
of N	200	x	x

This suggests an anova decomposition

	df
Fertilizer	3
N	1
K	1
N × K	1
Hybrid	2
Fertilizer × Hybrid	6

Fill in the sums of squares for this table.

(c) The experimenter is particularly interested in the nitrogen response and wants the anova table

	df
Fertilizer	3
Between K	1
Within K	2
Between N in K=0	1
Between N in K=60	1
Hybrid	2
Fertilizer × Hybrid	6

Fill in the sums of squares in this table and explain how they are related to the ones in (b).

1.21 A microarray experiment is to be conducted, and there are only funds for eight chips. The RNA is to come from mice, and there are enough subjects available. The design will be of the form

Treatment	Control
□□	□□
□□	□□

where each square represents a microarray. There are three options to consider:
(1) Use eight mice, one for each chip.
(2) Use four mice, splitting the RNA of each mouse onto two chips.
(3) Use 16 mice, pooling the RNA of 2 mice on each chip.
(a) Referring to (1.13), for $\sigma_B^2 = 1$ and $\sigma_W^2 = 1/4$, for each of the three options, calculate the power to detect a difference in means[3] of 1 at $\alpha = .05$.
(b) Comment on the results of (a), and how you would advise an experimenter.
(c) Verify (1.13).

1.22 Researchers are developing a more virulent strain of a virus that infects cabbage-eating insects. They have three strains under investigation: a wild type, a more virulent type (HOB), and a crippled type (1A). An important question for the

---

[3] It is common in microarray experiments to look for "twofold differences", which is interpreted as $\mu_2/\mu_1 = 2$. It is also typical to transform the expression-level data using the $\log_2$ scale, and since $\log_2(2) = 1$, we have $\delta = 1$.

researcher is: Even though the HOB strain kills more rapidly when it infects an insect, is it equally infective?

As a first step in answering this question, we must examine the virus/insect biology. This is an occluded virus; this means that many virions (infective particles) are enclosed in a protein body (occlusion body). The insects ingest occlusion bodies, which then dissolve in their gut, liberating virions that actually cause the infection. Other work has shown that the various strains of the virus have similar infectivity as virions. The question is: Do the different strains of the virus have similar numbers of virions/occlusion body?

Data were collected from the three types of virus. For each one, counts of virions on two replicates slides were obtained. The data are

Wild Slide		1A Slide		HOB Slide	
I	II	I	II	I	II
51	38	42	34	46	41
48	39	47	41	76	49
27	23	53	40	48	105
51	28	30	37	34	53
47	24	47	31	83	53
39	31	58	42	62	31
51		57	36	64	29
29		41	52	59	

(a) Construct a model and an anova, testing for a difference in virions. (The count data should probably be transformed with a log or square root.)

(b) Using your estimate of treatment variance from the data, perform a power analysis to find out how many counts of virions/occlusion bodies must the researcher collect to detect a difference between the wild and HOB types. Fill in the following table.

		Power desired	
		.80	.90
Minimum difference	15%		
they wish to detect	10%		
as % of mean	5%		
of wild type	1%		

Accompaniment

1.23 Referring to Section 1.8.1:

(a) Verify the expression for the likelihood function (1.14).

(b) Show that the likelihood estimators of $\mu + \tau_i$ are $\bar{y}_{i\cdot}$.

(c) Show that if we restrict $\sum_i \tau_i = c$, we can write the likelihood estimator of $\mu + c$ as $\bar{y}$ and $\hat{\tau}_i = \bar{y}_{i\cdot} - \bar{y} + c$.

1.24 For the general partitioning of sums of squares into contrasts:

(a) Verify that A^* of (1.18) is idempotent if the set of contrasts satisfy (1.17), and hence the uncorrected treatment sum of squares can be written as a sum of uncorrelated contrasts.

(b) Verify that A^{**} of (1.19) is idempotent if the set of contrasts satisfy (1.17), and hence the corrected treatment sum of squares can be written as a sum of uncorrelated contrasts. You will need the condition that $a_i'1 = 0$, $i \geq 2$.

1.25 Often in dose-response studies researchers expect to find a threshold or minimum effective dose. Consider a set of k doses plus a control dose that are equally spaced on some scale, for example on a log dose scale. One approach to locating the threshold data is to use a set of contrasts among the dose levels. Here we will look at three sets of contrasts, closely following the approach taken by Ruberg (1989).

A direct approach uses *Helmert contrasts*, as first seen in Exercise 1.10. These can be constructed by the following rule: We get the contrast coefficient a_{pq} for the qth dose in the pth contrast by

$$a_{pq} = \begin{cases} -1 & \text{if } q < p \\ p & \text{if } q = p \\ 0 & \text{if } q > p, \end{cases}$$

where

$$q = 0, 1, \ldots, k - 1 \text{ and } p = 1, 2, \ldots, k.$$

For example,

Contrast (p)	Coefficient (q) 0	1	2	3
1	-1	1	0	0
2	-1	-1	2	0
3	-1	-1	-1	3

Another approach uses *step contrasts*. These are constructed by the rule that set s_{pq}, the coefficient for the qth dose in the pth contrast, is given by

$$s_{pq} = \begin{cases} p - k - 1 & \text{if } q < p \\ p & \text{if } q \geq p, \end{cases}$$

where

$$q = 0, 1, \ldots, k - 1 \text{ and } p = 1, 2, \ldots, k.$$

For example,

Contrast (p)	Coefficient (q) 0	1	2	3
1	-3	1	1	1
2	-2	-2	2	2
3	-1	-1	-1	3

The third approach we will consider uses *Basin contrasts*. These are constructed using the rule

$$b_{pq} = \begin{cases} -2(k - p + 1)/(k - p + 2) & \text{if } q < p, \\ b_{p(q-1)} + k + 1 & \text{if } q \geq p, \end{cases}$$

where

$$q = 0, 1, \ldots, k - 1 \text{ and } p = 1, 2, \ldots, k.$$

For example:,

	Coefficient (q)			
Contrast (p)	0	1	2	3
1	-6	-2	2	6
2	-3	-3	1	5
3	-1	-1	-1	3

(a) For each of the three types of contrast, describe, in words, the effects tested by the first and second contrast $(p = 1, 2)$ in each set.

(b) In the three examples given, if the cell sizes are equal, are the contrasts orthogonal?

(c) Consider the hypotheses

$$H_0 : \mu_0 = \mu_1 = \cdots = \mu_k,$$
$$H_1 : \mu_0 \leq \mu_i, \quad 1 = 1, 2, \ldots, k,$$
$$H_2 : \mu_0 = \mu_1 = \cdots = \mu_i < \mu_j, \quad j = i + 1, \ldots, k,$$
$$H_3 : \mu_0 \leq \mu_1 \leq \cdots \leq \mu_k.$$

Which pair of hypotheses are most appropriate for the threshold dose problem. Would you suggest a different hypothesis?

(d) Referring to the hypotheses in (c), which contrast or set of contrasts is most appropriate for testing

(i) H_0 vs. H_1, (ii) H_0 vs. H_2, (iii) H_0 vs. H_3.

(e) Here is the data in Table 3 of Ruberg (1989),

Dose(mg/kg/day)	Mean $\pm$ Std. Dev.
Control	6.20 ± 3.08
10	6.14 ± 2.32
20	6.54 ± 2.77
30	7.67 ± 2.32
40	9.37 ± 1.87

Construct all three sets of contrasts and determine the threshold dose using each type of contrast. Use $\alpha = .05$ for a one-tailed test.

(f) Ruberg also provides cutoff points to control the *experimentwise* error rate (see Section 2.9.1). Redo part (e) using these experimentwise cutoff points, and compare the conclusions to those in part (e).

1.26 Referring to Theorem 2.21

(a) Fill in the details to show that $SS(\text{Trt})/\sigma^2 \sim \chi^2_{t-1}$, and that the ratio of mean squares has an F-distribution.

(b) Suppose we have a twoway anova

$$Y_{ijk} = \mu + \alpha_i + \beta_j + (\alpha\beta)_{ij} + \varepsilon_{ijk}$$

with $i = 1, \ldots, a$, $j = 1, \ldots b$, $k = 1, \ldots, r$ and $\varepsilon_{ijk} \sim N(0, \sigma^2)$, independent. Show that $SS(A)/\sigma^2$ and $SS(B)/\sigma^2$ are independent χ^2 random variables, and verify their degrees of freedom.

(c) Show that SS(Within)$/\sigma^2$ is also χ^2, independent of the Sums of squares in part (b), and hence establish the validity of the usual F-test on main effects.

Note also that SS(A $\times$ B)$/\sigma^2$, where SS(A $\times$ B) is the interaction sum of squares, is also χ^2, independent of everything else. See Exercise 3.26.

1.8 Technical Notes

1.8.1 Estimability

Although it is typical, especially in design books, to talk about estimability in terms of unbiasedness, as we have done, this is quite a limiting view. The essential point about a nonidentifiable model is that there is no unique likelihood estimator. Thus, from (1.1), the likelihood function is

$$L(\mu, \tau | \mathbf{y}) = \prod_{i=1}^{t} \prod_{j=1}^{r} \frac{1}{\sqrt{2\pi\sigma^2}} e^{-\frac{1}{2\sigma^2}[y_{ij}-(\mu-\tau_i)]^2}$$

$$(1.14) \qquad = \left(\frac{1}{\sqrt{2\pi\sigma^2}}\right)^{tr} e^{-\frac{1}{2\sigma^2}\sum_{ij}(y_{ij}-\bar{y}_{i.})^2} e^{-\frac{r}{2\sigma^2}\sum_i[\bar{y}_{i.}-(\mu+\tau_i)]^2},$$

showing that if we translate $\mu \to \mu+\delta$ and $\tau_i \to \tau_i-\delta$, the likelihood function remains constant. This is why the overparameterized model is nonidentifiable. If we restrict $\sum_i \tau_i$ to any fixed value, identifiability is restored.

Also, we note that "estimability" really has nothing to do with unbiasedness. Whether a parameter has an unbiased estimator has no bearing on its estimability. The only thing that matters is that the likelihood function is not overparameterized. In fact, there are many cases in which biased estimators are preferred over unbiased estimators, as they could trade a small amount of bias for a large variance reduction. (For an introduction to such ideas, see Casella and Berger 2001, Section 7.3.)

1.8.2 Orthogonality and Covariance

Definition 1.11, which defines orthogonal contrasts, is nothing more than a statement about trigonometry; that the cosine of the angle between two vectors is zero, and so the angle is 90°. For two vectors

$$\mathbf{a} = (a_1, a_2, \ldots, a_p)' \text{ and } \mathbf{b} = (b_1, b_2, \ldots, b_p)',$$

the *cosine* of the angle between the vectors $\mathbf{a}$ and $\mathbf{b}$ is

$$\cos(\mathbf{a}, \mathbf{b}) = \frac{\sum_i a_i b_i}{\sqrt{\sum_i a_i^2 \sum_i b_i^2}},$$

so the definition of orthogonality is that the cosine is equal to zero, which makes the angle between the vectors 90°. Realize that this is a geometric property, not a statistical property.

In statistics, we want to partition variation, and contrasts are a means of doing this. However, it is very important to understand the difference between *orthogonal* and *uncorrelated*. To partition variation, it turns out that what is important is for contrasts to be uncorrelated. This can become tricky because, unlike orthogonality, which is an inherent geometric property, correlation is a property of a model. So a pair of contrasts may be uncorrelated in one model and correlated in another model, all of the time being orthogonal!

Note: Perhaps the easiest way to understand the distinction between orthogonal and uncorrelated is realize that we should only use orthogonal (or nonorthogonal) for contrasts in *parameters*, and we should only use correlated (or uncorrelated) for contrasts in *statistics*.

When we are in the case of a oneway anova, with independent cell means and both equal variance *and* equal cell sizes, then uncorrelated and orthogonal are the same (see Section 1.4).

As was stated in Definition 1.12, two contrasts $\sum_{i=1}^{t} a_i \bar{y}_i$ and $\sum_{i=1}^{t} b_i \bar{y}_i$ are orthogonal if their defining vectors are at $90°$, so $\sum_i a_i b_i = 0$. The same two contrasts are *uncorrelated* if $\text{Cov}(\sum_i a_i Y_i, \sum_i b_i Y_i) = 0$. We can calculate

$$(1.15) \quad \text{Cov}\left(\sum_i a_i \bar{Y}_i, \sum_i b_i \bar{Y}_i\right) = \sum_i a_i b_i \text{Var}(\bar{Y}_i) + 2\sum_{i>i'} a_i b_{i'} \text{Cov}(\bar{Y}_i, \bar{Y}_{i'}),$$

so the contrasts are uncorrelated if $\text{Cov}(\bar{Y}_i, \bar{Y}_{i'}) = 0$ and $\sum_i a_i b_i \text{Var}(\bar{Y}_i) = 0$. The former occurs if the means are uncorrelated, and the latter can occur if $\text{Var}(\bar{Y}_i) = \sigma^2/r_i$ and $\sum_i a_i b_i/r_i = 0$, which is the case in the oneway anova. Orthogonal contrasts are important because they give us a means of partitioning variation, in that we can break up sums of squares in the following way.[4] Suppose we have a full set of p orthogonal vectors $\mathbf{a}_1, \mathbf{a}_2, \ldots, \mathbf{a}_p$ where $\mathbf{a}_i = (a_{i1}, a_{i2}, \ldots, a_{ip})$ and

$$\mathbf{a}_i'\mathbf{a}_i = 1 \text{ and } \mathbf{a}_i'\mathbf{a}_{i'} = 0 \text{ for all } i \text{ and } i'.$$

Define the matrix $A = \sum_i \mathbf{a}_i \mathbf{a}_i'$, the sum of the outer products of the vectors $\mathbf{a}_i$. Then direct matrix multiplication will verify that

$$A^2 = \left(\sum_i \mathbf{a}_i \mathbf{a}_i'\right)^2 = \sum_i \sum_{i'} \mathbf{a}_i \mathbf{a}_i' \mathbf{a}_{i'} \mathbf{a}_{i'}' = \sum_i \mathbf{a}_i \mathbf{a}_i' = A$$

because $\mathbf{a}_i'\mathbf{a}_{i'} = 1$ if $i = i'$ and 0 if $i \neq i'$. Recall that a matrix A satisfying the condition $A^2 = A$ is called *idempotent*[5]. For this matrix A, however, there is more. Since the set of contrasts are all orthogonal, the matrix A has full rank. Moreover, by construction it is symmetric, and the only full rank symmetric idempotent matrix is the identity matrix. Thus, $A = I$.

If we have data $\mathbf{y} = (y_1, y_2, \ldots, y_p)$, then the total variation in the data (total uncorrected sum of squares) is

$$\sum_i y_i^2 = \mathbf{y}'\mathbf{y} = \mathbf{y}'A\mathbf{y}$$

---

[4] Unfortunately, to adequately deal with this, we need to resort to the use of matrix algebra.

[5] An idempotent matrix also satisfies $\text{rank}(A) = \text{trace}(A)$

$$(1.16) \qquad = \mathbf{y}' \left(\sum_i \mathbf{a}_i \mathbf{a}_i' \right) \mathbf{y} = \sum_i \mathbf{y}' \mathbf{a}_i \mathbf{a}_i' \mathbf{y}$$

$$= \sum_i \left(\sum_j a_{ij} y_j \right)^2 ,$$

showing that the total sum of squares can be broken down into sums of squares of orthogonal contrasts.

We typically use "corrected" sums of squares, that is, we center at the mean. In matrix notation, define $\mathbf{1} = (1, 1, \ldots, 1)$, then the mean of the elements of $\mathbf{y}$ can be written $\bar{y} = (1/p)\mathbf{1}'\mathbf{y}$, and we form our matrix A from the outer products of

$$\frac{1}{\sqrt{p}} \mathbf{1}, \mathbf{a}_2, \ldots \mathbf{a}_p,$$

where all of the contrasts $\mathbf{a}_i$ are orthogonal to $\mathbf{1}$ (since their components sum to zero). We then have

$$\sum_i y_i^2 = \frac{1}{p}(\mathbf{1}'\mathbf{y})^2 + \sum_{i=2}^p (\mathbf{a}_i'\mathbf{y})^2 = p\bar{y}^2 + \sum_{i=2}^p (\mathbf{a}_i'\mathbf{y})^2,$$

or

$$\sum_i (y_i - \bar{y})^2 = \sum_{i=2}^p (\mathbf{a}_i'\mathbf{y})^2.$$

Finally, if we are partitioning a treatment sum of squares, each of the y_i will be means. If they are all based on the same number of observations, then the above algebra does not change. However, if the means have differing numbers of observations, then orthogonal and uncorrelated become two different things. So suppose that we have a vector of means $\bar{\mathbf{y}} = (\bar{y}_1, \bar{y}_2, \ldots, \bar{y}_p)$, where $\bar{y}_i$ is based on r_i observations. The uncorrected total sum of squares is now

$$\sum_i r_i \bar{y}_i^2 = \bar{\mathbf{y}}' D \bar{\mathbf{y}}.$$

where D is a $p \times p$ diagonal matrix $D = \mathrm{diag}(r_1, r_2, \ldots, r_p)$. The presence of the matrix D complicates the construction of the idempotent matrix of contrasts, as in (1.16). What we want to do is write $\bar{\mathbf{y}}' D \bar{\mathbf{y}} = (D^{1/2}\bar{\mathbf{y}})'(D^{1/2}\bar{\mathbf{y}})$, where the elements of $D^{1/2}$ are the square roots of the elements of D. However, if we use the same matrix as in (1.16), the presence of D will stop it from being idempotent. This is why, in the unequal r_i case, we need the condition that $\sum_i a_i b_i / r_i = 0$.

We construct our idempotent matrix as follows. Our set of p orthogonal vectors $\mathbf{a}_1, \mathbf{a}_2, \ldots, \mathbf{a}_p$ now must satisfy

$$\sum_j a_{ij} a_{i'j} / r_j = \mathbf{a}_i' D^{-1} \mathbf{a}_{i'} = 0$$

(1.17) and

$$\sum_j a_{ij}^2 / r_j = 1 \text{ (this is just a matter of scaling)}.$$

The matrix

(1.18)
$$A^* = \sum_i D^{-1/2} \mathbf{a}_i \mathbf{a}_i' D^{-1/2}$$

then satisfies

$$A^{*2} = \sum_i \sum_{i'} D^{-1/2} \mathbf{a}_i \left[\mathbf{a}_i' D^{-1/2} D^{-1/2} \mathbf{a}_{i'} \right] \mathbf{a}_{i'}' D^{-1/2} = \sum_i D^{-1/2} \mathbf{a}_i \mathbf{a}_i' D^{-1/2} = A^*,$$

because the term in square brackets is equal to zero if $i \neq i'$ and equal to 1 otherwise. Thus A^* is a full rank idempotent matrix and is equal to the identity. We then have

$$\bar{\mathbf{y}}' D \bar{\mathbf{y}} = \bar{\mathbf{y}}' D^{1/2} A^* D^{1/2} \bar{\mathbf{y}} = \sum_i \bar{\mathbf{y}}' D^{1/2} D^{-1/2} \mathbf{a}_i \mathbf{a}_i' D^{-1/2} D^{1/2} \bar{\mathbf{y}} = \sum_i (\mathbf{a}_i' \bar{\mathbf{y}})^2,$$

and the total sum of squares has been partitioned into nonoverlapping pieces. The $\mathbf{a}_i$s are not orthogonal, but they result in linear combinations of $\bar{\mathbf{y}}$ that are uncorrelated and partition the sum of squares.

The corrected treatment sum of squares is $\sum_i r_i(\bar{y}_i - \bar{\bar{y}})^2$, where $\bar{\bar{y}} = \sum_i r_i \bar{y}_i / \sum_i r_i$. We can partition this by using the vector $\mathbf{1}$, in a variation of what we did in the equal r_i case. Start with the vector $D\mathbf{1}/\sqrt{\mathbf{1}'D\mathbf{1}}$, and for $i \geq 2$, choose the remaining $p - 1$ vectors $\mathbf{a}_i$ so that the entire set satisfies (1.17). Note that this requires that $\mathbf{a}_i'\mathbf{1} = 0$, $i \geq 2$. We now form the matrix A^{**} analogous to (1.18), obtaining

(1.19)
$$A^{**} = \frac{D^{1/2} \mathbf{1} \mathbf{1}' D^{1/2}}{\mathbf{1}' D \mathbf{1}} + \sum_{i=2}^{p} D^{-1/2} \mathbf{a}_i \mathbf{a}_i' D^{-1/2}.$$

which is again idempotent and equal to the identity, and gives the decomposition

(1.20)
$$\sum_i r_i \bar{y}_i^2 = N \bar{\bar{y}}^2 + \sum_{i=2}^{p} (\mathbf{a}_i' \bar{\mathbf{y}})^2.$$

See Exercise 1.24.

1.9 Miscellanea

1.9.1 Microarray Design I

In a microarray experiment[6] RNA from a subject is placed on a microarray (a glass slide) that contains genetic material, which we refer to as genes, from an organism. Depending on the organism, there can be over $54,000$ genes on a microarray. The RNA is "hybridized" to the microarray, and gene expression level is measured through fluorescence. The more the gene is expressed (that is, the more it reacts to the subject RNA) the higher the level of fluorescence. Thus, for one subject (experimental unit) we can have up to $54,000$ data points.

---

[6] My knowledge of molecular biology is at the kindergarten level. This section represents my understanding of these processes, and the statistical consequences. Do not be surprised if I confuse DNA and RNA (although I do know that RNA is reversed-transcribed into cDNA for the spotted arrays, whatever that means!)

Although this data flood presents a large problem for inference, as there must be many guards for multiple testing, there is really no added problem from a design point. In designing a microarray experiment, we should concentrate on getting it right for one gene. As the other 53, 999 data points are measured on subsamples of the experimental unit(!), they have no bearing on constructing a good design.

One thing that has large implications in the statistical design is the fact that there are distinct kinds of microarray platforms. There is the complementary DNA (cDNA) platform, which is a two-dye system, as well as other two-dye chips made by companies such as Agilent. There is also the oligonucleotide microarray, a single-dye system made by Affymetrix and Nimblegen. The Affymetrix microarray, or "chip" has become the standard in human experiments, while the other arrays are more popular in plant and animal studies because of their greater flexibility in the design of the chip (what genes to include).

As mentioned, the measurement from a microarray is a fluorescence, or a count of stimulated pixels. As it is typically right skewed, the measurement is almost always log transformed. There are also heterogeneity of variance problems, with measurements both at the low end and high end having decreased variance. These problems can be mitigated through transformations.

In the two-dye system, two RNA samples (typically from two different treatments or experimental units) are hybridized to the same microarray. Each sample is labelled with a fluorescent dye (either Cy3 or Cy5), which gives a color to its fluorescence (red or green). What is then measured is the relative fluorescence of the red/green RNA samples.

It is the relative florescence that is important, for another problem with microarray data is that the amount of RNA that is hybridized cannot be tightly controlled, and the more RNA the more florescence. There is much pre-processing of microarray data to address this problem, as well as the other problems mentioned above.

Note that the microarray, the chip, is in fact a block. The fact that, from a statistical design view, it is no different from a plot of land in the field makes our job easy. In fact, procedures that were developed for field plot data can be applied to microarray data without much change.

For a single gene, we can model a microarray experiment as

$$y_{ij} = \mu + \tau_i + \beta_j + \varepsilon_{ij},$$

where y_{ij} is the log expression level, τ_i is the treatment, and β_j is the microarray (block). This can now be analyzed as a randomized complete block design.

The two dye system poses other problems, and there are other concerns if we model more that one gene at a time. We will return to both of these questions. The companies mentioned above, as well as many others, market (somewhat pricey) software for microarray analysis. There is also some fine software that is freely available: Array Tools is available as a free download from NIH, and Bioconductor is an R package available from the R webpage.

This section drew on the work of Kerr and Churchill (2001ab), Schulze and Downward (2001), and Simon et al. (2003), which are a good start for further reading.

2

Completely Randomized Designs

And so it was ... borne in upon me that very often, when the most elaborate statistical refinements possible could increase the precision by only a few percent, yet a different design involving little or no additional experimental labour might increase the precision two-fold, or five-fold or even more

R. A. Fisher (1962)

If the idea looked lousy, I said it looked lousy. If it looked good, I said it looked good. Simple proposition.

Richard P. Feynman
Surely You're Joking, Mr. Feynman

2.1 Introduction

The terminology *Completely Randomized Design (CRD)* refers to the experiment design, not the data layout. In a CRD the treatments can be either crossed or nested, but the key feature is that the randomization must be carried out throughout the data layout. One important implication of this is that all factors in a CRD must be fixed factors – as a random factor is necessarily a restriction on randomization, there are no random factors in a CRD.

> CRDs have only fixed factors

A theoretical consequence of the fact that all factors in a CRD must be fixed factors is that we can study the theory for *all* CRD simply by looking at the oneway CRD. This follows because of the simple error structure of the CRD, and the fact that any effect can be built up through contrasts. However, this fact is only useful as a theoretical tool, say when we are trying to develop some distributional properties, as the data layout and the treatment structure always play an important practical role. But the oneway CRD is the place to start.

Most importantly, the randomization structure of the CRD implies that there is only one error term, the within error, and all effects are tested against it.

2.1.1 A Oneway CRD Model

Example 2.1. ONEWAY CRD An experiment was done to assess *in vitro digestibility* (IVD) of dried forage samples for alfalfa grown at different temperatures. The object is to determine if growing temperatures affected the feeding quality of alfalfa. There are four observations for each temperature, with data layout:

<table>
<tr><td></td><td colspan="4">Growing Temperatures
(Celsius)</td></tr>
<tr><td></td><td>17°</td><td>22°</td><td>27°</td><td>32°</td></tr>
<tr><td></td><td>x</td><td>x</td><td>x</td><td>x</td></tr>
<tr><td>IVD</td><td>x</td><td>x</td><td>x</td><td>x</td></tr>
<tr><td>Values</td><td>x</td><td>x</td><td>x</td><td>x</td></tr>
<tr><td></td><td>x</td><td>x</td><td>x</td><td>x</td></tr>
</table>

The data are in dataset IVD. This experiment was done as a oneway CRD, which means that treatments were assigned at random to the different units. In particular, the four IVD values measured at 17° were *not* taken from four plots exposed together to the same 17° temperature. They were four separate trials, with independent exposure to the 17° temperature. That is, a treatment is selected at random, an the experimental unit is then subjected to the treatment. If 22° is selected twice in a row, the temperature must be reset!   ‖

The test on treatments is

$$F = \frac{\text{MS(Treatments)}}{\text{MS(Within)}},$$

which, under H_0 has an F-distribution. The validity of this test can be justified with Cochran's Theorem (see the discussion following Theorem 2.20).

2.1.2 CRD and the Two-sample *t*-test

The oneway anova is a generalization of the two sample t-test, and by building on the theory of contrasts, gives us an effective way of comparing many means. If we had just two growing conditions, say 17° and 22°, we could then do a two sample t-test to check if the responses are significantly different. If we denote the means, variances, and cell sizes by

Growing Temperatures
(Celsius)

	17°	22°	27°	32°
	$\bar{y}_1$	$\bar{y}_2$	$\bar{y}_3$	$\bar{y}_4$
IVD	s_1^2	s_2^2	s_3^2	s_4^2
Values	n_1	n_2	n_3	n_4

the t-test is

$$t_{\text{calc}} = \frac{\bar{y}_1 - \bar{y}_2}{s\sqrt{\frac{1}{n_1} + \frac{1}{n_2}}} = 1.399, \quad p = .211,$$

where $s^2 = [(n_1 - 1)s_1^2 + (n_2 - 1)s_2^2]/[n_1 + n_2 - 2]$ is the pooled variance. We see that there is no significant difference in response between the 17°C and 22°C growing temperature. We compare this t-test to a oneway anova using only treatments 17°C and 22°C.

Source	df	Sum Sq	Mean Sq	F	p
Trt	1	0.211	0.211	1.957	0.211
"Residuals"	6	0.647	0.107		

The anova table is taken from R output. Like all computer packages, R calls the last line of the anova "residuals". However, for us it is very important to actually keep track of what this is: within? interaction? technical replication? Here, in the oneway CRD, we have "within" error.

Within error, sometimes called "pure" error, is very different from a "residual". As the name implies, a residual is something that is left over. In statistics, the residual is left over from the model fit. | A digression on "residual" error

For example, if a higher-order interaction term is used as an error term, it is left over from a model fit, and its validity as an error estimate depends on the validity of the model. In contrast, the validity of a within error as an error estimate is not dependent on the fit of the model. Regardless of the model, each experimental unit within the cell receives exactly the same treatment, and hence any difference between the observations is not dependent on the model, and only reflects the inability of the experiment to replicate itself.

Example 2.2. IVD ONEWAY CRD CONTINUED

Returning to the IVD anova, we know that the oneway anova on two treatments is identical to the two sample t-test. We can continue, and if we (for example) use the simple contrasts of Exercise 1.10, we could run three t-tests and find the results in Table 2.1.

Although in this example the results are, qualitatively, about the same, it is important to realize that the anova uses the pooled MSE for its estimate of error, so the contrast tests have 12 degrees of freedom rather than 6. In the anova we get to use an error estimate based on pooling the within error

Table 2.1. Two-sample t tests and anova contrast t tests corresponding to IVD simple contrasts

Test	df	Two-Sample t t_{calc}	p	df	Anova contrasts t_{calc}	p
17°C vs. 22°C	6	1.399	0.211	12	1.181	.260
27°C vs. 32°C	6	−.160	.878	12	−.181	.858
17°C + 22°C vs. 27°C + 32°C	6	−5.851	.001	12	−6.102	< .0001

from all four treatments, even if those treatments are not being tested in the contrast.

> Orthogonal contrasts do not guarantee independent tests

It is also important to remember that even though we are using orthogonal contrasts, which are independent in this case, the tests are *not* independent because they share the same denominator.

If we apply the formula for contrast sums of squares (Definition 1.14) to Table 2.1, we see that the contrast sums of squares will add to the anova treatment sum of squares. Moreover, the contrast t-statistics are a breakdown of the anova F-test in that (referring to Table 2.1)

$$\frac{(1.181)^2 + (-0.181)^2 + (-6.102)^2}{3} = 12.89$$

which is the F-statistic from the complete anova (Exercise 2.3).   ‖

2.1.3 CRD Anova

The oneway anova F-test is summarized in Table 2.2, which reflects the partitioning of the sums of squares according to (1.8)

The CRD design, of course, does not stop at the oneway. As we will see, the treatment design in a CRD can take many forms, the defining feature being that the randomization is unrestricted throughout the table, and there in no correlation between any two observations.

Example 2.3. RED CLOVER TWOWAY CRD   To investigate the effect of sulfur and nitrogen on the growth of red clover, a plant scientist conducted a greenhouse experiment using a CRD with the treatments in a crossed layout. The sulfur levels were applied at rates of 0, 3, 6, and 9 pounds/acre, and the rate of nitrogen application was either 0 or 20 pounds/acre. Greenhouse pots were prepared with uniform soil, allowing for $r = 3$ pots per treatment combination. The data are given in Table 2.3   ‖

Table 2.2. Oneway anova table.

Source of Variation	Degrees of Freedom	Sum of Squares	Mean Square	F-Statistic
Treatments	$t-1$	$\text{SS(Trt)} =$ $\sum_i r(\bar{y}_{i\cdot} - \bar{\bar{y}})^2$	$\text{MS(Trt)} =$ $\frac{\text{SS(Trt)}}{t-1}$	$F = \dfrac{\text{MS(Trt)}}{\text{MS(Within)}}$
Within	$t(r-1)$	$\text{SS(Within)} =$ $\sum_i \sum_j (y_{ij} - \bar{y}_{i\cdot})^2$	$\text{MS(Within)} =$ $\frac{\text{SS(Within)}}{t(r-1)}$	
Total	$rt-1$	$\text{SS(Total)} =$ $\sum \sum (y_{ij} - \bar{\bar{y}})^2$		

Table 2.3. Dry matter yields, in grams/pot, of red clover.

		Sulfur			
		0	3	6	9
		4.48	4.70	5.21	5.88
	0	4.52	4.65	5.23	5.98
		4.63	4.57	5.28	5.88
Nitrogen		– – – – – –			
		5.76	7.01	5.88	6.26
	20	5.72	7.11	5.82	6.26
		5.78	7.02	5.73	6.37

The partitioning of the sums of squares for the twoway anova is a straightforward generalization of the oneway partition, giving

$$\text{SS(Total)} = \text{SS(Treatment A)} + \text{SS(Treatment B)} + \text{SS(A} \times \text{B)} + \text{SS(Within)},$$

where, here, the interaction term measures the nonadditivity of the treatments, and the within serves as the error for all estimation and tests. The details of the partitioning are

(2.1)
$$\sum_{i=1}^{t}\sum_{j=1}^{g}\sum_{k=1}^{r}(y_{ijk} - \bar{\bar{y}})^2 = \sum_{i=1}^{t} rb(\bar{y}_{i\cdot\cdot} - \bar{\bar{y}})^2 + \sum_{j=1}^{g} ra(\bar{y}_{\cdot j\cdot} - \bar{\bar{y}})^2$$
$$+ \sum_{i=1}^{t}\sum_{j=1}^{g}(\bar{y}_{ij\cdot} - \bar{y}_{i\cdot\cdot} - \bar{y}_{\cdot j\cdot} + \bar{\bar{y}})^2$$
$$+ \sum_{i=1}^{t}\sum_{j=1}^{g}\sum_{k=1}^{r}(y_{ijk} - \bar{y}_{ij\cdot})^2,$$

which results in the anova table given in Table 2.4.

Table 2.4. Twoway anova table

Source of Variation	Degrees of Freedom	Sum of Squares	Mean Square	F-statistic
T	$t-1$	$\text{SS(T)} =$ $\sum_i rg(\overline{y}_{i..} - \overline{\overline{y}})^2$	$\text{MS(T)} =$ $\frac{\text{SS(T)}}{t-1}$	$F = \dfrac{\text{MS(T)}}{\text{MS(Within)}}$
G	$g-1$	$\text{SS(G)} =$ $\sum_j rg(\overline{y}_{.j.} - \overline{\overline{y}})^2$	$\text{MS(G)} =$ $\frac{\text{SS(G)}}{g-1}$	$F = \dfrac{\text{MS(G)}}{\text{MS(Within)}}$
T × G	$(t-1)(g-1)$	$\text{SS(T} \times \text{G)} =$ $\sum_{ij} r(\overline{y}_{ij.} - \overline{y}_{i..}$ $-\overline{y}_{.j.} + \overline{\overline{y}})^2$	$\text{MS(T} \times \text{G)} =$ $\frac{\text{SS(T} \times \text{G)}}{(t-1)(g-1)}$	$F = \dfrac{\text{MS(T} \times \text{G)}}{\text{MS(Within)}}$
Within	$tg(r-1)$	$\text{SS(Within)} =$ $\sum_{ijk} (y_{ijk} - \overline{y}_{ij.})^2$	$\text{MS(Within)} =$ $\frac{\text{SS(Within)}}{tg(r-1)}$	
Total	$tgr-1$	$\text{SS(Total)} =$ $\sum_{ijk} (y_{ijk} - \overline{\overline{y}})^2$		

Remember that the treatment design, which is a twoway factorial, is all that is needed to determine all of the anova table except the last column. The F-ratios can only be properly formed with knowledge of the experiment design, and those in Table 2.4 reflect the fact that we have a CRD.

> The treatment design gives the anova table

The error term in the CRD anova is a true "within" error. It comes from true replication of the experimental unit within the smallest cell of the experiment. All of the units in the cell are subjected to exactly (we hope!) the same treatment combination, so any differences in the response is model-independent true error. In the CRD all tests are done against this term.

There is often discussion about how to perform tests on main effects, T and G, in the presence of interaction, and how to interpret these effects. We will address this topic later in Section 2.4, when we look at expected mean squares.

Example 2.4. TWOWAY CRD CONTINUED The anova for the data of Example 2.3 is given in Table 2.5. The data can be found in dataset `RedClover`. All tests are done against the within MSE, and all terms are wildly significant, showing that there are effects due to Sulphur and Nitrogen, and they also interact. To understand what is going on we need some further analysis. We will return to this example in Section 2.5. ‖

2.2 Model and Distribution Assumptions

The IVD experiment (Example 2.1) is a oneway CRD with model (1.1), that is

Table 2.5. Twoway CRD anova for the red clover data.

Source	df	Sum Sq	Mean Sq	F-statistic	p
Sulfur	3	3.06	1.02	285.53	$< .00001$
Nitrogen	1	7.83	7.83	2185.63	$< .00001$
Sulphur $\times$ Nitrogen	3	3.76	41.25	349.78	$< .00001$
Within	16	0.06	0.0036		

(2.2) $$Y_{ij} = \mu + \tau_i + \varepsilon_{ij}, \quad i = 1, \ldots, t; \quad j = 1, \ldots, r,$$

where Y_{ij} is the observed response, μ is an overall mean, τ_i is the treatment effect, and ε_{ij} is the error. For identifiability we can assume that $\bar{\tau} = 0$; otherwise we can consider the treatment effect to be $\tau_i - \bar{\tau}$ and the overall mean to be $\mu + \bar{\tau}$ (see the discussion in Section 1.1). We also assume

(i) The random variables $\varepsilon_{ij} \sim \mathrm{N}(0, \sigma^2)$ for $i = 1, \ldots, t$, and $j = 1, \ldots, r$ (normal errors with equal variances).

(ii) $\mathrm{Corr}(\varepsilon_{ij}, \varepsilon_{i'j'}) = 0$.

It is somewhat more common to assume that the ε_{ij} are iid, from which (ii) would follow. Under normality, this is equivalent to the assumption of zero correlation. We think it is most important to emphasize correlation structure, however, and thus present the assumptions in this way.

The mean and variance of Y_{ij}s are

(2.3) $$\mathrm{E}(Y_{ij}) = \mu + \tau_i, \quad \mathrm{Var}\,(Y_{ij}) = \sigma^2.$$

Building on the oneway CRD, more complicated CRD experiments will tend to have a factorial structure for the treatment design. We will mainly emphasize twoway factorials in our development, for most of the theoretical development is straightforward for higher-order factorials (see Exercise 2.23 and Section 2.5.2).

A model for the twoway CRD is

(2.4) $$Y_{ijk} = \mu + \tau_i + \gamma_j + (\tau\gamma)_{ij} + \varepsilon_{ijk},$$
$$i = 1, \ldots, t; \quad j = 1, \ldots, g, \quad k = 1, \ldots, r,$$

where Y_{ijk} is the observed response, μ is an overall mean, τ_i is one treatment effect, and γ_j is the other treatment effect. The term $(\tau\gamma)_{ij}$ represents the interaction of the two factors, a deviation from an additive response. Finally, ε_{ijk} is the error.

We further assume

(i) The random variables $\varepsilon_{ijk} \sim \mathrm{N}(0, \sigma^2)$ for $i = 1, \ldots, t$, and $j = 1, \ldots, g, k = 1, \ldots, r$ (normal errors with equal variances).

(ii) $\mathrm{Corr}(\varepsilon_{ijk}, \varepsilon_{i'j'k'}) = 0$.

Note that without loss of generality we can assume

(2.5) $$\bar{\tau} = \bar{\gamma} = \overline{(\tau\gamma)} = 0,$$

for this is equivalent to redefining the mean level μ as $\mu + \bar{\tau} + \bar{\gamma} + \overline{(\tau\gamma)}$ and has no effect on interpretation of the parameters. However, for identifiability, it is necessary to go further, and it is typical to also assume that

(2.6) $$\overline{(\tau\gamma)}_{i\cdot} = \overline{(\tau\gamma)}_{\cdot j} = 0 \text{ for all } i, j$$

> Restrictions
> for
> identifiability

This is because the data from such an experiment will be in an $g \times t$ table, with the parameter estimates coming from the cell means. Losing 1 degree of freedom for the overall mean, that leaves $tg - 1$ degrees of freedom to estimate the parameters. Since we have $g + t + tg$ effect parameters in model (2.4), the restrictions (2.6) ensure identifiability.

To estimate the parameters in the CRD, we use least squares. We give some details for the twoway CRD, and leave other cases for exercises. Under model (2.4), least squares is quite straightforward in the CRD design. The least squares estimates satisfy

(2.7) $$\min_{\mu, \tau_i, \gamma_j, (\tau\gamma)_{ij}} \sum_{i=1}^{t} \sum_{j=1}^{g} \sum_{k=1}^{r} (y_{ijk} - \mu - \tau_i - \gamma_j - (\tau\gamma)_{ij})^2.$$

Under (2.5), the solution (see Exercise 2.7) is given by

(2.8)
$$\bar{\bar{y}} = \hat{\mu}$$
$$\bar{y}_{ij\cdot} - \bar{\bar{y}} = \hat{\tau}_i + \hat{\gamma}_j + \widehat{(\tau\gamma)}_{ij}$$
$$\bar{y}_{i\cdot\cdot} - \bar{\bar{y}} = \hat{\tau}_i + \widehat{(\tau\gamma)}_{i\cdot}$$
$$\bar{y}_{\cdot j\cdot} - \bar{\bar{y}} = \hat{\gamma}_j + \widehat{(\tau\gamma)}_{\cdot j}$$

where we see that, due to the identifiability constraint, the parameter estimate contains two pieces - for example - $\hat{\tau}_i + \widehat{(\tau\gamma)}_{i\cdot}$, where τ_i is often interpreted as a "main effect", an effect that is constant across the levels of γ, while $\overline{(\tau\gamma)}_{i\cdot}$ is an average effect. That is, the effect may be different in the levels of γ, and we are just estimating the average.

It is important to understand that we *cannot* separate these effects, main and average, without an additional *assumption*, such as $\overline{(\tau\gamma)}_{i\cdot} = 0$.

> **Note:** It is very important to understand the difference between the identifiability constraint (2.6) and the *assumption* that $\overline{(\tau\gamma)}_{i\cdot} = 0$.

The identifiability constraint gets us a set of parameters to estimate. It does *not* imply that $(\tau\gamma)_{i\cdot} = 0$, but rather has the effect of redefining the parameters. That is, the effect of (2.6) is equivalent to redefining the parameters as

$$\tau_i' = \tau_i + (\overline{\tau\gamma})_{i\cdot},$$
$$\gamma_j' = \gamma_j + (\overline{\tau\gamma})_{\cdot j},$$
$$(\tau\gamma)_{ij}' = (\tau\gamma)_{ij} - (\overline{\tau\gamma})_{i\cdot} - (\overline{\tau\gamma})_{\cdot j}.$$

Thus, the average interaction effect does not go away, it just relocates.

In our calculations we will assume that $\bar\tau = \bar\gamma = (\overline{\tau\gamma}) = 0$ or, equivalently, the overall mean is $\mu + \bar\tau + \bar\gamma + (\overline{\tau\gamma})$. Also, to keep notation manageable, we will also assume that $(\overline{\tau\gamma})_{i\cdot} = (\overline{\tau\gamma})_{\cdot j} = 0$. Thus, a treatment effect will be written as τ_i, but this should always be interpreted as $\tau_i + (\overline{\tau\gamma})_{i\cdot}$.

2.3 Expected Squares and F-tests

The calculation of expected values of mean squares (EMS) is an important part of any design, as it indicates the correct denominators for F-tests in the anova and the correct error terms for testing and estimating contrasts. Moreover, it helps us to see which replication helps control different sources of variation, and thus helps us in setting up a better design.

Although the direct calculation of EMS can be painful, it is important to carry it our carefully, and not rely on so-called "EMS algorithms", which can sometimes give incorrect results unless used very carefully. Moreover, after doing a few of the calculations, the procedure becomes fairly transparent and the actual calculations can go quite smoothly.

We will do the calculations for the twoway CRD of model (2.4), leaving the other CRDs for exercises (see Exercises 2.11 and 2.23). Calculation of the EMS is the first step in justifying the F-tests in Table 2.4.

Starting with the first treatment sum of squares, we have

$$\mathrm{E}(\mathrm{SS(T)}) = \mathrm{E}\sum_i rg(\overline{Y}_{i\cdot\cdot} - \overline{\overline{Y}})^2$$
$$= \mathrm{E}\sum_i rg\left([\mu + \tau_i + (\overline{\tau\gamma})_{i\cdot} + \varepsilon_{i\cdot\cdot}] - [\mu + \varepsilon_{\cdot\cdot\cdot}]\right)^2$$
$$= \sum_i rg\tau_i^2 + \mathrm{E}\sum_i rg\left(\bar\varepsilon_{i\cdot\cdot} - \bar\varepsilon_{\cdot\cdot\cdot}\right)^2,$$

where the cross term in zero in the last line. From Lemma 2.16 it follows that

$$(2.9) \qquad \mathrm{E}\sum_i rg\left(\bar\varepsilon_{i\cdot\cdot} - \bar\varepsilon_{\cdot\cdot\cdot}\right)^2 = rg\left(1 - \frac{1}{t}\right)\sum_i \mathrm{Var}(\bar\varepsilon_{i\cdot\cdot}) = (t-1)\sigma^2$$

and thus

$$E(SS(T)) = (t-1)\sigma^2 + rg \sum_i \tau_i^2.$$

Continuing in this fashion, we can produce Table 2.6 (see Exercises 2.22 and 2.20).

Table 2.6. Expected Mean Squares for twoway CRD anova.

Source	df	EMS
Treatment T	$t-1$	$\sigma^2 + \frac{rg}{t-1}\sum_i \tau_i^2$
Treatment G	$g-1$	$\sigma^2 + \frac{rt}{g-1}\sum_j \gamma_i^2$
T × G	$(t-1)(g-1)$	$\sigma^2 + \frac{r}{(t-1)(g-1)}\sum_{ij}(\tau\gamma)_{ij}^2$
Within	$tg(r-1)$	σ^2

The null hypothesis of no effect of treatment T is

(2.10) $H_0 : \tau_i = 0$ for all i

and is tested by

$$F_{t-1,tg(r-1)} = \frac{MS(\text{Treatment T})}{MS(\text{Within})}.$$

The other tests are formed similarly. Note that all tests are against the within error – a very simple situation. Justification of the tests comes through Cochran's Theorem, which we relegate to Technical Note 2.8.3.

There is sometimes discussion about the interpretation of "main effects" in the presence of interaction. That is, if there is interaction, which means that one treatment acts differently depending on the levels of the other treatment, then there is sometimes concern about the interpretation of the "treatment effect".

However, upon close examination of the parameterization of the model, there is really no complication here. Without restricting the parameters, the interaction test is of the hypothesis

$$H_0 : (\tau\gamma)_{ij} - (\bar{\tau\gamma})_{i.} - (\bar{\tau\gamma})_{.j} = 0 \text{ for all } i, j,$$

and whether this is true has no bearing on the sizes of $(\bar{\tau\gamma})_{i.}$ and $(\bar{\tau\gamma})_{.j}$. Furthermore, as we cannot separate τ_i and $(\bar{\tau\gamma})_{i.}$ without further assumptions (that are unverifiable), the treatment effect is "always" an average effect. The treatment hypotheses, such as (2.10), are concerned with the sizes of the average effects.

Interpreting treatment effects

2.4 Estimating Contrasts

Under model (2.4) we have already derived the least squares estimates of
the parameters. Since contrasts in these parameters are the typical focus of
inference, we now develop those inferences.

Point Estimates

As least squares estimates are unbiased, it follows immediately that, for example,

$$(2.11) \qquad\qquad \mathrm{E}\left(\bar{Y}_{i\cdot\cdot} - \bar{\bar{Y}}\right) = \tau_i,$$

and a contrast $\sum_i a_i \left(\bar{Y}_{i\cdot\cdot} - \bar{\bar{Y}}\right)$ is an unbiased estimator of $\sum_i a_i \tau_i$ with variance

$$(2.12) \quad \mathrm{Var}\left(\sum_i a_i (\bar{Y}_{i\cdot\cdot} - \bar{\bar{Y}})\right) = \mathrm{Var}\left(\sum_i a_i (\bar{\varepsilon}_{i\cdot\cdot} - \bar{\bar{\varepsilon}})\right) = \frac{\sigma^2}{rg} \sum_i a_i^2,$$

with analogous formulas for the other effects (see Exercise 2.8).

For the oneway CRD it follows immediately that

$$(2.13) \qquad \mathrm{E}(\bar{Y}_{i\cdot} - \bar{Y}) = \hat{\tau}_i, \quad \mathrm{Var}\left(\sum_i a_i \hat{\tau}_i\right) = \frac{\sigma^2}{r} \sum_i a_i^2,$$

see Exercise 2.6.

Note that in the estimation of contrasts, the term involving $\bar{\bar{Y}}$ cancelled.
Thus, the variance calculation only involved independent terms. If we had
estimated τ_i alone, with $\bar{Y}_{i\cdot} - \bar{Y}$, we would have to deal with a covariance
term (see Exercise 2.8).

Variance Estimates

The residuals from the least squares fit of the model (2.4) are

$$y_{ijk} - \hat{\mu} - \hat{\tau}_i - \hat{\gamma}_j - (\hat{\tau\gamma})_{ij} = y_{ijk} - \bar{y}_{ij\cdot} = \varepsilon_{ijk} - \bar{\varepsilon}_{ij\cdot},$$

which represents the within variance. The within sum of squares has expected
value

$$\mathrm{E}[\mathrm{SS(Within)}] = \mathrm{E}\left(\sum_{ijk}(\varepsilon_{ijk} - \bar{\varepsilon}_{ij\cdot})^2\right) = rtg\left(1 - \frac{1}{r}\right)\sigma^2,$$

making $\mathrm{MS(Within)} = \frac{1}{tgr(r-1)} \sum_{ijk}(y_{ijk} - y_{ij\cdot})^2$ an unbiased estimate of σ^2,
and, for example, we can estimate a contrast variance

$$(2.14) \qquad\qquad \hat{\mathrm{Var}}\left(\sum_i a_i \hat{\tau}_i\right) = \frac{\mathrm{MS(Within)}}{rg} \sum_i a_i^2.$$

Inference in the CRD ———————————————————

The main objects of interest in an anova are treatment contrasts, which are usually tested following an anova. Inference for contrasts can be based on multiple comparison procedures (Miscellanea 2.9.1), or on individual tests and intervals. In either case, the inferential distribution is built up from the model (2.4)

Under model (2.4)

$$Y_{ijk} \sim N\left(\mu + \tau_i + \gamma_j + (\tau\gamma)_{ij}, \sigma^2\right), \quad \text{Cov}(Y_{ijk}, Y_{i'j'k'}) = 0,$$

and using results from Sections 1.8 and 2.4 we can build up the distribution of any contrast. For example,

$$(2.15) \qquad \sum_i a_i Y_{i\cdot\cdot} \sim N\left(\sum_i a_i \tau_i, \frac{\sigma^2}{rg} \sum_i a_i^2\right)$$

and hence

$$\frac{\sum_i a_i Y_{i\cdot\cdot} - \sum_i a_i \tau_i}{\sqrt{\frac{\sigma^2}{rg} \sum_i a_i^2}} \sim N(0,1).$$

We now can apply Theorem 2.23 to replace σ^2 with $\hat{\sigma}^2 = \text{MS(Within)}/(tg(r-1))$ to get

$$\frac{\sum_i a_i Y_{i\cdot\cdot} - \sum_i a_i \tau_i}{\sqrt{\frac{\hat{\sigma}^2}{rg} \sum_i a_i^2}} \sim t_{tg(r-1)}.$$

2.5 Deeper into Factorials ———————————————————

In this section we look a little closer at the interpretations of interaction in factorial experiments. There is often the urge to get as much as possible out of an experiment and, in doing so, to include many different treatments. However, interpretations of treatment effects, in the face of many interactions, can be difficult. We look at a number of examples of types of interactions that can be expected, and try to better understand them through contrasts.

2.5.1 Investigating Interactions ———————————————————

The existence of an interaction means that the effect of one treatment is dependent on the levels of another. This makes interpretation more difficult, and also can cause problems if the experimenter is looking to control future responses by setting treatment levels.

One overall distinction in interactions is between *qualitative* and *quantitative* interactions. In the first case, although the treatment response differs according to the levels of another factor, the response is only changed in quantity. In the second case there is a change in quality, that is, the "better" treatment changes according to the level of another treatment.[1]

Example 2.5. FISH MICROARRAYS REVISITED   As an example, we revisit Example 1.16, where the experimenter measured gene expression level in fish tissue as a function of two treatments. Figure 2.1 shows a plot of the cell means, often called an "interaction plot". There is can be seen that there is a qualitative interaction - the lines cross - and the effect of hCG is different at the different levels of tissue.   ‖

Fig. 2.1. Interaction plot for log expression-level data from the fish tissue experiment. The solid line corresponds to the absence of hCG, and the dashed line to the presence of hCG.

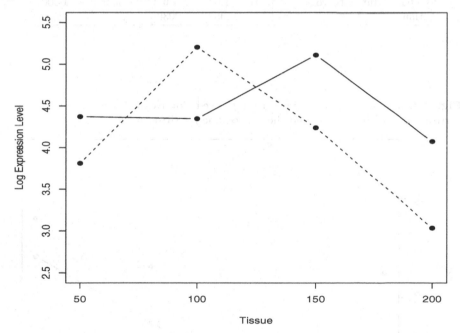

Example 2.6. RED CLOVER REVISITED   We next revisit another experiment seen in Examples 2.3 and 2.4. The interaction plot for the two treatments is shown in Figure 2.2, where there is a strong quantitative interaction.

---

[1] Procedures that test for qualitative interactions have been developed, especially in the biostatistics literature. See, for example, Gail and Simon (1985) or Piantidosi and Gail (1993).

Although the lines do not cross, there seems to be a strong elevation of yield at one combination of Sulfur and Nitrogen. The anova, summarized in Table 2.7 shows that all effects are significant.

The fact that all the polynomial trends are significant does not really tell us much - in terms of using this information we see that Nitrogen = 20 is preferred. Because everything is so significant we suspect that the peak at Sulfur = 3 is real, which seems to point to an optimal combination.   ‖

Table 2.7. Breakdown of interaction effect for the red clover data

Source	df	Sum Sq	Mean Sq	F-statistic	p
Sulfur × Nitrogen	3	3.76	1.25	349.78	< .00001
Linear Sulfur × Nitrogen	1	1.40	1.40	388.2	< .00001
Quadratic Sulfur × Nitrogen	1	0.72	0.72	199.33	< .00001
Cubic Sulfur × Nitrogen	1	1.64	1.64	456.95	< .00001
Within	16	0.057	0.0036		

Fig. 2.2. Interaction plot for yield from the red clover experiment. The solid line corresponds to Nitrogen=0, and the dashed line to Nitrogen=20.

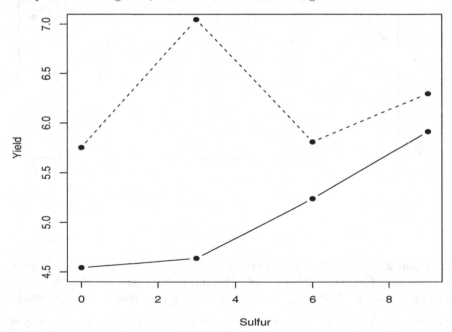

Before leaving the subject of interpretation of interactions, we look at one more example that arose in a QTL experiment[2], where the experimenter was investigating the interaction of two QTL, known in the genetics literature as *epistasis*.

Example 2.7. QTL EPISTASIS Data were collected in an experiment to examine the main effects and interactions of two QTL. The data look like

$$
\begin{array}{c}
\text{QTL1}
\end{array}
$$

	AA	Aa	aa
BB	x x	x x	x x
QTL2 Bb	x x	x x	x x
bb	x x	x x	x x

In QTL experiments there are a number of interaction patterns that are expected, and one researcher at the University of Florida wondered if, simply due to the interaction, could it happen that the mere presence of QTL2 in the experiment could increase the power to detect an effect due to QTL1. That is, for typical interaction patterns such as those shown in Figure 2.3, does increasing the difference in the means of QTL1 result in an increase in the means of QTL2?

To answer this question we examine the EMS of the effects, which are given in Table 2.6 for the twoway analysis. The EMS are important because they are the noncentrality parameters corresponding to the treatment hypothesis tests (Technical Note 2.8.4).

If $H_0 : \tau_i = 0$ for all i is true, then the usual ratio of mean squares has a central F distribution and, under the alternative, the distribution is noncentral F with noncentrality parameter $\sum_i \tau_i^2$. The F-statistic is stochastically increasing in its noncentrality parameter, which implies that the ratio of mean squares will tend to be large if this parameter is larger. Thus, the larger this value is, the more power we have to reject H_0 and detect QTL differences. Now we look more closely at the pattern of means in Figure 2.3, and show that by increasing the marginal mean difference in QTL1, we automatically increase the marginal mean difference in QTL2.

First, we show that when there is no interaction the values of one factor cannot influence the other, but when there is any interaction whatsoever, the levels of one factor can cause higher values in the other.

If there is no interaction, then $(\tau\gamma)_{ij} = 0$, so the margins control the cells, and when computing any marginal mean the effect of the other parameter disappears because of the restrictions (2.5). So in this case the effects are independent of one another, and no matter what happens in the means of one parameter, it can have no effect on the marginal means of the other parameter. This case is shown in the left panel of Figure 2.3.

---

[2] *Quantitative Trait Loci (QTL)* are regions on the genome that can be linked to quantitative traits. For example, one may find a certain region on the corn genome to be linked with yield.

Fig. 2.3. The left panel shows no interaction between two QTL, while the right panel shows a typical interaction pattern between two QTL.

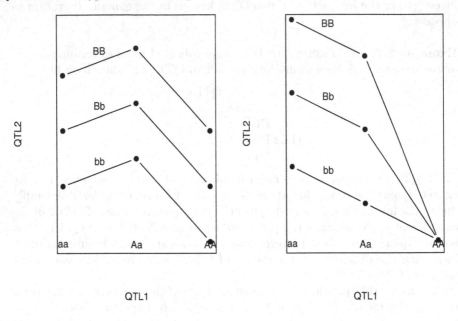

We now illustrate a pattern of cell means that look like the right panel. The cell entries are the cell mean parameters and the marginal entries are the marginal mean parameters. We use two free parameters, a and b, to arrive at the pattern in Table 2.8.

Table 2.8. Cell means that correspond to the pattern displayed in the right panel of Figure 2.3.

	QTL1			Average
	$a + \frac{4}{3}b$	$\frac{4}{3}b$	$-a - \frac{2}{3}b$	$\frac{2}{3}b$
QTL2	$a + \frac{1}{3}b$	$\frac{1}{3}b$	$-a - \frac{2}{3}b$	0
	$a - \frac{2}{3}b$	$-\frac{2}{3}b$	$-a - \frac{2}{3}b$	$-\frac{2}{3}b$
Average	$a + \frac{1}{3}b$	$\frac{1}{3}b$	$-a - \frac{2}{3}b$	0

Since Table 2.8 adds to zero for any choice of a and b, any values represent a legitimate set of QTL parameters.

Now the main point is that if the main effect of QTL2 increases, that is, if b increases, because of the interaction this has the effect of increasing the main effect of QTL1. That is, as b increases, the marginal means of QTL1 separate more, giving us more power to detect a difference in QTL1, solely due to the interaction with QTL2.

This is really an interesting occurrence

||

Thus, the presence of interaction can make one marginal factor seem significant simply due to the presence of the other factor!

2.5.2 Higher-Order Factorials

Optimizing resources in experimentation would suggest that putting more treatments in one experiment is a good thing – we can argue that:

(1) With many treatments in one experiment, we get simultaneous information. This allows better comparisons.
(2) With more treatments to explain variation, the error will be reduced.
(3) Running one experiment saves resources and time.

These are all excellent arguments, but in practice big factorials, except in certain situations, are not a good thing to do. What we saw in the previous section is that in the presence of interaction, effects can get muddled together and precise inferences become problematic. And this problem worsens when we move to higher-order factorial; threeway, fourway, ten-way! (Just take a look at the EMS from the threeway in Exercise 2.23, and the formal conclusions from the hypothesis tests.)

Having said this, we need to understand how to deal with higher-order factorials for at least two reasons: (i) Experimenters do them, and (ii) They are good for exploration.

Although many-way designs can be run with treatments having many levels, in such designs we run into the problem of interpretation of factors with many degrees of freedom. A significant threeway interaction with 18 degrees of freedom is difficult to make sense of - there are 18 contrasts to think about!

Higher order exploratory factorials are best run with treatments having two levels – the levels are typically chosen to span the range of possible treatment levels, representing a "high" and "low" setting. This gives the greatest chance of finding an effect. (See Section 6.3.) Also, since the treatment has only two levels and 1 degree of freedom, it is represented by one contrast, making interpretation easy. Moreover, all interactions between two-level treatments have 1 degree of freedom, making their interpretation easy

Another circumstance that sometimes arises in higher-order factorials is the lack of within error. Although sometimes the experimental conditions are replicated to get a within error, sometimes they are not. In such cases it is typical to use the higher-order interactions (threeway and higher) as error terms. The thinking is that these higher-order factors do not carry much main effect information and are mostly error anyway. Whether this is true, to use these are errors means that we assume that their contribution to the EMS is only through σ^2, and there is no contribution from the interaction. This leads to conservative tests, meaning that it is harder to reject H_0, so if we do reject we have some confidence that a true effect has been found.

Example 2.8. 2 × 2 FACTORIAL   A 2 × 2 factorial can be represented by the four cell means

Factor A

Factor B	θ_{11}	θ_{12}
	θ_{21}	θ_{22}

and each effect can be represented by a contrast

A effect B effect $A \times B$

1	−1
1	−1

1	1
−1	−1

1	−1
−1	1

where the contrasts merely account for the mean differences. Note, in particular, that the interaction contrast is the "difference of the differences".   ‖

Example 2.9. INTERACTION CONTRASTS   If factor A has four levels and factor B has three levels, then there are 6 interaction contrasts for $A \times B$, each with 12 coefficients. These can be obtained as all products of three orthogonal contrasts in A and two orthogonal contrasts in B:

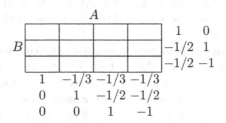

				1	0
B				−1/2	1
				−1/2	−1
1	−1/3	−1/3	−1/3		
0	1	−1/2	−1/2		
0	0	1	−1		

The six interaction contrasts are now obtained as the products of the coefficients, and the interpretation is a bit more straightforward. For example, the first B contrast compares the first level of B with the average of the other two, and the first A contrast compares the first level of A with the average of the other three. The product of these contrasts is an interaction contrast.

A

B	1	−1/3	−1/3	−1/3
	−1/2	1/6	1/6	1/6
	−1/2	1/6	1/6	1/6

This interaction contrast compares the B contrast applied to level 1 of A with the average B contrast in the other levels of A. In other words, it is as if we collapsed the above table into a 2×2 table

		A	
		Level 1	Average of Levels 2-4
B	Level 1		
	Average of Levels 2-3		

and did an interaction contrast here.

So we see that it is possible to interpret interactions using contrasts, but the task is not simple and the interpretations may not be meaningful. And we have only done twoway interactions – adding a third factor complicates things even more (see Exercise 2.15). ‖

Example 2.10. A SOCIAL SCIENCE EXPERIMENT   To examine the effect of different factors on a citizenship test performance, a factorial experiment was set up with the following factors:

(1) Education: Three levels (less than high school, high school, greater than high school),
(2) Home Environment: Four levels,
(3) Country of Origin: Seven countries.

The response variable was $Y =$ score on a citizenship aptitude test.

The twoway interaction of Education $\times$ Home Environment would be of interest, but the other twoway interactions are of less interest. Moreover, the threeway interaction, with 36 degrees of freedom, would be very difficult to interpret. ‖

Finally, we look at the classical higher-order factorial design (see Section 6.3 for another treatment of these designs), the 2^n design, where we have n factors, each at two levels. Things are actually quite simple here because everything has two levels – all effects are contrasts and all contrasts are easily obtainable as products.

It is common to specify the two levels of each factor, low and high, in the following way. The high level of A is denoted by a, and the low level is denoted by the absence of a. Thus the treatment combination ab has A and B at their high level and C at the low level, while ac has A and C high and B low. The treatment combination with all factors at their low level is denoted by (1).

Example 2.11. A 2^4 FACTORIAL   A 2^4 factorial design would have $2^4 = 16$ treatment combinations, and the full anova would specify 15 factors: 4 main effects, $\binom{4}{2} = 6$ twoway interactions, $\binom{4}{3} = 4$ threeway interactions, and 1 fourway interaction. Each effect has one degree of freedom, and all contrasts are obtainable as products. This makes for easy interpretation. A is the difference, AB is the difference of the differences, ABC is the difference of the AB differences, etc.

The following table shows some of the effects in the 2^4 design. The low level is -1 in the contrast and the high level is 1. The interaction contrasts are all products; for example the AD contrast is obtained by multiplying the A contrast by the D contrast, and the ABD contrast comes from $AB \times D$ (or $AD \times B$ or $A \times B \times D$).

Treatment	A	B	D	AD	ABD
(1)	-1	-1	-1	1	-1
d	-1	-1	1	-1	1
c	-1	-1	1	-1	1
⋮	⋮	⋮	⋮	⋮	⋮
abc	1	1	-1	-1	-1
abcd	1	1	1	1	1

Unless there was express interest in the threeway and higher interactions, it would be typical to pool these terms for an error estimate (Exercise 2.16)   ‖

2.6 Adjusting for Covariates

The analysis of covariance (ancova) in some sense has no business being in a design book, as it is more of an analysis tool than a design tool. However, its function is to reduce variance of the treatment means, so from that view it is a design strategy.

Here we look at a few examples of ancova, and do some variance calculations to understand how ancova works, and how it may benefit (on not!) a design.

Example 2.12. SOME ANCOVA EXAMPLES

(1) In agriculture, an experiment is done to examine the effect of treatments on yield. However, it is known that the density/plot of the plants could affect yield, and this variable is used as a covariate.
(2) In a nutrition experiment where the growth of laboratory rats is tracked, the initial weight of the rats influences their growth rate, and could be used as a covariate.
(3) In a microarray experiment with a spotted array, the florescence of a spot is related to the size of the spot, which could vary from gene to gene. Thus, spot size is a possible covariate.

‖

In each of the above examples the candidate covariate satisfies two conditions:

(1) The covariate is related to the response, and can account for variation in the response.
(2) The covariate is *not* related to the treatment.

> **Note:** We will see in (2.23) that the second condition is extremely important, and if it is not satisfied it can erase all the benefits of using a covariate.

A covariate functions somewhat like a block, in that it removes variation. However, it may not be a planned part of the experiment (it may be noticed after the design is set up) and it is typically continuous, so it really cannot be blocked over. If the covariate is recognized during the design phase, we could stratify over it, effectively using it to balance the observations by having a range of the covariate in each treatment.

Example 2.13. CORN YIELD COVARIANCE   When trying to compare yields of different varieties of corn, the response is confounded due to the fact that the number of plants in a plot may vary. Since the number of plants clearly influences yield, and should not be related to any treatment, it is an ideal covariate. The following data are the yield (Y) (dry weight) of four varieties of corn, along with a covariate (X) measuring the number of plants per plot.

	Cornell M-4		Robson 360		Ohio K-24		Ohio M-15	
Obs.	X	Y	X	Y	X	Y	X	Y
1	20	12.8	20	12.2	20	14.1	13	8.6
2	17	11.0	20	10.0	20	13.1	18	10.2
3	20	10.9	16	9.8	20	12.8	17	8.7
4	15	9.1	20	9.8	20	11.8	14	7.3
5	20	9.6	19	9.8	20	10.8	15	9.3
6	15	9.3	20	12.1	13	7.8	11	8.2

Examining the data suggests that, regardless of the variety, the yield is increased if the number of plants per plot is increased (Figure 2.4). ||

Here we will concentrate of the design effect of the covariate, in particular looking at the effect of the covariate on the variance of a treatment contrast, and attempting to document when the use of a covariate will be beneficial.

We will only look at models with one covariate, and stay with CRDs. The design effect in more complicated situations should be clear, and for details of more complex ancova models there are data analysis books that treat this. (A more advanced treatment can be found in Mead 1988; see also Federer and Meredith 1992.)

Recall the oneway anova model (2.2)

$$Y_{ij} = \mu + \tau_i + \varepsilon_{ij}, \quad i = 1, \dots t, \quad j = 1, \dots, r.$$

If there is a covariate, x_{ij}, we can use an ancova model

$$(2.16) \quad Y_{ij} = \mu + \tau_i + \beta(x_{ij} - \bar{x}) + \varepsilon_{ij}, \quad i = 1, \dots t, \quad j = 1, \dots, r,$$

where $\varepsilon_{ij} \sim N(0, \sigma^2)$, independent, and we add to the anova model a "regression" piece that adjusts for the covariate. Although it is not immediately clear how to estimate the treatment means in this model, we can appeal to least squares to obtain

$$(2.17) \quad \widehat{\mu + \tau_i} = \bar{y}_i - \hat{\beta}_1(\bar{x}_i - \bar{x}),$$

where

$$\hat{\beta}_1 = \frac{\sum_{ij}(x_{ij} - \bar{x}_{i.})(y_{ij} - \bar{y}_{i.})}{\sum_{ij}(x_{ij} - \bar{x}_{i.})^2}.$$

If we restrict $\sum \tau_i = 0$, then $\hat{\mu} = \bar{\bar{y}}$ and we can estimate the effects τ_i. However, this breaks down if the cell sizes are unequal (Exercise 2.29).

Fig. 2.4. Least squares line fit to the four varieties of corn in Example 2.13. Regardless of the treatment, there is a positive relationship between yield and number of plants per plot.

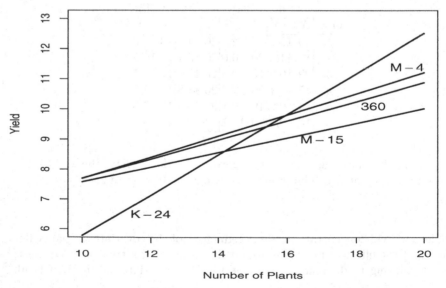

Note that the means are adjusted to by covariate mean with the assumption that, in each group, the slope is the same. This assumption is both crucial and bothersome. Without it, the mean comparisons could be very sensitive to where they are adjusted – look at Figure 2.4. There it seems that the equal slope assumption may be violated, but if we fit different slopes then, depending on where we adjust the means, they may be far apart or close together. It seems best to try to live with the equal slope assumption if covariance analysis is to be performed, unless there is a good reason to adjust each mean by a specific covariate value.

Since the least squares estimate is unbiased, we have $E(\widehat{\mu + \tau_i}) = \mu + \tau_i$ with variance

$$
\begin{aligned}
\text{Var}(\widehat{\mu + \tau_i}) &= \text{Var}(\bar{Y}_i - \hat{\beta}_1(\bar{x}_i - \bar{x})) \\
&= \text{Var}(\bar{Y}_i) + (\bar{x}_i - \bar{x})^2 \text{Var}(\hat{\beta}_1) \\
&= \sigma^2 \left(\frac{1}{r} + \frac{(\bar{x}_i - \bar{x})^2}{\sum_{ij}(x_{ij} - \bar{x}_{i\cdot})^2} \right),
\end{aligned}
$$

(2.18)

where we use the fact that $\bar{Y}_i$ and $\hat{\beta}_1$ are independent.

There are two ways we can *interpret* (2.16), neither of which is formally correct as an ancova model, but each of which lends some insight:

(1) If we write (2.16) as

$$
Y_{ij} - \mu - \tau_i = \beta(x_{ij} - \bar{x}) + \varepsilon_{ij},
$$

then the left side of the equation are the anova residuals, so we can think of ancova as doing a regression on the anova residuals. Note that in this interpretation we see that the variation due to regression will come out of the anova residual, and thus should decrease experimental error.

(2) If we write (2.16) as

$$
Y_{ij} - \beta(x_{ij} - \bar{x}) = \mu + \tau_i + \varepsilon_{ij},
$$

then we can think of ancova as an anova on the regression residuals. In this case the observations have been adjusted, and their variance has been decreased, so the anova error should be decreased.

Although both of these interpretations are useful, the ancova actually does things simultaneously, and fits the model

(2.19)
$$
Y_{ij} - \bar{Y}_{i\cdot} = \beta(x_{ij} - \bar{x}_{i\cdot}) + \varepsilon_{ij},
$$

where we adjust the means in each treatment. Notice that the total sum of squares for this regression is, in fact, the within sum of squares from the oneway anova, that is,

$$\text{SS(Regression Total)} = \sum_{ij}(y_{ij} - \bar{y}_{i\cdot})^2 = \text{SS(Within from anova)},$$

and thus the ancova adjustment does come out of the anova "residual".

Partitioning the Sums of Squares

Model (2.19) actually leads to two different partitioning of the sums of squares. Since the regression comes out of the anova within, the ancova table can be written

Source	df	SS
Treatments	$t-1$	$r\sum_i(\bar{y}_i - \bar{y})^2$
Within Treatments	$t(r-1)$	$\sum_{ij}(y_{ij} - \bar{y}_{i\cdot})^2$
Regression (Covariate after Trt)	1	$\hat{\beta}_1\sum_{ij}(x_{ij} - \bar{x}_{i\cdot})(y_{ij} - \bar{y}_{i\cdot})$
Residual	$t(r-1)-1$	$\sum_{ij}[(y_{ij} - \bar{y}_{i\cdot}) - \hat{\beta}_1(x_{ij} - \bar{x}_{i\cdot})]^2$

where $\hat{\beta}_1$ is given at (2.17). Here we have partitioned the sum of squares by first fitting the anova, and then pulling out the effect of the covariate from the regression, as in interpretation (1) above. This table allows us to do the usual anova test, and see if the regression is significant by forming the F-ratio MS(Regression)/MS(Residual).

Although this test may be interesting, it does not get at the heart of the ancova rationale, in that we want to see if the *adjusted yields* are significantly different. To do this, we first partition the sums of squares by doing the regression, and then removing the treatment variability from the residuals, as in interpretation (2) above. The resulting ancova table is

Source	df	SS
Regression	1	$\hat{\beta}_0\sum_{ij}(x_{ij} - \bar{x})(y_{ij} - \bar{y})$
Residual from Regression	$tr-2$	$\sum_{ij}[(y_{ij} - \bar{y}) - \hat{\beta}_0(x_{ij} - \bar{x})]^2$
Treatment (after Covariate)	1	SS(Residual from Regression) $-$SS(Residual)
Residual	$t(r-1)-1$	$\sum_{ij}[(y_{ij} - \hat{\beta}_1 x_{ij}) - (\bar{y}_{i\cdot} - \hat{\beta}_1\bar{x}_{i\cdot})]^2$

where

$$\hat{\beta}_0 = \frac{\sum_{ij}(x_{ij} - \bar{x})(y_{ij} - \bar{y})}{\sum_{ij}(x_{ij} - \bar{x})^2},$$

and it is easiest to get the adjusted treatment sum of squares by subtraction. The SS(Residual from Regression) becomes the total sum of squares for the adjusted treatment anova, and SS(Residual) is actually a within sum of squares of the adjusted data $y_{ij} - \hat{\beta}_1 x_{ij}$; we have written it this way in the second table. However, realize that the bottom-line residual is the same in both tables.

Example 2.14. CORN YIELD COVARIANCE CONTINUED The two anova tables for the corn data are

Covariate After Treatment					Treatment After Covariate			

Source	df	SS	MS
Varieties	3	27.955	9.318
Within	20	46.765	2.338
Plants (after Var.)	1	21.729	21.729
Residual	19	25.036	1.318

Source	df	SS	MS
Plants	1	43.916	43.916
Residual (from Reg.)	22	30.804	1.400
Varieties (after Plants)	3	5.768	1.923
Residual	19	25.036	1.318

where the test statistic for varieties, after adjusting for the covariate, is

$$F_{3,19} = \frac{1.923}{1.318} = 1.459,$$

which is not significant. ||

Testing Treatments _____

The ancova test on treatments, done in the above example, is a test of the adjusted treatments. Formally, we test the hypotheses

$$H_0 : Y_{ij} = \mu + \beta(x_{ij} - \bar{x}) + \varepsilon_{ij} \text{ vs. } H_1 : Y_{ij} = \mu + \tau_i + \beta(x_{ij} - \bar{x}) + \varepsilon_{ij},$$

where the null hypothesis specifies only a regression relationship, and the alternative specifies a treatment effect in addition to the regression. To test these hypotheses note that the H_0 model is a special case of the H_1 model (we also say that the H_0 model is nested in the H_1 model), and thus the H_1 model will always provide a better fit (since it has more parameters). To measure if this fit is better, we can use the F-ratio

$$(2.20) \quad F = \frac{[SS(\text{Residual from } H_0) - SS(\text{Residual from } H_1)]/(df_{H_0} - df_{H_1})}{SS(\text{Residual from } H_1)/df_{H_1}},$$

where df_{H_0} and df_{H_1} are the residual degrees of freedom under the respective models. This is the test done in Example 2.14. Specifically, from the above ancova tables we have

$$F_{t-1,\, t(r-1)-1}$$
$$= \frac{\left(\sum_{ij} [(y_{ij} - \bar{y}) - \hat{\beta}_0(x_{ij} - \bar{x})]^2 - \sum_{ij} [(y_{ij} - \bar{y}_{i\cdot}) - \hat{\beta}_1(x_{ij} - \bar{x}_{i\cdot})]^2 \right) / (t-1)}{\left(\sum_{ij} [(y_{ij} - \bar{y}_{i\cdot}) - \hat{\beta}_1(x_{ij} - \bar{x}_{i\cdot})]^2 \right) / (t(r-1)-1)}.$$

Fig. 2.5. Adjusted and unadjusted treatment means for the data of Example 2.13. The fitted lines are from model (2.16). The ancova adjusts the means to their covariate mean (left panel), and the anova (unadjusted) is equivalent to adjusting the means to the overall covariate mean (right panel).

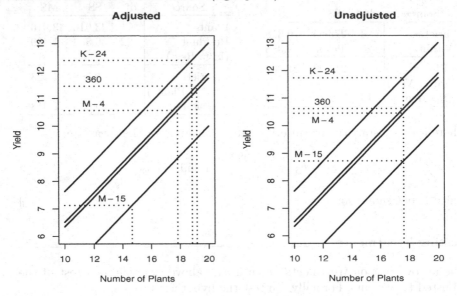

Note: The strategy to get the F-statistic in (2.20) is a general one for linear models.

If the H_0 model is nested in the H_1 model, then (2.20) provides a valid test of whether the richer model H_1 is a significant improvement.

Estimating Contrasts

Building on (2.18), the variance of a contrast is given by

$$\mathrm{Var}\left(\sum_i a_i(\widehat{\mu + \tau_i})\right) = \mathrm{Var}\left(\sum_i a_i \bar{Y}_i\right) + \mathrm{Var}(\hat{\beta}_1)\left(\sum_i a_i(\bar{x}_i - \bar{x})\right)^2$$

$$(2.21) \qquad = \frac{\sigma^2}{r}\sum_i a_i^2 + \frac{\sigma^2}{\sum_{ij}(x_{ij} - \bar{x}_{i\cdot})^2}\left(\sum_i a_i \bar{x}_i\right)^2,$$

where we use the independence of $\bar{Y}_i$ and $\hat{\beta}_1$ and the fact that $\sum_i a_i = 0$. We then estimate σ^2 with the MS(Residual) from the ancova.

Example 2.15. CONTRASTS FOR CORN YIELD EXPERIMENT Returning to Example 2.13, we calculate the estimated treatment means and their standard errors. Figure 2.5 illustrates the covariance adjustment of the means. There it can be seen that the adjustment increases the mean estimate as the number of plants increases, and results in a spreading out of the treatment means. The ancova also reduces the treatment variances, as the following table shows.

	Varieties			
	360	K-24	M-15	M-4
Unadjusted Mean	10.617	11.733	8.717	10.450
Std. Error	0.624	0.624	0.624	0.624
Adjusted Mean	11.447	12.384	7.124	10.562
Std. Error	0.496	0.486	0.563	0.469

Note, in particular, the adjustment for M-15, which had fewer plants per plot. The ancova tells us that the yield of M-15 is, in fact, lower than the others even after adjusting for the number of plants.

If we test the contrast $H_0 : \tau_{M-15} = \frac{1}{3}(\tau_{360} + \tau_{K-24} + \tau_{M-4})$, we find

$$\text{Unadjusted:}\quad t_{20} = -4.100,$$
$$\text{Adjusted:}\quad t_{19} = -6.659,$$

both wildly significant. Thus, even though the ancova gave greater separation, this differences was large enough so that the original anova would find it. See Exercise 2.13 for more. ∥

Variance of a Treatment Difference _____

We close this section with a deeper look at the variance of a treatment difference, which may add some insight as to when the ancova can be expected to improve things. From (2.21), the estimated variance of the difference of two adjusted means is

$$\text{Var}\left((\widehat{\mu + \tau_i}) - (\widehat{\mu + \tau_{i'}})\right) = \frac{2\hat{\sigma}^2}{r} + \frac{\hat{\sigma}^2}{\sum_{ij}(x_{ij} - \bar{x}_{i\cdot})^2}(\bar{x}_i - \bar{x}_{i'})^2,$$

where $\hat{\sigma}^2$ is the MS(Residual) from the ancova and we recognize the sum in the denominator as the within sum of squares from an anova of X on Treatments; we denote it by SS(Within$_x$). An idea that goes back to Finney (1946) is to replace this variance with a common one that can be used for all of the differences. To do this, we can replace the term $(\bar{x}_i - \bar{x}_{i'})^2$ with its average over all pairs.

First, it can be verified that for any numbers $b_1, b_2, \ldots, b_n$ we have

(2.22) $$\sum_{i=1}^{n}\sum_{i'=1}^{n}(b_i - b_{i'})^2 = 2n\sum_{i=1}^{n}(b_i - \bar{b})^2.$$

There are $t(t-1)$ nonzero terms in the $\bar{x}_i - \bar{x}_{i'}$ pairs, and applying (2.22) we obtain

$$\text{Var}\left((\widehat{\mu+\tau_i}) - (\widehat{\mu+\tau_{i'}})\right) \approx \frac{2\hat{\sigma}^2}{r} + \frac{\hat{\sigma}^2}{\sum_{ij}(x_{ij} - \bar{x}_{i\cdot})^2}\left[\frac{2}{t-1}\sum_i(\bar{x}_i - \bar{x})^2\right]$$

$$(2.23) \qquad = \frac{2\hat{\sigma}^2}{r}\left(1 + \frac{1}{t-1}\frac{\text{SS}(\text{Trt}_x)}{\text{SS}(\text{Within}_x)}\right),$$

where $\text{SS}(\text{Trt}_x)$ is the treatment sum of squares from the anova of X on Treatments. From (2.23) we see two things very clearly

(1) The variance of a treatment difference is reduced as the regression of Y on X in (2.19) improves, as $\hat{\sigma}^2$ will decrease.
(2) The variance of a treatment difference is *increased* if the covariate X is related to the treatment, that is, if $\text{SS}(\text{Trt}_x)/\text{SS}(\text{Within}_x)$ increases. This is a clear message that it is never advantageous to use a covariate that is related to the treatment.

2.7 Exercises

Essential

2.1 Referring to Example 2.1
 (a) For this experiment to be a CRD, explain how the data would need to be collected.
 (b) Complete the anova table:

Source	df	SS	MS	F
Temp	-	5.86	-	-
Within	-	-	-	
Total	-	7.67		

 (c) The experimenter is trying to decide which of two different temperature contrasts to test (i) $(-3, 1, 1, 1)$ or (ii) $(-3, -1, 1, 3)$. Give the experimenter an interpretation of each of these contrasts.
 (d) For contrast (i) of part (c), find two others that are orthogonal.
 (e) Calculate the t-statistic for testing H_0: $-3\tau_1 - \tau_2 - \tau_3 + 3\tau_4 = 0$, and find the proportion of variation in Temperature that is *not* explained by the contrast $(-3, -1, 1, 3)$.

2.2 Referring to Section 1.2 of Chapter 1:
 (a) Write $\bar{Y}_{i\cdot} - \bar{\bar{Y}}$ as a contrast in Y_{ij}.
 (b) Use the contrast formulas to find the mean and variance of $\bar{Y}_{i\cdot} - \bar{\bar{Y}}$. Check that they agree with those in Section 1.2
 (c) Write $Y_{ij} - \bar{Y}_{i\cdot}$ as a contrast in Y_{ij} and then verify (1.6) and (1.6).

2.3 Referring to the IVD data of Example 2.1:
 (a) Describe, in words, the conclusions that you would draw from the results of Table 2.1.
 (b) Produce the complete anova table for the IVD data.

      (*i*) Break down the treatment sum of squares into the contrast sum of squares corresponding to Table 2.1.

      (*ii*) Reproduce the *t*-statistics for the contrasts.

      (*iii*) Verify that the average squared *t*-statistic is the anova *F*-ratio.

    (c) Use another set of orthogonal contrasts, those that correspond to the linear, quadratic and cubic trend, to break down the treatment sum of squares. Test the significance of each contrast and describe the conclusions that can be made.

2.4 Here we look more closely at parameter restrictions.

    (a) Referring to model (2.4), show that the restriction $\bar{\tau} = \bar{\gamma} = \overline{(\tau\gamma)}_{i\cdot\cdot} = \overline{(\tau\gamma)}_{\cdot j\cdot} = 0$ results in exactly $tg - 1$ parameters to be estimated.

    (b) Extend the results of (a) to a threeway CRD with all interactions.

    (c) For a general factorial with T different treatments, each at t_i levels, $i = 1, \ldots T$, show that there are $\prod_i t_i - 1$ degrees of freedom to estimate effect parameters. Describe a restriction on the parameters that will result in identifiability, that is, that will reduce the parameters space to $\prod_i t_i - 1$ elements.

2.5 For the oneway model (2.2),

    (a) Show that under the assumption that $\bar{\tau} = 0$, which makes the τ_i estimable, the least squares estimates in the oneway CRD are given by

$$\hat{\tau}_i = \bar{y}_{i\cdot} - \bar{\bar{y}}, \quad \hat{\mu} = \bar{\bar{y}}.$$

    Derive the variances of $\hat{\tau}_i$ and $\hat{\tau}_i - \hat{\tau}_{i'}$.

    (b) For the following oneway CRD, provide the anova table and estimates of the parameter effects and their variances. To determine diet quality, male weanling rats were fed diets with various protein levels. Fifteen rats were randomly assigned to one of three diets, and their weight gain in grams was recorded. The data are (also in dataset **Protein**).

Diet Protein Level		
Low	Medium	High
3.89	8.54	20.39
3.87	9.32	24.22
3.26	8.76	30.91
2.70	9.30	22.78
3.82	10.45	26.33

2.6 Show the following for the oneway CRD (1.1):

    (a) $E(\bar{Y}_{i\cdot} - \bar{Y}) = \hat{\tau}_i$.

    (b) A contrast $\sum_i a_i \left(\bar{Y}_{i\cdot\cdot} - \bar{\bar{Y}}\right)$ is an unbiased estimator of $\sum_i a_i \tau_i$ with variance $\frac{\sigma^2}{r} \sum_i a_i^2$.

    (c) If there are r_i observations per treatment, then (a) and (b) hold with variance $\sigma^2 \sum_i a_i^2 / r_i$.

2.7 For the twoway CRD (2.4):

    (a) Verify the derivation of the least squares estimators and their variances. (*Hint:* First add $\pm y_{ij\cdot}$ and show that the estimates will only depend on $y_{ij\cdot\cdot}$. Then write $\tau_i + \gamma_j + (\tau\gamma)_{ij} = \gamma_{ij}$ and show that the least squares estimate of γ_{ij} is $y_{ij\cdot\cdot}$.)

(b) For the data of Example 2.4, found in dataset `RedClover`, estimate all of the treatment effects and their variances.

2.8 Calculation of variances in the twoway CRD.

(a) Show that the estimator $\bar{Y}_{i\cdot\cdot} - \bar{\bar{Y}}$ has variance

$$\text{Var}\left(\bar{Y}_{i\cdot\cdot} - \bar{\bar{Y}}\right) = \text{Var}\left(\varepsilon_{i\cdot\cdot} - \bar{\varepsilon}\right) = \left(1 - \frac{1}{t}\right)\frac{\sigma^2}{rg}.$$

Note that $\varepsilon_{i\cdot\cdot}$ and $\bar{\varepsilon}$ are correlated, which has to be accounted for in the variance calculation. Justify the application of Lemma 2.16 with $W_i = \varepsilon_{i\cdot\cdot}$.

(b) Verify (2.12) by showing

$$\sum_i a_i(\bar{Y}_{i\cdot\cdot} - \bar{\bar{Y}}) = \sum_i a_i(\varepsilon_{i\cdot\cdot} - \bar{\varepsilon}) = \sum_i a_i\varepsilon_{i\cdot\cdot}$$

and then use the fact that $\varepsilon_{i\cdot\cdot}$ are independent with variance σ^2/rg.

(c) Find the expectation and variance for contrasts in the other effect estimates in (2.8).

(d) Repeat (b) and (c) for the case of unequal cell sizes, where there are r_{ij} observations per cell. As a guide, first verify that

$$\text{Var}\left(\sum_i a_i(\bar{Y}_{i\cdot} - \bar{Y})\right) = \sigma^2 \sum_i \frac{a_i^2}{\sum_j r_{ij}}$$

2.9 For the following experiment

(*i*) specify the model equation

(*ii*) set up the anova table (source, df and EMS)

(*iii*) specify two hypotheses and how they would be tested

The cathode warm-up time in seconds was determined for three different tube types using eight observations on each tube type. The order of the experiment was completely randomized. The results were

	Tube Type		
	A	B	C
Warm-up	19 20	20 40	16 19
Time	23 20	20 24	15 17
(Seconds)	26 18	32 22	18 19
	18 35	27 18	26 18

2.10 (Finding an Optimal Region)[3]

If we are looking for maximum yield it would be wasteful to examine regions of low yield. Typically, the main features of such regions are *first-order*, that is, they are main effects rather than interactions. Thus, in a first search for such regions, we would be willing to use interaction terms as a denominator. Moreover, we consider this the first of a series of experiments; after we approximately locate a region of optimal response we will continue with a more precise experiment.

---

[3] Methods for finding optimal regions are called *Response Surface* methods. Although we are not treating these methods in detail, they are an important topic. The textbooks of Mead (1988) and Dean and Voss (1999) contain chapter length treatments; a thorough introduction is the book by Khuri and Cornell (1996).

As a particular example, an experimenter is rearing beneficial insects and is trying to find the conditions that produce insects most rapidly. The researcher has four large environmental chambers that can be used to control temperature, photoperiod, and humidity. There are other factors, for example, diet, but these will not be considered at this time.

We consider four possible designs, where photoperiod is hours of light and dark in a 24-hour period; for example, 16:8 is 16 hours of light and 8 hours of dark. The designs are given below.

(1)	Chamber	Temperature (°C)	Relative Humidity %	Photoperiod
	1	20	60	16:8
One factor	2	25	60	16:8
at a time	3	30	60	16:8
	4	35	60	16:8

(2)	Chamber	Temperature (°C)	Relative Humidity %	Photoperiod
	1	20	60	16:8
One factor	2	20	60	16:8
at a time	3	25	60	16:8
	4	25	60	16:8

(3)	Chamber	Temperature (°C)	Relative Humidity %	Photoperiod
	1	20	60	16:8
Factorial,	2	20	60	14:10
ignore	3	25	60	16:8
humidity	4	25	60	14:10

(4)	Chamber	Temperature (°C)	Relative Humidity %	Photoperiod
	1	20	60	14:10
Fraction	2	25	60	16:8
	3	20	80	14:10
	4	25	80	16:8

(Experiment (4) is an example of a *fractional factorial.* See Section 6.3.)

(a) Construct the anova table (just source and df) for each design.

(b) Recall that the major goal of these experiments is to locate the condition for optimal growth. List at least one advantage and one disadvantage of each design.

(c) Which design would you use for locating the region with the fastest insect growth?

2.11 Here we calculate expected mean squares to justify the F-test in Table 2.2.

(a) Show that

$$E(SS(Trts)) = rE \sum_{i=1}^{t} [\bar{y}_{i\cdot} - \bar{y}]^2 = rE \sum_{i=1}^{t} [(\tau_i - \bar{\tau}) + (\bar{\varepsilon}_{i\cdot} - \bar{\varepsilon})]^2$$

$$= r \sum_{i=1}^{t} \tau_i^2 + r \sum_{i=1}^{t} E(\bar{\varepsilon}_{i \cdot} - \bar{\varepsilon})^2 \quad \text{[cross term is zero]}$$

$$= r \sum_{i=1}^{t} \tau_i^2 + (t-1)\sigma^2 \quad \text{[Lemma 2.16]}.$$

(b) Calculate the other EMS and produce the anova table

Source	df	EMS
Treatments	$t-1$	$\sigma^2 + \frac{r}{t-1}\sum_i \tau_i^2$
Within	$r(t-1)$	σ^2

(c) For model (2.2), suppose $j = 1, \ldots r_i$, so there are unequal numbers of experimental units per treatment. Produce the anova table analogous to that in part (b).

2.12 Kuehl (1994) reported data on weight gain (mg) of shrimp cultured in aquaria, subjected to different levels of temperature (T), density of shrimp populations (D), and water salinity (S). The factors were crossed, and the experiment was run as a threeway CRD. Here is a schematic of the data, given in dataset Shrimp.

	$T = 25°C$				$T = 35°C$	
	D				D	
	80	160			80	160
10%	x, x, x	x, x, x		10%	x, x, x	x, x, x
S 25%	x, x, x	x, x, x		S 25%	x, x, x	x, x, x
40%	x, x, x	x, x, x		40%	x, x, x	x, x, x

(a) Produce the anova table and do the appropriate F-tests.
(b) You should find that, at the .05 level, the $T \times D$ and $S \times D$ interactions are not significant. Provide interpretations of this.
(c) You should find that, at the .05 level, the $T \times S$ and $T \times S \times D$ interactions are significant. Provide interpretations of this.
(d) Explore the interactions in part (c) further. Look at interaction with the linear and quadratic contrasts of S. Can you refine your conclusions from part (c)?
(e) Suppose you were asked to design a second experiment to further explore the treatments that result in large weight gain. Based on what you have learned in parts (a)-(d), what would you suggest?

2.13 Referring to Example 2.5, look further into the interaction term to see if there is any significant effects.
(a) Get the contrast coefficients for the three orthogonal (in the parameters) interaction contrasts for linear, quadratic and cubic tissue $\times$ hCG.
(b) Calculate the contrast sums of squares for each contrast, and perform the anova F-tests. Note that the estimated contrasts are correlated, and the sums of squares do not add to that of the 3 df interaction.
(c) Perform the same linear, quadratic, and cubic breakdowns for the main effect of tissue.

(d) What can you conclude about the effects and interactions?

2.14 Referring to Example 2.7:

(a) Produce a table like Table 2.7 for the left panel of Figure 2.3, where there is no interaction. Show that the marginal effect parameters act independently.

(b) The following figure has two more possible QTL interactions patterns.

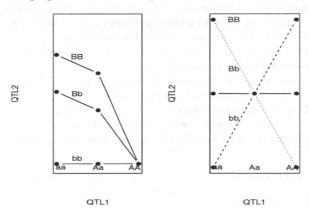

Show that, by constructing tables similar to Table 2.8, the left panel results in the marginal means of one QTL affecting the other, while for the second panel (where there is interaction) the marginal QTLs act independently (in fact, one has no effect).

2.15 Referring to Example 2.9:

(a) Fill in the table with the six interaction contrasts for $A \times B$.

(b) Make a similar table using polynomial contrasts.

(c) Verify, using the cell means, that the contrast from the collapsed 2×2 table is exactly the same as the contrast with all twelve cells.

(d) Suppose that a third factor C, with three levels, was crossed with A and B. Obtain the coefficients of the $A \times B \times C$ interaction contrast that compares the first level of B with the average of the other two, the first level of A with the average of the other three, and a linear trend in C. Also, give the collapsed table, and attempt to interpret the contrast in words.

2.16 Referring to Example 2.11:

(a) Fill in the table with all of the effect contrasts.

(b) Give the anova table, source, df, and EMS, where we pool the threeway and fourway interactions to get an error estimate. Specify the assumptions needed for this pooling.

2.17 Referring to Example 2.13:

(a) Verify the analysis of covariance, the adjusted treatment means and their standard errors.

(b) Test the significance of the pairwise differences, using both adjusted and unadjusted means. Do the conclusions change from the anova to the ancova?

(c) Redo part (b) using the average standard error of the difference give in (2.23). How do the results compare to part (b)? Based on your assessment of $SS(\text{Trt}_x)$ was the ancova worthwhile?

2.18 An experiment described in Snedecor and Cochran (1989) (same Cochran[4]) concerned treatment of leprosy in the Philippines. On each patient, six sites were examined for leprosy bacilli, with the abundance of the bacilli being measured. The covariate, X, is a score representing the abundance before treatment, and the response Y is the same score after several months of treatment with antibiotics A and B or a control C. The data are in dataset `Antibiotic`.

(a) Run both an anova on Y and an ancova with covariate X. Explain the results with respect to the significance of the treatments.

(b) The treatment design suggests testing the two orthogonal contrasts $A + B - 2C$ and $A - B$. Do this both for the adjusted and unadjusted means. Describe your conclusions.

(c) Is the initial measurement X a good covariate? Explain. You may want to look at the anova of X on treatment.

(d) Plot the individual regression lines for each drug, along with the overall regression (from the covariance analysis) and the unadjusted means. Explain the plot in terms of the results of the analysis.

Accompaniment

2.19 *A Useful Identity:*

(a) Show that for numbers x_{ij},

$$\sum_{i=1}^{t}\sum_{j=1}^{t}(x_{ij} - \bar{x}_i - \bar{x}_j + \bar{x})^2 = \sum_{ij} x_{ij}^2 - b\sum_{i}\bar{x}_i^2 - t\sum_{j}\bar{x}_j^2 + bt\bar{x}^2.$$

(b) If X_{ij} are random variables with $EX_{ij} = 0$, show that

$$E\left[\sum_{i=1}^{t}\sum_{j=1}^{t}(X_{ij} - \bar{X}_i - \bar{X}_j + \bar{X})^2\right]$$
$$= \sum_{ij}\mathrm{Var}X_{ij} - b\sum_{i}\mathrm{Var}\bar{X}_i - t\sum_{j}\mathrm{Var}\bar{X}_j + bt\mathrm{Var}\bar{X}.$$

2.20 (a) Show how to apply Lemma 2.16 to establish (2.9).

(b) Finish the calculations to fill in Table 2.6. To calculate the EMS for the interaction term you might first want to establish that

$$\bar{\varepsilon}_{ij\cdot} - \bar{\varepsilon}_{i\cdot\cdot} - \bar{\varepsilon}_{\cdot j\cdot} + \bar{\varepsilon}_{\cdots} \sim N\left(0, \frac{(t-1)(g-1)}{r}\sigma^2\right).$$

2.21 (Some anova theory: the t-F relationship)

(a) Referring to Example 2.2, there it was demonstrated numerically that the oneway CRD anova F-statistic is equal to the average squared t-statistics from uncorrelated contrasts. Show, analytically, that this is true for any set of orthogonal contrasts.

---

[4] This is really a famous pair. You know some of what Cochran did. Among the accomplishments of Snedecor is the derivation of the anova F-statistic, which has sometimes been called "Snedecor's F". Snedecor named it "F" in honor of Fisher.

(b) Another way to derive the anova F-statistics is as a *maximum* of t-statistics. In a oneway CRD with t treatments and n_i observations per treatment, we test contrast hypotheses by

$$\text{reject } H_0^{(a)} : \sum_i a_i \mu_i = 0 \text{ if } |t^{(a)}| = \frac{\left|\sum_i a_i \bar{y}_i\right|}{\sqrt{\text{MSE} \sum_i a_i^2/n_i}} > t_{\alpha/2}.$$

(i) The *union-intersection* principle (see Casella and Berger 2001, Sections 8.3.3 and 11.2) tells us that $H_0 : \mu_1 = \mu_2 = \cdots = \mu_t$ is true if and only if $H_0^{(a)}$ is true for *every* set of contrasts $(a_1, a_2, \ldots, a_k)$. Prove this. (Thus, H_0 is the intersection of $H_0^{(a)}$.)

(ii) Building on (i), show that H_0 will be rejected if at least one $H_0^{(a)}$ is rejected. (Thus, the rejection region of H_0 is a union of the individual rejection regions.)

(iii) Finally, argue that H_0 will be rejected if and only if $\max_a |t^{(a)}|$ exceeds its critical point, and establish that

$$\max_{(a_1,a_2,\ldots,a_k):\sum_i a_i=0} \frac{\left(\sum_i a_i \bar{y}_i\right)^2}{\sum_i a_i^2/n_i} = \sum_i n_i (\bar{y}_i - \bar{\bar{y}})^2,$$

showing that the maximum of the squared t-statistics is the anova F. (This is a difficult maximization.)

2.22 Complete the calculations to produce the EMS in Table 2.6. In particular, apply the identity in Exercise 2.19 to show

$$E \sum_{ij} (\bar{Y}_{ij} - \bar{Y}_i - \bar{Y}_j + \bar{\bar{Y}})^2 = bt \left[\text{Var}\bar{\varepsilon}_{ij} - \text{Var}\bar{\varepsilon}_i - \text{Var}\bar{\varepsilon}_j + \text{Var}\bar{\varepsilon} \right].$$

2.23 For a threeway CRD, all factors fixed, a model is

$$y_{ijkl} = \mu + \alpha_i + \delta_j + \gamma_k + (\alpha\delta)_{ij} + (\alpha\gamma)_{ik} + (\delta\gamma)_{jk} + (\alpha\delta\gamma)_{ijk} + \varepsilon_{ijkl},$$

where $i = 1, \ldots, t, j = 1, \ldots, b$, $k = 1, \ldots, c$, $l = 1, \ldots, r$, and $\varepsilon_{ijkl} \sim N(0, \sigma^2)$, independent, and we have the identifiability restrictions

$$\bar{\alpha} = \bar{\delta} = \bar{\gamma} = (\bar{\alpha\delta}) = (\bar{\alpha\gamma}) = (\bar{\delta\gamma}) = (\bar{\alpha\delta\gamma}) = 0,$$

$$(\bar{\alpha\delta})_i = (\bar{\alpha\delta})_j = (\bar{\alpha\gamma})_i = (\bar{\alpha\gamma})_k = (\bar{\delta\gamma})_j = (\bar{\delta\gamma})_k = 0,$$

$$(\bar{\alpha\delta\gamma})_i = (\bar{\alpha\delta\gamma})_j = (\bar{\alpha\delta\gamma})_k = (\bar{\alpha\delta\gamma})_{ij} = (\bar{\alpha\delta\gamma})_{ik} = (\bar{\alpha\delta\gamma})_{jk} = 0.$$

(a) Verify that the EMS for factor A is given by

$$E(MS(A)) = \sigma^2 + \frac{1}{t-1} \sum_{i=1}^{t} \alpha_i^2.$$

(b) Verify that the EMS for the $A \times B$ interaction is given by

$$E(MS(A \times B)) = \sigma^2 + \frac{1}{(t-1)(b-1)} \sum_{i=1}^{t} \sum_{j=1}^{b} (\bar{\alpha\delta})_{ij}^2.$$

(c) Based on what you learned in (a) and (b) fill in the complete threeway anova table (Source, df, EMS). (You do not need to do any more calculations – you should be able to deduce what the other EMS will be.)

(d) An experimenter will typically state a hypothesis in words. For the two hypothesis

$$H_0 : \text{No B effect and } H_0 : \text{No BC effect}$$

state each hypothesis in terms of the appropriate parameter values.

2.24 Refer to Miscellanea 2.9.2. If the treatment is applied, we assume that either $\delta_{ij} \sim$ iid $N(0, \sigma_\delta^2)$, or $\delta_i \sim$ iid $N(0, \sigma_\delta^2)$, in both cases independent of $\varepsilon_{ij} \sim N(0, \sigma_\varepsilon^2)$.

(a) Show that, if we have the case where $\delta_i \sim$ iidN$(0, \sigma_\delta^2)$, then $\text{Cov}(Y_{ij}, Y_{ij'}) = \sigma_\delta^2$, and in the other cases the Y_{ij} are all independent.

(b) Show that the following table summarizes the EMS for the case of modeling the error due to applying the treatment, and that if $\delta_{ij} = \delta_i$ there is no test on treatments.

EMS for Attributes

Source	df	EMS $\delta_{ij} = 0$	EMS $\delta_i \sim$ iid $N(0, \sigma_\delta^2)$	EMS $\delta_{ij} \sim$ iid $N(0, \sigma_\delta^2)$
Treatments	$t-1$	$\sigma_\varepsilon^2 + \frac{r}{t-1}\sum_i \tau_i^2$	$\sigma_\varepsilon^2 + r\sigma_\delta^2 + \frac{r}{t-1}\sum_i \tau_i^2$	$\sigma_\varepsilon^2 + \sigma_\delta^2 + \frac{r}{t-1}\sum_i \tau_i^2$
Within	$t(r-1)$	σ_ε^2	σ_ε^2	$\sigma_\varepsilon^2 + \sigma_\delta^2$

(This table once again illustrates the importance of true replication. In the middle EMS column the treatment is not truly replicated, and the result is no test on treatments!)

2.25 Here we see that a single contrast has a chi-squared distribution.

(a) Verify that the matrix $\mathbf{aa'}$ in (2.24) is idempotent.

(b) Show that (a) implies that for $Y_i \sim N(0, \sigma^2)$ and any contrast, we have

$$\frac{\left(\sum_i a_i Y_i\right)^2}{\sigma^2 \sum_i a_i^2} \sim \chi_1^2$$

2.26 Referring to Technical Note 2.8.3:

(a) Show that A_1 is idempotent and SS(Within) $= \mathbf{Y}'A_1\mathbf{Y}$.

(b) Show that A_2 is idempotent and SS(Within) $= \mathbf{Y}'A_2\mathbf{Y}$.

(c) Show that $A_1 A_2 = A_2 A_1 = 0$, and hence $A_1 + A_2$ is idempotent and the assumptions of Cochran's Theorem are satisfied.

(d) Prove Theorem 2.21.

2.27 Referring to Technical Note 2.8.3, here we will prove Theorem 2.22.

(a) Establish (1)-(3) of Theorem 2.22, using arguments similar to those in Exercise 2.26.

(b) To prove (4), we need to find the matrix A_3 that satisfies $\mathbf{Y}A_3\mathbf{Y} =$ SS(T $\times$ G), and show that A_3 is idempotent and satisfies $A_1 A_3 = A_3 A_1 = 0$, where A_1 satisfies $\mathbf{Y}A_1\mathbf{Y} =$ SS(Within). Cochran's Theorem can then be applied to get the F-test. (The matrix A_3 is constructed in a manner similar to the interaction matrix in Technical Note 3.8.2.)

2.28 Prove Theorem 2.23 by showing that, under model (2.4):

(a) $\text{Cov}(\bar{Y}_i, Y_{i'jk} - \bar{Y}_{i'j}) = 0$ for all i, i', j, k.

(b) $\mathrm{Cov}(\bar{Y}_j, Y_{ij'k} - \bar{Y}_{ij'}) = 0$ for all i, j, j', k.
(c) $\mathrm{Cov}(\bar{Y}_{ij} - \bar{Y}_i - \bar{Y}_j, Y_{i'j'k} - \bar{Y}_{i'j'}) = 0$ for all i, i', j, j', k.
(d) Use the normality assumption to go from uncorrelated to independence in (a)-(c), and use the properties of the t-distribution (Section 2.8.2) to complete the proof.

2.29 Referring to Section 2.6:
(a) Use model (2.16) to derive the least squares estimators (2.17) and show that the least squares estimator of $\mu + \tau_i$ is unbiased.
(b) Show that, in (a), if $\sum_i \tau_i = 0$, then we can estimate τ_i. If the cell sizes are unequal (r_i instead of r), derive the least squares estimates. How can we now estimate τ_i?
(c) Derive the variance of the least squares estimate of $\mu + \tau_i$ in the case of unequal r_i.

2.30 For the variance of an ancova contrast:
(a) Verify the variance (2.21).
(b) Prove the identity (2.22) and use it to verify the variance approximation (2.23). (Add $\pm\bar{b}$ on the left side and expand.)
(c) Suppose that, in (2.22), each b_i is a mean based on r_i observations. Formulate and prove an analogous identity to (2.22), and then establish an analogous average variance to (2.23).

2.31 The following data are measurements on the strength index of three varieties of cotton, where the treatments are pounds of potassium oxide per acre (dataset **Imbalance**). Here we want to see the effect of imbalance on the anova.

Strength Index of Cotton
Varieties

Treatment	1	2	3
36	7.06	7.75, 8.22	7.95, 8.59
60	7.51, 7.5	8.08, 8.18	8.69, 8.39
84	7.27, 7.49	7.9	8.04
108	6.55, 6.47	7.26, 7.06	7.35
132	6.97, 7.14	7.52, 7.83	7.63, 7.2

(a) Verify the anova tables in Miscellanea 2.9.3.
(b) Produce one anova table that contains the partial sums of squares. (You can do this by combining the two tables from part (a) or by using the R command `drop1`.)
(c) In terms of the analysis of these data, explain why the table in part (b) is most appropriate.
(d) Calculate the contrast sum of squares for the linear effect of potassium oxide and test its significance.
(e) Estimate the contrast and its standard error, and give a 95% confidence interval.

2.8 Technical Notes

2.8.1 Helpful Lemma I

The following lemma is helpful in calculating expected sums of squares in CRD designs.

Lemma 2.16. *Suppose that $W_1, W_2, \ldots, W_n$ satisfy $\mathrm{E}W = 0$, $\mathrm{Var}W = \sigma^2$, and $\mathrm{Cov}(W_i, W_{i'}) = 0$. Then*

(1) $\mathrm{E}\sum_{i=1}^{n}(W_i - \bar{W})^2 = \sum_{i=1}^{n}\mathrm{Var}W_i - n\mathrm{Var}\bar{W} = \left(1 - \frac{1}{n}\right)\sum_{i=1}^{n}\mathrm{Var}W_i$,

(2) $\mathrm{Cov}(W_i - \bar{W}, W_{i'} - \bar{W}) = -\frac{1}{n}\sigma^2$,

(3) $\mathrm{Var}(W_i - \bar{W}) = \mathrm{E}(W_i - \bar{W})^2 = \left(1 - \frac{1}{n}\right)\sigma^2$.

If the W_i are normal, then

(4) $W_i - \bar{W} \sim N\left(0, \left(1 - \frac{1}{n}\right)\sigma^2\right)$.

2.8.2 Anova Theory

In this section we outline the statistical theory behind anova, which leads us to t-tests and F-tests, as well as confidence intervals. Many of the details that we suppress here can be found in Casella and Berger (2001), especially in Section 5.3.

Theorem 2.17 (Independence of Linear Combinations). *Let $Y_i \sim N(\mu_i, \sigma_i^2)$, $i = 1, 2, \ldots, r$, be jointly normal, with $\mathrm{Cov}(Y_i, Y_{i'}) = \phi_{ii'}$, and let $W_j = \sum_{i=1}^{r} a_{ij}Y_i$, $j = 1, 2, \ldots, m$ be any collection of linear combinations. Then*

(1) The random variables W_j, $j = 1, 2, \ldots, m$, are jointly normal, with

$$W_j \sim N\left(\sum_{i=1}^{r} a_{ij}\mu_i, \sum_{i=1}^{r} a_{ij}^2\sigma_i^2 + 2\sum_{i>i'} a_{ij}a_{i'j}\phi_{ii'}\right).$$

(2) If $\mathrm{Cov}(W_j, W_{j'}) = 0$, W_j and $W_{j'}$ are independent.

(3) If $\mathrm{Cov}(W_j, W_{j'}) = 0$ for all $j \neq j'$, then the W_j are all independent.

The proof follows from the properties of the normal distribution and some algebra. It uses the fact that, because of the special form of the normal distribution, if the covariances are zero the density function factors into products.

Theorem 2.18 (From normals to χ^2 random variables). *Let χ_p^2 denote a chi squared random variable with p degrees of freedom.*

(1) If $Z \sim N(0, 1)$, then $Z^2 \sim \chi_1^2$.

(2) If W_i, $i = 1, 2, \ldots, r$ are independent, each distributed as $\chi_{p_i}^2$, then $\sum W_i \sim \chi_{\sum p_i}^2$.

So the degrees of freedom of independent χ^2 random variables sum. Of course, W_i could be the square of a standard normal or, more generally, if $Y_i \sim N(\mu_i, \sigma_i^2)$, $i = 1, 2, \ldots, r$ are independent, then

$$\sum_{i=1}^{r}\left(\frac{Y_i - \mu_i}{\sqrt{\sigma_i^2}}\right)^2 \sim \chi_r^2.$$

Now, we take the final step to the t and F distributions.

Theorem 2.19 (t and F distributions).
(1) If $Z \sim N(0,1)$ and $V \sim \chi_p^2$, independent of Z, then

$$\frac{Z}{\sqrt{V/p}} \sim t_p,$$

where t_p is Student's t distribution with p degrees of freedom.
(2) If $W \sim \chi_q^2$ and $V \sim \chi_p^2$, independent, then

$$\frac{W/q}{V/p} \sim F_{q,p},$$

where $F_{q,p}$ has an F distribution with q numerator degrees of freedom and p denominator degrees of freedom.

2.8.3 Cochran's Theorem for CRDs

Thus far, we have reached the t and F distributions when starting from independent normal random variables. However, we need more, as we will often be dealing with variables of the form $Y_i - \bar{Y}$, and these are not independent. More generally, in anova, we are typically interested in the distributions of sums of squares that are built up from linear combinations (actually contrasts) of normal variables. (For more details about anova theory see Graybill and Hultquist 1961, Albert 1976, Khuri 1982. Speed (1987) gives a very detailed derivation of this decomposition.)

For example, from Theorem 2.17, consider a linear combination $\sum_i a_i Y_i$ with $\sum_i a_i^2 = 1$. Write $\sum_i a_i Y_i = \mathbf{a'Y}$, where $\mathbf{a} = (a_1, a_2, \dots, a_p)$ and $\mathbf{Y} = (Y_1, Y_2, \dots, Y_p)$. Then

$$(2.24) \qquad \left(\sum_i a_i Y_i \right)^2 = (\mathbf{a'Y})^2 = \mathbf{Y'(aa')Y},$$

where the matrix $(\mathbf{aa'})$ satisfies $(\mathbf{aa'})^2 = (\mathbf{aa'})$. Recall that a matrix A that satisfies the condition $A^2 = A$ is called *idempotent*, and there is a strong connection between idempotent matrices and the chi squared distribution. The major result that we need is known as Cochran's Theorem, which dates back to Cochran (1934)[5].

Theorem 2.20 (Cochran's Theorem). *Let $\mathbf{Y} \sim N(0, \Sigma)$, and let A_k, $k = 1, 2, \dots, m$ satisfy $\sum_{k=1}^m A_k = A$, where $A\Sigma$ is idempotent. If*

$$A_k \Sigma \text{ is idempotent for every } k \text{ and } A_k \Sigma A_{k'} = 0, \quad k \neq k',$$

then
(1) $\mathbf{Y}'A_k\mathbf{Y} \sim \chi^2_{tr(A_k\Sigma)}$ for every k,
(2) $\mathbf{Y}'A_k\mathbf{Y}$ and $\mathbf{Y}'A_{k'}\mathbf{Y}$ are independent for $k \neq k'$,
(3) $\mathbf{Y}'A\mathbf{Y} \sim \chi^2_{tr(A\Sigma)}$.

---

[5] A complete treatment of Cochran's Theorem can also be found in Stuart and Ord 1987, Chapter 15 or Scheffé1959 Appendix VI.

If $\Sigma = \sigma^2 I$, a scalar multiple of the identity matrix, it follows from the previous discussion that $\mathbf{Y}'\mathbf{Y}/\sigma^2 = \sum Y_i^2/\sigma^2 \sim \chi_p^2$. Using Cochran's theorem, if A is an idempotent matrix then we can write

$$\mathbf{Y}'\mathbf{Y} = \mathbf{Y}'A\mathbf{Y} + \mathbf{Y}'(I - A)\mathbf{Y},$$

and, when divided by σ^2, both quadratic forms on the right side are independent χ^2 random variables, with degrees of freedom equal to the trace of the matrix in the quadratic form. That is, for $Y \sim N(0, \sigma^2 I)$ and idempotent A we have

$$\mathbf{Y}'A\mathbf{Y}/\sigma^2 \sim \chi_{tr(A)}^2 \text{ and } \mathbf{Y}'(I - A)\mathbf{Y}/\sigma^2 \sim \chi_{tr(I-A)}^2,$$

where $tr(A)$ is the sum of the diagonal elements. Since $A(I - A) = 0$, we have partitioned $\chi_p^2 = \chi_{tr(A)}^2 + \chi_{tr(I-A)}^2$.

As we have seen, many of the sums of squares that we deal with in anova are built up from contrasts in normal random variables. We now look at the oneway CRD anova

$$Y_{ij} = \mu + \tau_i + \varepsilon_{ij}, \quad \varepsilon_{ij} \sim N(0, \sigma^2), \text{ independent}, \quad i = 1, \ldots t, \quad j = 1, \ldots, r,$$

and work our way to the F distribution for the test on treatments.

Within cell i, we have observations $Y_{i1}, Y_{i2}, \ldots, Y_{ir}$ and we can write the deviations from the mean as

$$\begin{pmatrix} Y_{i1} - \bar{Y}_{i\cdot} \\ Y_{i2} - \bar{Y}_{i\cdot} \\ \vdots \\ Y_{ir} - \bar{Y}_{i\cdot} \end{pmatrix} = \left(I - \frac{1}{r}J\right) \begin{pmatrix} Y_{i1} \\ Y_{i2} \\ \vdots \\ Y_{ir} \end{pmatrix},$$

where I is the identity matrix and J is a matrix of ones. Now we note that

$$\left(I - \frac{1}{r}J\right)\left(I - \frac{1}{r}J\right) = \left(I - \frac{1}{r}J\right)$$

is an idempotent matrix. The entire set of deviations can be written as

$$\begin{pmatrix} Y_{11} - \bar{Y}_{1\cdot} \\ Y_{12} - \bar{Y}_{1\cdot} \\ \vdots \\ Y_{tr} - \bar{Y}_{t\cdot} \end{pmatrix} = \begin{pmatrix} \left(I - \frac{1}{r}J\right) & 0 & \cdots & 0 \\ 0 & \left(I - \frac{1}{r}J\right) & \cdots & 0 \\ \vdots & \vdots & \vdots & \vdots \\ 0 & 0 & & \left(I - \frac{1}{r}J\right) \end{pmatrix}_{tr \times tr} \begin{pmatrix} Y_{11} \\ Y_{12} \\ \vdots \\ Y_{tr} \end{pmatrix}$$

$$\stackrel{\text{def}}{=} A_1 \mathbf{Y}.$$

It can be shown that A_1 is idempotent and $SS(\text{Within}) = \mathbf{Y}'A_1\mathbf{Y}$. Similarly, we can write the deviations for the treatment sum of squares as

$$\begin{pmatrix} \bar{Y}_{1\cdot} - \bar{\bar{Y}} \\ \bar{Y}_{2\cdot} - \bar{\bar{Y}} \\ \vdots \\ \bar{Y}_{t\cdot} - \bar{\bar{Y}} \end{pmatrix} = \left(I - \frac{1}{t}J\right) \begin{pmatrix} \bar{Y}_{1\cdot} \\ \bar{Y}_{2\cdot} \\ \vdots \\ \bar{Y}_{t\cdot} \end{pmatrix} = \left(I - \frac{1}{t}J\right)\frac{1}{r} \begin{pmatrix} \mathbf{1}_{r \times 1} & 0 & \cdots & 0 \\ 0 & \mathbf{1}_{r \times 1} & \cdots & 0 \\ \vdots & \vdots & \vdots & \vdots \\ 0 & 0 & & \mathbf{1}_{r \times 1} \end{pmatrix}_{t \times tr} \mathbf{Y},$$

however, the matrix multiplying $\mathbf{Y}$ is not square and cannot be idempotent. To remedy this we take t copies of it (this is why there is a t multiplying the treatment sum of squares) and define

$$A_2 = \begin{pmatrix} \left(I - \frac{1}{t}J\right) \\ \left(I - \frac{1}{t}J\right) \\ \vdots \\ \left(I - \frac{1}{t}J\right) \end{pmatrix}_{tr \times t} \frac{1}{r} \begin{pmatrix} 1_{r \times 1} & 0 & \cdots & 0 \\ 0 & 1_{r \times 1} & \cdots & 0 \\ \vdots & \vdots & \vdots & \vdots \\ 0 & 0 & & 1_{r \times 1} \end{pmatrix}_{t \times tr}$$

and then A_2 is idempotent and SS(Treatments) $= \mathbf{Y}' A_2 \mathbf{Y}$. We are now in the position to apply Cochran's Theorem to get the following, whose proof is left to Exercise 2.26.

Theorem 2.21 (Oneway CRD F-distribution). *For the oneway CRD*

$$Y_{ij} = \mu + \tau_i + \varepsilon_{ij}, \quad \varepsilon_{ij} \sim \mathrm{N}(0, \sigma^2), \ independent, \quad i = 1, \ldots t, \quad j = 1, \ldots, r,$$

(1) SS(Within)$/\sigma^2 \sim \chi^2_{t(r-1)}$.
(2) Under $H_0 : \tau_i = 0$ for all i, SS(Trt)$/\sigma^2 \sim \chi^2_{t-1}$, independent of SS(Within).
(3) Under $H_0 : \tau_i = 0$ for all i, $F = \frac{\mathrm{SS(Trt)}/(t-1)}{\mathrm{SS(Within)}/(t(r-1))} \sim F_{t-1, t(r-1)}$.

A similar theorem holds for the twoway (and higher) CRDs, with the complication that we have to deal with the matrices for the interaction terms. We leave that pleasant task to Exercise 2.27, and simply state the theorem here.

Theorem 2.22 (Twoway CRD F-distributions). *For the twoway CRD of model (2.4),*
(1) SS(Within)$/\sigma^2 \sim \chi^2_{tg(r-1)}$.
(2) Under $H_0 : \tau_i = 0$ for all i, SS(T)$/\sigma^2 \sim \chi^2_{t-1}$, independent of SS(Within) and $F = \frac{\mathrm{SS(T)}/(t-1)}{\mathrm{SS(Within)}/(tg(r-1))} \sim F_{t-1, tg(r-1)}$.
(3) Under $H_0 : \gamma_j = 0$ for all j, SS(G)$/\sigma^2 \sim \chi^2_{g-1}$, independent of SS(Within) and $F = \frac{\mathrm{SS(G)}/(g-1)}{\mathrm{SS(Within)}/(tg(r-1))} \sim F_{g-1, tg(r-1)}$.
(4) Under $H_0 : (\tau\gamma)_{ij} = 0$ for all i,j, SS(T $\times$ G)$/\sigma^2 \sim \chi^2_{(t-1)(g-1)}$, independent of SS(Within) and $F = \frac{\mathrm{SS(TxG)}/((t-1)(g-1))}{\mathrm{SS(Within)}/(tg(r-1))} \sim F_{(t-1)(g-1), tg(r-1)}$.

Finally, the distribution of contrasts in the CRD is also straightforward. We state the theorem here and leave the proof to Exercise 2.28.

Theorem 2.23 (Contrasts in the twoway CRD). *For the twoway CRD of model (2.4), let $\hat{\sigma}^2 = \mathrm{MS(Within)}/(tg(r-1))$. Then for any contrast*
(1) $\sqrt{gr} \sum_i a_i(\bar{Y}_i - \tau_i)/\hat{\sigma} \sim t_{tg(r-1)}$.
(2) $\sqrt{tr} \sum_j a_j(\bar{Y}_j - \gamma_i j)/\hat{\sigma} \sim t_{tg(r-1)}$.
(3) $\sqrt{r} \sum_{ij} a_{ij}(\bar{Y}_{ij} - \bar{Y}_i - \bar{Y}_j + \bar{\bar{Y}} - (\tau\gamma)_{ij})/\hat{\sigma} \sim t_{tg(r-1)}$.

There has been much research in answering the question "When is a sum of squares an analysis of variance", the title of Albert (1976). What is meant is, under what conditions can we partition the total sum of squares into orthogonal pieces, and get out the usual F-tests. This was first addressed by Graybill and Hultquist (1961), who investigated the connections between such a decomposition and minimal sufficiency, which implies that the inference would then be based on the likelihood function. The most general treatment of this problem can be found in Speed (1987). On a related note, Bradley (1973) gives conditions for the equivalence of likelihood and least squares estimates in linear models (see Berger and Casella 1992 for a more elementary treatment of this equivalence).

2.8.4 Noncentral Distributions

In the anova, if we set up the ratio of mean squares correctly, then under the appropriate H_0 their ratio will have an F distribution. More precisely, it is a *central* F distribution as described in Theorem 2.19. If H_0 is false, and the appropriate parameter is nonzero, then the ratio of mean squares has a *noncentral* F distribution. This noncentral distribution, and its properties, is related to the power of the anova tests, and a good understanding of the behavior of the F-statistic under the noncentral distribution will be helpful in better evaluating anova strategies. Here we discuss the oneway and twoway CRD; extensions to other designs should be clear.

The central F distribution (we usually omit the adjective unless we need to be very clear) is obtained from the ratio of two independent central χ^2 random variables. The noncentral F comes from the noncentral χ^2 distribution.

Definition 2.24. If $X_1, X_2, \ldots, X_p$ are independent normal random variables, $X_i \sim N(\theta_i, \sigma^2)$, then

$$W = \frac{\sum_i X_i^2}{\sigma^2} \sim \chi_p^2(\lambda),$$

a *noncentral* χ^2 random variable with p degrees of freedom and noncentrality parameter $\lambda = \sum_i \theta_i^2$. If $V \sim \chi_q^2$ (central), independent of W, then

$$\frac{W/p}{V/q} \sim F_{p,q}(\lambda),$$

a *noncentral* F with degrees of freedom p and q and noncentrality parameter $\lambda = \sum_i \theta_i^2 / \sigma^2$.

The density function of the noncentral χ^2 is quite imposing, being an infinite sum of weighted central χ^2 densities. We will not display it here, but rather share a more useful characterization of the noncentral χ^2: If

$$K \sim \text{Poisson}(\lambda) \text{ and } W|K \sim \chi_{p+2K}^2, \text{ then } W \sim \chi_p^2(\lambda),$$

which also tells us that the weights in that infinite sum are Poisson probabilities. The important property about these noncentral distributions is that they are *stochastically increasing* in the noncentrality parameter. That is, for a fixed value a

$$W \sim \chi_p^2(\lambda) \Rightarrow P_\lambda(W > a) \text{ is an increasing function of } \lambda$$
$$T \sim F_{p,q}(\lambda) \Rightarrow P_\lambda(T > a) \text{ is an increasing function of } \lambda.$$

Thus, if W or T are test statistics, and we reject H_0 for large values of the statistic, then the power of the test increases with λ, since the probability of the statistics exceeding the cutoff point increases in λ.

Oneway CRD

In the oneway CRD of model (2.2), with EMS given in Exercise 2.11, we test $H_0 : \sum_i \tau_i^2 = 0$ with $F = \mathrm{MS(Treatments)/MS(Within)}$, calculated as in Table 2.2. If H_0 is violated, then $\mathrm{E}(\bar{Y}_i - \bar{\bar{Y}}) = \tau_i$, the numerator sum of squares is noncentral χ^2 with noncentrality parameter $\sum_i \tau_i^2/\sigma^2$, and the F-statistic has the corresponding noncentral F distribution. The power of the F-test increases with $\sum_i \tau_i^2$. Thus the power increases no matter which τ_i increases, and the power is constant on spheres where $\sum_i \tau_i^2$ is constant. Note also that the denominator of the F-statistic remains an independent central χ^2.

Twoway CRD

Similar conclusions hold for the twoway CRD of model (2.4) with respect to the tests of the main effects and the interactions, with denominator mean square MS(Within). That is, the power of the respective tests is increasing in $\sum_i \tau_i^2$, $\sum_i \gamma_i^2$, and $\sum_i (\tau\gamma)_i^2$ (see Table 2.6). There is an interesting occurrence, however, if we look at the practice of pooling error terms.

In Definition 2.24, if V is an independent *noncentral* χ^2, say $V \sim \chi_q^2(\delta)$, then $T = (W/p)/V/q)$ has a *doubly noncentral* F distribution, $F_{p,q}(\lambda, \delta)$. Moreover, it should be clear that $P_{\lambda,\delta}(T > a)$ is increasing in λ for fixed δ, and decreasing in δ for fixed λ (see Scheffé1959, Section 4.8).

Now, in the twoway CRD, suppose that we, for example, test $H_0 : \sum_i \tau_i^2 = 0$ using a denominator that is obtained by pooling the interaction and within sum of squares. (Possibly based on first accepting the null hypothesis that there is no interaction.) If, in fact there is an interaction, so $\sum_i (\tau\gamma)_i^2 > 0$, then we have increased the noncentrality parameter in the denominator of the F-ratio, which has a doubly noncentral F distribution. Thus, we have decreased the probability of the test statistic exceeding the cutoff or, in other words, we have decreased the power of the test. This is what is called a *conservative* test, where it is more difficult to reject H_0 because the distribution of the denominator of the statistic has been inflated. This procedure is more prone to making the Type I error, and a rejection is usually greeted with enthusiasm.

2.9 Miscellanea

2.9.1 Multiple Comparisons and Error Rates

Post-anova analysis will often involve many inferences, typically on contrasts or pairwise differences. Of course, we know that the simultaneous inference from many α level tests is not necessarily at level α, so the question is how do we control this in a meaningful way.

This is actually a difficult and important question, especially in light of microarray experiments where we might be faced with thousands of hypothesis tests. In such cases, we must intelligently balance power and false detection rates.

Classical Error Rate Control

First, there are many definitions of Type I Error for multiple comparisons, so exactly what is meant by "α level" is not always clear. Some of the types of error rates considered are *experimentwise error rate* or *comparisonwise error rate*. Miller (1981) and Hsu (1996) are good references for this topic.

> *Experimentwise error rate*: Controls the error rate of the entire experiment, so a 5% experimentwise α level means that if all nulls are true, there will be a false rejection in 5% of experiments, no matter how many tests are done. This is a very conservative criterion and is virtually useless for exploratory studies (no power)

> *Comparisonwise error rate*: Controls the error rate of comparisons, so a 5% comparisonwise α level means that if all nulls are true, there will be a false rejection in 5% of comparisons. So if you do 1000 comparisons you can expect 50 false rejections. This is a very liberal criterion with high power, and is good for exploratory studies. It will produce a lot of false rejections, however.

There is also a *familywise error rate*, which is the experimentwise error applied to a smaller group of comparisons, a "family".

Note that, if we are combining results of *independent* tests, then things are simple in the following sense: If we do m tests, each at level α, and if all of the null hypotheses are true, then the overall Type I error rate (experimentwise) error is $1 - (1 - \alpha)^m$, so 20 independent .05 tests have an overall Type I error rate of .64 and 20 independent .01 tests have an overall Type I error rate of .18. Simultaneous coverage of the 20 intervals would be .36 and .82, respectively. However, this is not a realistic situation, as we will seldom find independent tests in any experiment. For example, if we have a set of uncorrelated contrasts under normality, and we *know* the variance, or we use an independent variance estimate for each contrast, then we have independent tests. In almost any other situation, such as correlated contrasts or the use of pooled variances, the tests are dependent.

As a more realistic example, if we want to make a simultaneous $1-\alpha$ statement about the coverage of m confidence sets, then, from the Bonferroni Inequality, we can construct each confidence set to be of level $1 - \frac{\alpha}{m}$. If we do each test at level α/m then we are controlling the experimentwise error at level α. If we do each test at level α, then we are controlling the comparisonwise error at level α. In an anova with k treatments, simultaneous inference on all $k(k-1)/2$ pairwise differences can be made with confidence $1 - \alpha$ if each t interval has confidence $1 - 2\alpha/[k(k-1)]$.

An alternative to Bonferroni, which also controls the experimentwise error, is the Scheffé procedure, which provides simultaneous confidence intervals on *all* contrasts. The procedure states that simultaneously for all contrasts $(a_1, \ldots, a_k)$,

$$\sum_{i=1}^{k} a_i \bar{Y}_{i\cdot} - \mathbf{M}\sqrt{\hat{\sigma}^2 \sum_{i=1}^{k} a_i^2/n_i} \;\le\; \sum_{i=1}^{k} a_i \theta_i \;\le\; \sum_{i=1}^{k} a_i \bar{Y}_{i\cdot} + \mathbf{M}\sqrt{\hat{\sigma}^2 \sum_{i=1}^{k} a_i^2/n_i},$$

where $\mathbf{M} = \sqrt{(k-1)F_{k-1,n-k,\alpha}}$ and $\hat{\sigma}^2$ has $n - k$ degrees of freedom.

One of the real strengths of the Scheffé procedure is that it allows legitimate "data snooping", and we can test hypotheses that have been suggested by the data. Suggested hypotheses can bias the results and, hence, invalidate the inference, but intervals or tests that are valid for *all* contrasts, whether they have been suggested by the data, have already have been taken care of by the Scheffé procedure. So the bias is eliminated. Of course, we pay for this data snooping privilege because the confidence intervals are very wide, and the tests are not very powerful.

There is a plethora of other simultaneous inference procedures, most concerned with pairwise comparisons. A method due to Tukey gives simultaneous confidence intervals on all pairwise differences, not all contrasts.

Other types of multiple comparison procedures, which deal with pairwise differences, are more powerful than the Scheffé method or the Tukey method. Some procedures are the LSD (Least Significant Difference) Procedure, Protected LSD, Duncan's Procedure, and Student–Neumann–Keuls' Procedure. These last two are *multiple range* procedures, where the cutoff point to which comparisons are made changes between comparisons. (These should be avoided, as they can lead to contradictory inferences.) │ Avoid multiple range procedures

False Discovery Rate Control

There are many other error rate definitions, but we will only mention one more here, one that has become very popular in recent years, and has seen much use in microarray analysis. Table 2.9 is adapted from Benjamini and Hochberg (1995).

Table 2.9. Number of errors committed when testing m hypotheses, where m_0 nulls are true

	Declared Significant	Declared Nonsignificant	Total
True Null Hypothese	V	$m_0 - V$	m_0
False Null Hypotheses	S	$m - m_0 - S$	$m - m_0$
	R	$m - R$	m

Note that the only numbers that we know in Table 2.9 are m and R, but error rates can be obtained as probabilities and expected values.

$$\mathrm{E}\left(\frac{V}{m}\right) = \text{Comparisonwise Error Rate}$$
$$P(V \geq 1) = \text{Familywise Error Rate}$$

and testing each hypothesis at α/m guarantees that $P(V \geq 1) \leq \alpha$, while testing each hypothesis at α guarantees that $\mathrm{E}(V/m) \leq \alpha$.

Benjamini and Hochberg (1995) suggest that we should control another error rate

$$\mathrm{E}\left(\frac{V}{R}\right) = \text{False Discovery Rate,}$$

which is the proportion of discoveries (rejections of H_0) that are false (H_0 is true). They note that if all the nulls are true, this is equivalent to the familywise error rate, but is a smaller number once $m_0 < m$.

A procedure for controlling the false discovery rate (FDR) at a set value q^* is the following:

(1) For testing m hypotheses, obtain the p-values and *order* them $p_{(1)} < p_{(2)} < \cdots < p_{(m)}$. (Note that $p_{(i)}$ is not the p-value for the ith hypothesis.)

(2) Let k be the largest i for which $p_{(i)} < (i/m)q^*$.

(3) Reject all nulls with a p-value less than $p_{(k)}$.

Controlling the FDR has almost become a standard in microarray analysis, partially due to the popularity of a procedure known as SAM (Statistical Analysis of Microarrays, Tusher *et al.* 2001). The SAM procedure is essentially multiple t-tests processed using the FDR control.

There has been much other work done in estimating the FDR (Storey and Tibshirani 2003) and on other variations of FDR. A large literature on FDR is developing, and good entries are Storey (2002, 2003), Genovese and Wasserman (2002, 2004), and Genovese *et al.* (2006).

2.9.2 Application or Attribute?

The question has sometimes come up about whether a treatment that is *applied* is different from a treatment that is an *attribute*. For example, a treatment could be a certain level of nitrogen in a fertilizer that is applied to a plot of land, or a treatment could be an age group, which is an attribute. Statistically, is there a difference?

This discussion could be about the difference between designed experiments and observational studies, but it is not. We are here dealing with designed studies, but could include, as treatments, factors that are attributable to the experimental units rather than applied to the experimental units. A possible way to understand the difference is with the following model, a variation on the oneway CRD (2.2):

$$Y_{ij} = \mu + \tau_i + \delta_{ij} + \varepsilon_{ij},$$

where we add δ_{ij} as the error in applying treatment i to the jth unit in that treatment group. The error δ_{ij} can have one of three forms

$$\delta_{ij} = \begin{cases} 0 & \text{if the treatment is an attribute} \\ \delta_i & \text{if the treatment is applied once to the group} \\ \delta_{ij} & \text{if the treatment is applied independently to each unit.} \end{cases}$$

This model says that there is no error in the application of the treatment if the treatment is an attribute. So, for example, if τ_1 is the application of a fertilizer with 10% nitrogen, and the batch is mixed with 12%, that error is reflected in δ. (This is in contrast to ε, which would reflect the different responses in yield to the same treatment of 10% nitrogen.)

If we assume that the errors are independent, then there is no discernable difference between the first and third cases, that is, whether the treatment is an attribute, or applied independently to each unit, results in the same analysis. The only difference arises is if $\delta_{ij} = \delta_i$, for then a correlation is introduced and the analysis cannot proceed as a oneway CRD. In fact, in a formal sense the analysis cannot proceed at all (Exercise 2.24).

2.9.3 Imbalance

In this book we are dealing almost exclusively with balanced designs. When looking at the experiment from the design viewpoint, it would be silly to design an unbalanced experiment, as too many things are given up. Dealing with imbalance in data is a topic that is better treated in a book devoted to analysis rather than design (see, for example, Rawlings *et al.* 1998).

A design is *balanced* if there are the same number of observations in each cell. Imbalance only becomes a serious problem in twoway and higher designs, as the oneway CRD can handle imbalance pretty well (although it wreaks havoc with contrasts, as we have seen in Chapter 1; see Exercise 1.18).

Imbalance is a fact of life. Rats or plants can die, microarray wells may not be correctly read, data may be lost. Sometimes we have the luxury of having extra observations in each cell to plan for this- which might be one of the only valid reasons for have within observations in an RCB. However, discarding data for any reason is typically a bad idea.[6]

For example, if every cell but one in a twoway design has 5 observations, and one has four observations, balance can be achieved by discarding, at random, one observation from each of the cells with 5 observations. Never, never do this. Never throw away data unless there is a really good reason to do so. Each data point has information, and discarding information is never good.

> Data are sacred!

There is a fundamental difference between unbalanced data and missing data. We take the approach that our analysis should always be based on the likelihood for the data. In the balanced case this agrees with the anova-based analysis, but in the unbalanced case things differ.

- In the unbalanced (but not missing) case, we can write down a model and a likelihood, and estimate and test based on the likelihood. In fact, this is regression analysis! There are problems, however, with identifiability constraints and estimability, but these can be handled.
- In the case where there is missing data, so we have empty cells, things get a bit more complicated. If the parameter for the cell is in the model then, formally, the likelihood function is calculated by averaging over all possible values that could have been in that cell. This is not a simple calculation, but can be handled by modern missing data techniques such as the EM algorithm (see, for example, Little and Rubin (2002). Alternatively, we can do an *observed case* analysis using the likelihood for the observed data, but this can sometimes be a difficult calculation, and may sometimes require discarding data.

The main problem caused by imbalance is that, under the usual models, effects are no longer orthogonal. For example, consider model (2.4) with unequal cell sizes where, for simplicity, we assume no interaction:

$$Y_{ijk} = \mu + \tau_i + \gamma_j + \varepsilon_{ijk}, \quad i = 1, \ldots, t; \quad j = 1, \ldots, g, \quad k = 1, \ldots, r_{ij},$$

and the r_{ij} are not necessarily equal. If we denote the total number of observations by $n = \sum_{ij} r_{ij}$, and the marginal totals by $r_{i\cdot} = \sum_j r_{ij}$ and $r_{\cdot j} = \sum_i r_{ij}$,

---

[6] If there is a nonstatistical reason to believe that a data point is an outlier, an analysis with and without it can show if it is unduly affecting the inference.

then

$$E(\bar{Y}_i - \bar{\bar{Y}}) = \tau_i - \frac{1}{n}\sum_i r_{i.}\tau_i + \frac{1}{r_{i.}}\sum_j r_{ij}\gamma_j - \frac{1}{n}\sum r_{.j}\gamma_j.$$

The identifiability constraints $\sum_i \tau_i = \sum_j \gamma_j = 0$ do not do much here, and thus there are "pieces" of extraneous parameters in this expected value no matter what we do. This means that the sums of squares will not be orthogonal – they will not add to the total – and we must be more careful if we do an anova. Specifically, we must decide between *sequential* or *partial* sums of squares.

The data in Exercise 2.31 is an unbalanced twoway. We can produce two different anova tables depending on which factor we fit first:

	Treatments Before Varieties				Varieties Before Treatments		
Source	df	SS		Source	df	SS
Treatments	4	4.253		Varieties	2	3.563
Varieties	2	3.004		Treatments	4	3.694
Residual	19	1.023		Residual	19	1.023

Since the sums of squares are not orthogonal, fitting one will account for some variation in the other, so the order matters. In each of these tables we see the sequential sums of squares, which add to the total. Note that in the second table, where varieties are fit first, the sum of squares for treatments is smaller.

The second line of each table gives the partial sum of squares for that factor and this is the sum of squares that should be used for evaluating the effect. For example, if we want to evaluate the effect of treatment, it should be done after the variation due to varieties is removed, resulting in SS(Treatments After Varieties) = 3.694. These partial sums of squares can be presented in one anova table, where the sums of squares will not add to the total sum of squares. Also note that in each table, since the residual is fit last, its sum of squares is the same.

So, in general, what to do? The nonorthogonality problem, which leads to problems with the identifiability constraints, does not go away even if we calculate least squares means. However, it should be clear that this problem grows worse if the imbalance grows worse. Thus, if the imbalance is minimal, we can forge ahead with partial sums of squares and, for the most part, ignore the imbalance in our effect estimates. If the imbalance is great, or if there are cells without data, then it is probably best to turn to a cell means model, which we first saw in (1.2), and here would be

$$Y_{ijk} = \theta_{ij} + \varepsilon_{ijk}, \quad i = 1,\ldots,t; \quad j = 1,\ldots,g, \quad k = 1,\ldots,r_{ij},$$

where θ_{ij} is the mean of cell (i,j), and $E\bar{Y}_{ij} = \theta_{ij}$. We can then estimate treatment effects with contrasts in the $\bar{Y}_{ij}$.

3

Complete Block Designs

We shall need to judge of the magnitude of the differences introduced by testing our treatments upon the different plots by the discrepancies between the performances of the same treatment in different blocks.

R. A. Fisher
The Design of Experiments, Section 26

I thanked him for the explanation; now I understood it. I have to understand the world, you see.

Richard P. Feynman
Surely You're Joking, Mr. Feynman

3.1 Introduction

Just as a oneway anova is a generalization of a two-sample t-test, a randomized complete block (RCB) design is a generalization of a paired t-test. In this first section we review some basics and do a small example, and show how to build up an RCB from pairwise t-tests.

In this book we discuss two types of block effects, fixed and random. In most textbooks blocks are treated as a random effect without much discussion of options, but there are clear instances where blocks are not random (see Example 4.1). However, in such cases these factors are still blocks because of the randomization pattern they induce and, in particular, the covariance structure they induce. We focus on this, and look very carefully at how to model the covariance, which we find is the overwhelmingly important concern. Whether the block is fixed or random is a function of the particular experiment, as long as the covariance is correctly accounted for then valid inferences can be drawn.

In this chapter we will mainly concentrate on the classical approach with the blocks considered as random, leaving details of fixed blocks models and implications to Chapter 4.

3.1.1 An RCB Model

A *block* (or *blocking factor*) is a categorization that is inherently different from a treatment in that a block is usually in an experiment for the express purpose of removing variation, not because there is any interest in finding block differences. The practice of blocking originated in agriculture, where experimenters took advantage of similar growing conditions to control experimental variances. For example, blocks could represent field plots in an agricultural experiment; while in an experiment with human subjects, the subjects themselves can be blocks.

Example 3.1. STRAWBERRY BLOCKS REVISITED   Recall Example 1.8, the field experiment about the adaptability of three varieties of strawberries to Venezuelan soil. The data are repeated in Table 3.1, which just gives the data layout, and does not tell us about the randomization or the field layout. Since this is an RCB, a possible field layout is

A	C	B
10.1	8.4	6.3

B	C	A
6.9	9.4	10.8

C	A	B
9.0	9.8	5.3

A	C	B
10.5	9.2	6.2

‖

The blocks are called *complete* blocks if every treatment appears in every block, so that the data are in a rectangular array. In the classical RCB there is one observation for each treatment – block combination and the data are observed according to the additive model

$$(3.1) \qquad Y_{ij} = \mu + \tau_i + \beta_j + \varepsilon_{ij}, \quad i = 1, \ldots, t, \quad j = 1, \ldots, b,$$

where μ is an overall mean, τ_i are treatment effects, β_j are block effects, and ε_{ij} are error random variables.

Table 3.1. Yields in kilograms from four blocks of land over a two-week period.

		Blocks			
		1	2	3	4
	A	10.1	10.8	9.8	10.5
Variety of					
Strawberry	B	6.3	6.9	5.3	6.2
	C	8.4	9.4	9.0	9.2

Table 3.2. Data from an RCB anova with one observation/cell

		Blocks				
		1	2	3	...	b
Treatments	1	y_{11}	y_{12}	y_{13}	$\cdots$	y_{1b}
	2	y_{21}	y_{22}	y_{23}	$\cdots$	y_{2b}
	$\vdots$	$\vdots$	$\vdots$	$\vdots$	$\vdots$	$\vdots$
	k	y_{k1}	y_{k2}	y_{k3}	$\cdots$	y_{kb}

Notice that we are again using an overparameterized model, as discussed in Section 1.1 (see also Section 2.4 for CRD). As was previously mentioned, when there is more than one factor, the overparameterized model seems easier to understand. Remember that with the overparameterized model the treatment and block effects represent deviations from an overall mean level.

Schematically, the data, y_{ij}, from an RCB anova is shown in Table 3.2 Note that there is only one observation for each treatment – block combination so, unlike the oneway anova, no observations were taken under the same experimental conditions. However, this version of an RCB is a most efficient design, and provides all of the necessary information to give valid inferences.

The remaining term in RCB to be defined is *randomized*. This term refers to the way that the observations are taken in each block, and is perhaps the most important term in the name, especially from a design standpoint. In each block, the treatments are run in a completely random manner, using a randomization restricted to take place within blocks. By way of contrast, the anovas of Chapter 2 are *completely randomized designs*, since the observations are taken in a manner that is random throughout the data, with no blocks to restrict randomization.

Example 3.2. BLOCKS AS A FIXED FACTOR   A new development in cake mixes is microwavable mixes. An experiment was done to assess the texture of microwave brownies when compared to traditional oven-baked brownies. Three brands of brownie mix (A, B, C) were used, with three cooking times (T), (2.5, 3 and 3.5 minutes) and two power settings (P), (600 and 800 watts) on the microwave.

The factors P and T were crossed, and for each brownie mix 12 packages were used. The data layout is

Brand A	Brand B	Brand C
T	T	T

```
  Brand A        Brand B        Brand C
     T              T              T
   ┌─┬─┬─┐        ┌─┬─┬─┐        ┌─┬─┬─┐
   │x│x│x│        │x│x│x│        │x│x│x│
 P │x│x│x│      P │x│x│x│      P │x│x│x│
   │x│x│x│        │x│x│x│        │x│x│x│
   │x│x│x│        │x│x│x│        │x│x│x│
   └─┴─┴─┘        └─┴─┴─┘        └─┴─┴─┘
```

Here, the three brands represent the brands of interest. Thus, all three factors are fixed. However, the factor "brands" is a block in the sense that it imposes a correlation structure. The packages from the same brand are correlated, and this must be accounted for in the analysis. ‖

In contrast to Example 3.1, here we have a case of fixed blocks. In the previous example we can argue that the four blocks (plot of land) are a random sample from all plots on which strawberries will be plants. But in this example there are only three brands of cakes, so we cannot argue that the three brands are a sample from a larger lot. This is what we will discuss in Chapter 4; see also Exercise 4.3.

3.1.2 RCB and the Paired t-test

Just as the CRD anova can be viewed as an extension of a two-sample t-test (Section 2.1.2), the RCB anova is a direct extension of the paired t-test. We look at this connection in detail, with a main purpose of better understanding the appropriate error terms.

Example 3.3. STRAWBERRY BLOCKS – RCB ANALYSIS   We can apply the twoway anova sums of squares to the RCB design (see (1.10) and (1.11)) to arrive at the anova table

Source	df	Sum Sq	Mean Sq	F	p
Block	3	1.722	0.574		
Trt	2	35.582	17.791	147.235	< .0001
T × B	6	0.725	0.121		

Note that the test on treatments is

$$(3.2) \qquad F = \frac{\text{MS(Treatments)}}{\text{MS(T × B Interaction)}},$$

and it is important to note that this is *always* the correct test for treatments in an RCB. As Fisher said at the beginning of this chapter, we test the treatment variation by measuring the performance of the treatments on different blocks, which is measured by the interaction.

We will continue, in the spirit of Fisher, and reconstruct the RCB analysis from t-tests, and then it will become crystal clear that the interaction is the correct error term.

First, suppose that there we only have treatments A and B in the data of Table 3.1. Since the observations are then paired by the blocks, the correct analysis is clearly a paired t-test. We take the differences $d = A - B$ (that is, $d_j = y_{1j} - y_{2j}$ and calculate

$$t_{\text{calc}} = \frac{\bar{d}}{\sqrt{\hat{\sigma}_d^2/4}} = 24.9694, \quad p = 0.00014,$$

where $\hat{\sigma}_d^2 = (1/3)\sum_j (d_j - \bar{d})^2$. We compare this paired t-test to an RCB using only treatments A and B:

Source	df	Sum Sq	Mean Sq	F	p
Block	3	1.724	0.575		
Trt	1	34.031	34.031	623.473	0.00014
T × B	3	0.164	0.055		

and note that $t_{calc}^2 = 623.473$, which is the F-statistic from the anova. This is, of course, no coincidence, but look what this implies:

$$t_{calc}^2 = \frac{\bar{d}^2}{\hat{\sigma}_d^2/4} = \frac{2(\bar{y}_{1\cdot} - \bar{y}_{2\cdot})^2}{\hat{\sigma}_d^2/2} = \frac{SS(Trt)}{MS(TxB)},$$

where SS(Trt) and MS(T × B) are from the RCB using only treatments A and B. Note that this shows that MS(T × B)$= \hat{\sigma}_d^2/2$, illustrating that the interaction term is the correct variance of a treatment contrast in an RCB.

To go a bit further, we can also do an RCB anova on the contrast orthogonal to $A - B$, namely $\frac{1}{2}A + \frac{1}{2}B - C$. To do this we create variables $y_{1j}^* = \frac{1}{2}y_{1j} + \frac{1}{2}y_{2j}$ and $y_{2j}^* = y_{3j}$ and produce the anova

Source	df	Sum Sq	Mean Sq	F	p
Block	3	1.001	0.334		
Trt	1	1.163	1.163	8.287	.0636
T × B	3	0.421	0.140		

and the paired t-test on $y_{1j}^* - y_{2j}^*$ gives $t_{calc}^2 = 8.287$.

Thus, we have broken up the full RCB table into two orthogonal t-tests, where is was clear how to calculate the error, and it was clear that the interaction provides the proper error term for tests on treatment contrasts.   ‖

Although Example 3.3 illustrates the proper tests for the case of two treatments, it may not be clear how to extend the example to the case of an arbitrary number of treatments in an RCB. The answer, which is explained by Fisher, is that the error term for treatments is formed from all of the error contrasts. What we have actually done here is to exploit the partitioning of sums of squares into contrast sums of squares (recall Definition 1.14) where we have

$$SS(Trt) = b\frac{(\bar{y}_{1\cdot} - \bar{y}_{2\cdot})^2}{2} + b\frac{(\frac{1}{2}\bar{y}_{1\cdot} + \frac{1}{2}\bar{y}_{2\cdot} - \bar{y}_{3\cdot})^2}{3/2},$$

(3.3)

$$SS(T \times B) = \frac{1}{2}\sum_j [(y_{1j} - y_{2j}) - (\bar{y}_{1\cdot} - \bar{y}_{2\cdot})]^2$$

$$+ \frac{2}{3}\sum_j \left[\left(\frac{1}{2}y_{1j} + \frac{1}{2}y_{2j} - y_{3j}\right) - \left(\frac{1}{2}\bar{y}_{1\cdot} + \frac{1}{2}\bar{y}_{2\cdot} - \bar{y}_{3\cdot}\right)\right]^2.$$

The t-tests come from using the components as numerator and denominator, but it is also the case that each component of SS(T × B) has the same expectation, and they all can be pooled for a common error term, which is the MS(T × B) in the RCB anova. (See Exercise 3.4 for a numerical illustration, and Exercise 3.27 for the full blown algebra.)

Thus, the "interaction" term in the RCB anova can be understood as a sum of error terms for all treatment contrasts, and is the correct pooled error estimate for any treatment contrast. As Fisher and Wishart (1930) remarked when analyzing an RCB with 6 blocks and 5 treatments, and hence 20 df for $T \times B$ (my italics)

> There are, therefore, 20 degrees of freedom which represent the differences among the different blocks of the comparisons between treatments. *These may be regarded as the discrepancies, or errors, of our experiment,* and if the treatments have been assigned their places in the different blocks wholly at random it may be shown that these 20 degrees of freedom do in fact supply a valid estimate of error for the treatment comparisons which the experiment was designed to make.

3.1.3 The RCB Anova

In the RCB the error comes from the variation of treatment contrasts across blocks, not from within a cell. It is possible to have a within error term in an RCB anova, which would happen if replications were taken within the treatment – block combinations. The model would look like (compare to (3.1))

$$(3.4) \qquad Y_{ijk} = \mu + \tau_i + \beta_j + (\tau\beta)_{ij} + \varepsilon_{ijk},$$
$$i = 1, \ldots, t, \quad j = 1, \ldots, b, \quad k = 1, \ldots, r.$$

In such a case there would be a "within" row added to the anova tables in Section 3.1.2, and the RCB anova table would look like Table 3.3.

We refer to Table 3.3 as an RCB anova table with *subsampling*, to indicate that the extra samples taken are typically subsamples of the experimental unit, and are not contributing degrees of freedom to the main contrasts of interest. That is, the test on treatments is exactly the same regardless of the presence of the within sum of squares term, in that the error degrees of freedom are the same. In this sense, the extra $bt(r-1)$ observations taken in Table 3.3 are a waste of effort with respect to the test on treatments.

The term

$$y_{ij} - \overline{y}_{i\cdot} - \overline{y}_{\cdot j} + \overline{\overline{y}}$$

is a "residual" in the sense that it is variation in the cell that is unexplained by the marginal means. However, this is *exactly* the definition of interaction, and this is the name that should be used. So, we do not refer to the error term as "residual," but rather as the "T × B" interaction. Although it is fine to call it "residual", we want to constantly emphasize that the correct error term comes from the T × B interaction.

Moreover, although this term can be called "residual", it is never a "within error", as in the CRD anova.

Example 3.4. ALFALFA BLOCKS - RCB ANALYSIS   Four varieties of alfalfa (Ladak, Narragansett, DuPuits and Flamand) were tested in an RCB with four blocks. The response variable was yield, in tons of dry hay per acre. For each Variety × Block cell there were three subsamples. The data are in `Alfalfa`, and are schematically given in Table 3.4 with anova table

Source	df	Sum Sq	Mean Sq	F	p
Block	3	3.982	1.327		
Variety	3	37.201	12.400	26.068	.000
Variety × Block	9	4.281	0.476	1.880	.092
Within	32	8.100	0.253		

Note that the test on treatments (Variety) is still against the Variety × Block interaction. The fact that there are three observations in each cell does nothing to improve the treatment test (well, almost nothing).

We sometimes gain the ability to test the interaction term, but this test is really of lesser interest. This is because whether or not there is an interaction between Treatments and Blocks is somewhat of an academic question since,

Table 3.3. RCB anova table with subsampling.

Source of Variation	Degrees of Freedom	Sum of Squares	Mean Square	F-statistic
Blocks	$b-1$	$\text{SS(Blocks)} = \sum_j t(\bar{y}_{.j.} - \bar{\bar{y}})^2$		
Trts.	$t-1$	$\text{SS(Trt)} = \sum_i b(\bar{y}_{i..} - \bar{\bar{y}})^2$	$\text{MS(Trt)} = \text{SS(Trt)}/(t-1)$	$F = \dfrac{\text{MS(Trt)}}{\text{MS(T} \times \text{B)}}$
T × B	$(b-1)(t-1)$	$\text{SS(T} \times \text{B)} = \sum_i \sum_j (\bar{y}_{ij.} - \bar{y}_{i..} - \bar{y}_{.j.} + \bar{\bar{y}})^2$	$\text{MS(T} \times \text{B)} = \frac{\text{SS(T} \times \text{B)}}{(b-1)(t-1)}$	
Sub-sampling (Within)	$bt(r-1)$	$\text{SS(Within)} = \sum_i \sum_j \sum_k (y_{ijk} - \bar{y}_{ij.})^2$	$\text{MS(Within)} = \frac{\text{SS(Within)}}{bt(r-1)}$	
Total	$rbt-1$	$\text{SS(Total)} = \sum \sum (y_{ijk} - \bar{\bar{y}})^2$		

Table 3.4. Alfalfa RCB data with three observations/cell

		Blocks			
		1	2	3	4
	1	y_{111}	y_{121}	y_{131}	y_{141}
		y_{112}	y_{122}	y_{132}	y_{142}
		y_{113}	y_{123}	y_{133}	y_{143}
Treatments	$\vdots$	$\vdots$	$\vdots$	$\vdots$	$\vdots$
	4	y_{411}	y_{421}	y_{431}	y_{441}
		y_{412}	y_{422}	y_{432}	y_{442}
		y_{413}	y_{423}	y_{433}	y_{443}

by their very nature, we cannot control blocks (see Section 3.5). So, if there is an interaction *and* a significant treatment effect, we know that this is an average effect, and we cannot expect the same treatment ordering in each block.

‖

Take note!

The design lesson to be learned from Example 3.4 is that subsampling in an RCB can be a waste of time and effort. If more observations can be added, if at all possible the number of blocks should be increased. This is the one surefire way to increase the error degrees of freedom for the treatment variance estimate.

3.2 Model and Distribution Assumptions ⎯⎯⎯⎯⎯⎯⎯

Blocking serves many purposes. Within a block there is homogeneity, so treatment comparisons are very precise. Between blocks there is heterogeneity, so treatment comparisons are made across a wide variety of situations, and thus, if we see treatment differences we can have some assurance that the differences are significant even in the face of block variability.

There is a distinction between *fixed* and *random* effects. With a fixed effect, all of the treatments (or levels of a treatment) of interest are included in the experiment, so what we designate as "Treatments" is always a fixed effect. (This is the case in a CRD.) Indeed, any factor for which the inference on means is of interest is a fixed effect.

In contrast, with a random effect, all of the levels of interest are *not* in the experiment. For example, in Example 3.3, although we are interested in inferring about treatment differences regardless of the type of soil (block), we certainly cannot have all types of soil in the experiment. Therefore, we regard the blocks in the experiment as representative of a population of blocks (for example, all soil types). If the blocks are a random sample from a population of blocks, they are a particular case of a *random effect* (or random factor).

Definition 3.5. A *factor* is a variable defining a categorization, in particular in an anova. A factor is a *fixed factor* if all the levels of interest are included in the experiment. A factor is a *random factor* if all the levels of interest are *not* included in the experiment and those that are can be considered to be randomly chosen from all the levels of interest.

In an anova, the treatments are always a fixed factor, as all of the levels of interest are in the experiment. Blocks are typically a random factor since not all levels of interest can be in the experiment. However, blocks are a special type of random factor. There is really no interest in random blocks; they are there only because we know that the treatments will behave differently on different blocks.

> **Note:** Realize that blocks are only effective if they span a wide variety of situations and result in a large sum of squares.

Look back at (1.11), but interpret it like this:

$$SS(\text{Total}) - SS(\text{Treatments})) = SS(\text{Blocks}) + SS(T \times B).$$

So the more variation that blocks can remove, the greater the efficiency of the design.

The right side of the above equation is leftover from the treatments. The more of this variation that blocks can remove, the smaller the unexplained - residual - error (interaction) will be.

The classical analysis on the RCB model given in (3.1) goes as follows. If we assume that the blocks are random, then the actual block means in the experiment, $\beta = (\beta_1, \ldots, \beta_b)$, are a realization of a random variable, and random variables Y_{ij} are observed according to the model

(3.5) $Y_{ij} = \mu + \tau_i + \beta_j + \varepsilon_{ij}, \quad i = 1, \ldots, t, \; j = 1, \ldots, b,$ $\qquad$ The classical RCB model

where

(1) The random variables $\varepsilon_{ij} \sim$ iid $N(0, \sigma_\varepsilon^2)$ for $i = 1, \ldots, t$, and $j = 1, \ldots, b$ (normal errors with equal variances).
(2) The random variables $\beta_1, \ldots, \beta_b$, are iid $N(0, \sigma_\beta^2)$ and are independent of ε_{ij} for all i, j.

The mean and variance of Y_{ij}, conditional on the β_js are

$$(3.6) \qquad \mathrm{E}(Y_{ij}) = \mu + \tau_i + \beta_j, \quad \mathrm{Var}\,(Y_{ij}) = \sigma_\varepsilon^2.$$

If we now take expectations over both the ε_{ij}s and the β_js we have

$$(3.7) \qquad \mathrm{E}Y_{ij} = \mu + \tau_i, \quad \mathrm{Var}\,Y_{ij} = \sigma_\beta^2 + \sigma_\varepsilon^2.$$

(See Exercise 3.21 for details.)

Remember that (3.5) describes the Y_{ij} conditional on the blocks, and conditionally they are independent. From (3.5),

$$
\begin{aligned}
\mathrm{Cov}(Y_{ij}, Y_{i'j'}|\beta_j, \beta_{j'}) &= \mathrm{Cov}(\mu + \tau_i + \beta_j + \varepsilon_{ij}, \mu + \tau_{i'} + \beta_{j'} + \varepsilon_{i'j'}) \\
(3.8) \qquad &= \mathrm{Cov}(\varepsilon_{ij}, \varepsilon_{i'j'}) = 0, \qquad [\text{property of covariance}]
\end{aligned}
$$

because here we are conditioning on the β_js. However, the important calculation is the unconditional covariance between observations, that is, when we also take expectations over the blocks. We then have, for Y_{ij} and $Y_{i'j}$ in block j, with $i \neq i'$,

$$
\begin{aligned}
\mathrm{Cov}(Y_{ij}, Y_{i'j}) &= \mathrm{Cov}(\mu + \tau_i + \beta_j + \varepsilon_{ij}, \mu + \tau_{i'} + \beta_j + \varepsilon_{i'j}) \\
&= \mathrm{Cov}(\beta_j + \varepsilon_{ij}, \beta_j + \varepsilon_{i'j}). \qquad [\text{property of covariance}]
\end{aligned}
$$

Now we use the fact that the β_js and ε_{ij}s are independent to write

$$(3.9) \qquad \mathrm{Cov}(\beta_j + \varepsilon_{ij}, \beta_j + \varepsilon_{i'j}) = \mathrm{Cov}(\beta_j, \beta_j) = \sigma_\beta^2$$

showing that not only does the model imply that there is correlation in the blocks, but that it is *positive* correlation. This is a consequence of the additive model (3.5) and the assumption that the εs and βs are independent. (See Exercise 3.21.)

| Conditional vs. unconditional inference | Note the difference between (3.8) and (3.9). If we decide to make inferences conditional on the observed blocks, the Y_{ij}s are uncorrelated, and the inference is only to those blocks in the experiment. If we infer to the population of blocks, using the unconditional model, we have a wider inference. [1] |

In addition, the unconditional correlation between Y_{ij} and $Y_{i'j}$ is

$$(3.10) \qquad \mathrm{Corr}(Y_{ij}, Y_{i'j}) = \frac{\mathrm{Cov}(Y_{ij}, Y_{i'j})}{\sqrt{(\mathrm{Var}\,Y_{ij})(\mathrm{Var}\,Y_{i'j})}} = \frac{\sigma_\beta^2}{\sigma_\beta^2 + \sigma_\varepsilon^2},$$

a quantity called the *intraclass correlation* (see Exercise 3.6). Note also that there is no correlation between observations in different blocks, as those observations are independent from the model assumptions.

---

[1] This distinction is examined by McLean *et al.* (1991), who characterize "narrow" and "broad" inference space.

Models such as (3.5), which can be referred to as *mixed models* (since there are both fixed and random factors) is somewhat problematic to address in terms of parameter estimates. This is because, by the very nature of a random factor, we are not really interested in estimating the levels of the factor that are in the experiment. Why? Because if the factor is truly random, the levels in the experiment are nuisance parameters, and only the variance of the factor is meaningful for inference. There are methods of estimating these random effects, however, that are typically referred to as *prediction* rather than estimation. See Technical Note 3.8.4 for some details.

Example 3.6. ESTIMATION OF RANDOM EFFECTS   The effectiveness of three anticoagulant drugs in dissolving blood clots was studied. Each of five subjects (blocks) received all three drugs (in random order with adequate washout time in between), and the length of time (in seconds) required for a cut of specified size to stop bleeding was recorded. The data are given in Table 3.5. This is clearly a case in which blocks must be modeled as random.

Table 3.5. Time (seconds) for bleeding to stop after the administration of an anticoagulant drug.

		A	B	C
			Anticoagulant drug	
	1	127.5	129.0	135.5
Person	2	130.6	129.1	138.0
(Block)	3	118.3	111.7	110.1
	4	155.5	144.3	162.3
	5	180.7	174.4	181.8

The five subjects in the experiment are to represent a random sample of all subjects; that is where we want to inference to apply. To model the blocks as fixed, where the inference only applies to the five subjects in the experiment, does not make sense.   ‖

3.3 Expected Squares and F-tests _____

We first look at EMS calculations for the case of one observation per treatment-block combination, considering the blocks to be random. Contrast the results here with those in Section 4.3. Although the overall conclusions and inferences are virtually the same, there are important differences in the details.

Table 3.6. Expected Mean Squares for RCB anova with random blocks and no subsampling.

Source	df	EMS
Blocks	$b-1$	$\sigma_\varepsilon^2 + t\sigma_\beta^2$
Treatments	$t-1$	$\sigma_\varepsilon^2 + \frac{b}{t-1}\sum_i [\tau_i - \bar{\tau}]^2$
TxB	$(t-1)(b-1)$	σ_ε^2

Here we calculate expected means squares, and look at potential F-tests, for the RCB model with blocks random. We note that there are many ways of specifying an RCB with random blocks, and many names have been attached to different models (see Miscellanea 3.9.1). However, here we will stay with model (3.5), one of the more commonly used models (see, for example, Dean and Voss 1999, Section 17.9).

As an illustration of an EMS calculation, consider

$$\text{ESS(Blocks)} = \text{E}\sum_j t(\bar{Y}_j - \bar{Y})^2 = t\text{E}\sum_j \left[\beta_j - \bar{\beta} + \bar{\varepsilon}_j - \bar{\varepsilon}\right]^2$$

$$= t\text{E}\sum_j \left[\beta_j - \bar{\beta}\right]^2 + t\text{E}\sum_j (\bar{\varepsilon}_j - \bar{\varepsilon})^2,$$

where we have used the fact that the β_j and ε_{ij} are independent of one another. Since they are also *iid*, each term above is the respective variance component, and we have from Lemma 3.16,

(3.11) $$\text{ESS(Blocks)} = (b-1)\left(t\sigma_\beta^2 + \sigma_\varepsilon^2\right).$$

Continuing in this fashion we can produce the anova in Table 3.6

The test on treatments is of the null hypothesis

(3.12) $$H_0 : \tau_i - \bar{\tau} = 0 \text{ for all } i .$$

We see that under H_0, MS(Trts) and MS(T × B) have the same expectation, and the F-test comes to us from Cochran's Theorem

Theorem 3.7 (F-test for RCB anova).   *Under model (3.5) with hypothesis (3.12),*

$$\frac{\text{SS}(\textit{Trts})}{\sigma_\varepsilon^2} \sim \chi_{t-1}^2, \qquad \frac{\text{SS}(T \times B)}{\sigma_\varepsilon^2} \sim \chi_{(b-1)(t-1)}^2,$$

independently, and thus

$$\frac{\text{MS}(\textit{Trts})}{\text{MS}(T \times B)} \sim F_{t-1,(b-1)(t-1)}.$$

Proof: See Technical Note 3.8.2   □

   The theory here is rather simple, only complicated by the fact that the Y_{ij} are correlated, so the application of Cochran's Theorem takes a little bit of work. But the important point is that there is a valid *F*-test against the "residual", with no need for a within sum of squares.

   What is sometimes of direct interest is to assess the treatment-block interaction, that is, to see whether treatment effect patterns are different in different blocks. From Table 3.6 there is no obvious way to do this, but we will revisit this question in Section 3.5.

Example 3.8. RCB INTERACTION PLOTS   The anticoagulant data in Table 3.5 is plotted in Figure 3.1. The plot suggests an interaction between the Subjects (Blocks) and the treatment. Treatments *A* and *C* swap places as taking the longest time for bleeding to stop, and *B* sometimes takes longer than *A*. The implications of such an interaction are that different treatments may be more effective for different subjects. However, the treatment effects are not significant (Example 3.9), so there is no significant difference in the average effects   ‖

Fig. 3.1. Interaction plot of the anticoagulant data of Table 3.5.

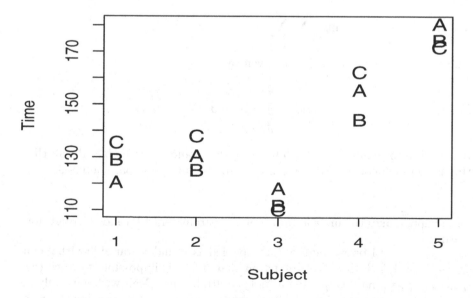

3.4 Estimating Contrasts

We next turn to estimating treatment contrasts and their variances. To start, we estimate the treatment effects using least squares, which provides us with unbiased estimates. It is important to realize, however, that the actual technique of least squares is a mathematical minimization, and does not take into account any error structure. Thus least squares estimates are the same whether blocks are fixed or random.

> **Note:** The technique of *ordinary least squares* derives estimates based only on the treatment design of an experiment, and ignores the experiment design.

Although it is typical to use ordinary least squares to get these estimates, blindly applying least squares can vary from problematic to nonoptimal. Estimates of error variances can change depending on whether we have fixed or random effects, and whether we have covariances. Alternatives such as *generalized* or *weighted* least squares may be reasonable. Here we will outline a typical estimation strategy based on ordinary least squares. More details are given in Technical Note 3.8.4.

Point Estimates

Using the treatment design of (3.5), the least squares estimates of μ, τ_i and β_j satisfy

$$(3.13) \qquad \min_{\mu, \tau_i, \beta_j} \sum_{i=1}^{t} \sum_{j=1}^{b} (y_{ij} - \mu - \tau_i - \beta_j)^2.$$

The solution (see Exercise 3.2) is given by

$$(3.14) \qquad \begin{aligned} \bar{y}_{i\cdot} - \bar{\bar{y}} &= \tau_i - \bar{\tau}, \\ \bar{y}_{\cdot j} - \bar{\bar{y}} &= \beta_j - \bar{\beta}, \\ \bar{\bar{y}} &= \mu + \bar{\tau} + \bar{\beta}. \end{aligned}$$

It is typical to require $\bar{\tau} = \bar{\beta} = 0$ to make the τ_i and β_j estimable. Recall that this is merely a convenience, amounting to redefining the parameters as

$$\tau_i' = \tau_i - \bar{\tau}, \qquad \beta_j' = \beta_j - \bar{\beta},$$

and simply makes the interpretation of the parameters a bit more straightforward.

A consequence of random blocks	However, if the β_j is modeled as a random variable with mean $\bar{\beta}$, then the restriction that $\bar{\beta} = 0$ is impossible to force (we have $E\beta_j = 0$, but not $\bar{\beta} = 0$), but just dealt with naturally in calculations. In fact, as we have previously mentioned, since β_j is random, there is usually little interest in the effect estimate $\hat{\beta}_j$, rather the interest is in the variance estimate.

   Least squares estimation does not take into account the fact that there is correlation in the blocks. To do so we could turn to generalized least squares. Such estimates, in theory, have smaller variance than the ordinary least squares estimators of (3.14), but accounting for the covariance in these point estimates can bring along another set of problems. Thus, we will use the least squares estimates of (3.14), which are unbiased, and examine their variances and the variance of contrast estimates. (See Exercise 3.2 and Technical Note 3.8.4 for details.)

   As least squares estimators are unbiased, they are so under model (3.5), and we have

$$(3.15) \qquad \begin{aligned} E(\bar{Y}_{i.} - \bar{Y}) &= E\,\hat{\tau}_i = \tau_i, \\ E(\bar{Y}_{.j} - \bar{Y}) &= E(\hat{\beta}_j - \bar{\beta}) = 0. \end{aligned}$$

   Thus a contrast estimate $\sum_i a_i \hat{\tau}_i$ is an unbiased estimate of the contrast $\sum_i a_i \tau_i$ and has variance

$$\begin{aligned} \mathrm{Var}\left(\sum_i a_i \hat{\tau}_i\right) &= \mathrm{Var}\left(\sum_i a_i(\bar{Y}_{i.} - \bar{\bar{Y}})\right) \\ &= \mathrm{Var}\left(\sum_i a_i(\bar{\varepsilon}_i - \bar{\varepsilon})\right) \qquad \text{[from (3.5)]} \\ &= \mathrm{Var}\left(\sum_i a_i\bar{\varepsilon}_i\right) \qquad [\textstyle\sum_i a_i\bar{\varepsilon} = 0]. \end{aligned}$$

Notice that the term involving $\bar{\beta}$, which occurs in both $\bar{Y}_{i.}$ and $\bar{\bar{Y}}$ cancels out – a result of the balance of the design. Finally, using the fact that the $\bar{\varepsilon}_i$ are all independent with variance σ_ε^2/b,

$$(3.16) \qquad \mathrm{Var}\left(\sum_i a_i \hat{\tau}_i\right) = \frac{\sigma_\varepsilon^2}{b} \sum_i a_i^2.$$

Variance Estimates

From (3.5) and (3.14) the residuals from the model are

$$y_{ij} - \hat{\mu} - \hat{\tau}_i - \hat{\beta}_j = y_{ij} - \bar{y}_{i.} - \bar{y}_{.j} + \bar{\bar{y}} = \varepsilon_{ij} - \bar{\varepsilon}_{i.} - \bar{\varepsilon}_{.j} + \bar{\bar{\varepsilon}} \stackrel{\mathrm{def}}{=} \hat{\varepsilon}_{ij},$$

and we would typically base our variance estimate on the residual mean square,

$$(3.17) \qquad \mathrm{SS(Res)} = \sum_{ij}(\varepsilon_{ij} - \bar{\varepsilon}_{i.} - \bar{\varepsilon}_{.j} + \bar{\bar{\varepsilon}})^2 = \sum_{ij}\hat{\varepsilon}_{ij}^2.$$

This should be the basis of our variance estimate, and we now look at its expectation. A relatively straightforward calculation (see Exercise 3.23) will show that
$$\text{ESS(Res)} = (b-1)(t-1)(1-\rho)\sigma_\varepsilon^2.$$

Of course, we realize that SS(Res) is exactly SS(T × B) from the anova in Table 3.3. Thus, the mean square $\hat{\sigma}_\varepsilon^2 = \text{MS(T × B)}$ is an unbiased estimator of the treatment variance, and for any contrast we can estimate its variance with

(3.18)
$$\hat{\text{Var}}\left(\sum_i a_i \hat{\tau}_i\right) = \frac{\text{MS(T × B)}}{b}\sum_i a_i^2.$$

Example 3.9. PAIRWISE COMPARISONS   Continuing with Example 3.6, the anova is

Source	df	SS	MS	F	p-value
Subjects	4	7162.9	1790.7	43.55	
Treatment	2	112.9	56.5	1.38	0.31
S × T	8	328.9	41.1		

giving us a variance estimate of $\hat{\sigma}_\varepsilon^2 = 41.1$. To compare the difference of any two treatment means, the contrast would have variance $\frac{41.1}{5}(1^2 + (-1)^2) = 16.44$.   ‖

Inference ───

The most important inference from a RCB anova concerns the treatments and, in particular, the estimation of contrasts between the treatments. We do all calculations under the RCB model (3.5). The parameter of interest is the treatment contrast $\sum_{i=1}^k a_i \tau_i$, whose estimator $\sum_{i=1}^k a_i \overline{Y}_{i\cdot}$ satisfies (Technical Note 3.8.4)

(3.19) $$E\left(\sum_{i=1}^t a_i \overline{Y}_{i\cdot}\right) = \sum_{i=1}^k a_i \tau_i \text{ and } \text{Var}\left(\sum_{i=1}^t a_i \overline{Y}_{i\cdot}\right) = \frac{\sigma_\varepsilon^2}{b}\sum_i a_i^2.$$

Since the Y_{ij}s are normal, we have

$$\sum_{i=1}^k a_i \overline{Y}_{i\cdot} \sim N\left(\sum_{i=1}^k a_i \tau_i, \frac{\sigma_\varepsilon^2}{b}\sum_i a_i^2\right)$$

and, therefore,

(3.20) $$\frac{\sum_{i=1}^k a_i \overline{Y}_{i\cdot} - \sum_{i=1}^k a_i \tau_i}{\sqrt{\frac{\sigma_\varepsilon^2}{b}\sum_i a_i^2}} \sim N(0, 1).$$

From (3.18) we can estimate σ_ε^2 with MS(T $\times$ B), and it remains to establish the distribution of

$$(3.21) \qquad \frac{\sum_{i=1}^k a_i \overline{Y}_{i\cdot} - \sum_{i=1}^k a_i \tau_i}{\sqrt{\frac{\text{MS(TxB)}}{b} \sum_i a_i^2}}.$$

That this, in fact, has Student's t-distribution with $(t-1)(b-1)$ degrees of freedom follows from Cochran's Theorem (Section 1.8) and a bunch of algebra with SS(T $\times$ B). We get the following theorem, whose proof is left to Exercise 3.26.

Theorem 3.10 (Distribution of Contrasts). *For the RCB model (3.5):*

(1) The quantities $\overline{Y}_{i'\cdot}$ and $Y_{ij} - \overline{Y}_{i\cdot} - \overline{Y}_{\cdot j} + \overline{\overline{Y}}$ are independent for all i and i'

(2) SS(T $\times$ B) = SS(Res) of (4.16) satisfies

$$\frac{\text{SS}(T \times B)}{\sigma_\varepsilon^2} \sim \chi^2_{(t-1)(b-1)},$$

(3)

$$\frac{\sum_{i=1}^k a_i \overline{Y}_{i\cdot} - \sum_{i=1}^k a_i \tau_i}{\sqrt{\frac{\text{MS(TxB)}}{b} \sum_i a_i^2}} \sim t_{(t-1)(b-1)}.$$

Thus, using Theorem 3.10, to test

$$H_0: \sum_{i=1}^t a_i \tau_i = 0 \qquad \text{vs.} \qquad H_1: \sum_{i=1}^t a_i \tau_i \neq 0$$

at level α, we have

$$(3.22) \qquad \text{reject } H_0 \text{ if } \left| \frac{\sum_{i=1}^t a_i \overline{Y}_{i\cdot}}{\sqrt{\frac{\text{MS(TxB)}}{b} \sum_{i=1}^t a_i^2}} \right| > t_{(b-1)(t-1),\alpha/2}.$$

More importantly, we get an interval estimator of $\sum a_i \tau_i$. With probability $1 - \alpha$,

$$(3.23) \qquad \sum_{i=1}^t a_i \overline{Y}_{i\cdot} - t_{(b-1)(t-1),\alpha/2} \sqrt{\frac{\text{MS}(T \times B)}{b} \sum_{i=1}^t a_i^2}$$

$$\leq \sum_{i=1}^t a_i \tau_i \leq \sum_{i=1}^t a_i \overline{Y}_{i\cdot} + t_{(b-1)(t-1),\alpha/2} \sqrt{\frac{\text{MS}(T \times B)}{b} \sum_{i=1}^t a_i^2}.$$

Implications of Complete Blocking ——————————————————————

Realize that the fact that the blocks are *complete* plays an important role in freeing treatment contrasts from block effects. If blocks are *incomplete*, that is, if not every block contains every treatment, then the treatment contrasts will not, in general, be independent of block effects. In complicated situations, an incomplete block design (Chapter 6) may be preferred (or necessary) and the resulting anova is more complicated than those considered here. Furthermore, even if an incomplete design is not preferred, it may be dictated by data-gathering problems.

A reasonable question to wonder about is what inference can be made about blocks, that is, can block effects be tested or estimated? This is an area where statisticians do not generally agree – it is almost a matter of taste. The formal mathematical statistics can be done in different, correct, ways and different, correct, answers can be obtained. In particular, the complete answer to this question is tied to the parameterization used for the model (see Miscellanea 3.9.1).

Looking at (1.11), also see the note above equation (3.5), it is clear that if the block sum of squares is very small, then blocking has increased the error estimate (since the degrees of freedom used for calculating the error mean square will have decreased). In other words, blocking pays off only if the blocks are significant.

> This is why we block

Thus, if we are in a situation where the blocks can be chosen, it makes sense to choose them as disparate as possible. This also makes good common sense, in that we want to verify our treatment comparisons on as wide a variety of situations as possible. And it also makes statistical sense, in that by increasing the block sum of squares we reduce the error sum of squares.

If the variation in blocks is not controllable, as in Example 3.6 where we block on subjects, but their use is dictated by the inherent design, then we just hope that the variation removed due to blocking is a large piece (and it typically is).

3.5 Modeling the Interaction ——————————————————————

In the preceding section the RCB model did not explicitly contain an interaction term, indeed, the interaction $(\tau\beta)_{ij}$ is confounded with the error ε_{ij}. Or, stated in another way, in the preceding section we could everywhere replace ε_{ij} with $(\tau\beta)_{ij} + \varepsilon_{ij}$ and the analyses would be unchanged.

However, if we want to learn about the block-treatment interaction, we must somehow replicate it. One way to do this is to have a model of the form

$$y_{ijk} = \mu + \tau_i + \beta_j + (\tau\beta)_{ij} + \varepsilon_{ijk},$$

where the term $(\tau\beta)_{ij}$ represents an interaction, that is, an effect in the (i,j) cell that is beyond the effect of $\mu + \tau_i + \beta_j$. Another way of replicating the interaction is to replicate the entire experiment; we will address that approach in Section 3.6.1.

We first look a little closer at the error term ε_{ijk}, recalling from Section 1.6 the difference between "true" replication and "technical" replication (the latter also going under the names "subsampling" or "pseudo-replication"). The important difference is that a true replication is the independent replication of the experimental unit, while the technical replication is subsampling within the same experimental unit. So, for example:

(1) In a microarray experiment, if RNA from the *same* subject is used in two different microarrays, this is a technical replication. A true replication would have RNA from different subjects on each microarray.

(2) In a block, if the treatment is variety of plant, and we have independent replicates of each variety, then we have true replication. If the treatment is fertilizer applied to a subplot with 5 plants of the same variety, then the 5 plants are a technical replication.

The important difference comes out in how we model the covariance (as always!). Technical replication (Section 1.6) means that the treatment was applied in a common way, which induces correlation, while true replication results in independent application of the treatment to different experimental units, and no correlation is induced. Thus, we model for $k \neq k'$

The action is in the covariance	$\mathrm{Corr}(\varepsilon_{ijk}, \varepsilon_{i'jk'}) = \begin{cases} \rho_\varepsilon & \text{for technical replication} \\ 0 & \text{for true replication.} \end{cases}$

A similar distinction is made by Gates (1995), and he allows having both true and technical replication simultaneously in the same experiment. Here we will mainly be concerned with whether the within error term arises from true or technical replication, and we will do our calculations with correlation ρ_ε, and can set $\rho_\varepsilon = 0$ for true replications. (Although we maintain that in most cases technical replication is a waste of time, so replications should only be done if they are true replications.)

In the RCB model with blocks random, the natural extension of model (3.5) is somewhat elusive, as it is not clear how to categorize the interaction, which will now be a mix of a fixed effect (treatment) and a random effect (blocks). There has been much written on such models; see the discussion in Miscellanea 3.9.1.

Here we will use the following model, which is a *variation* of the so-called Model II. (This can be found in, for example, Hocking 1973, 1985 Chapter 15; Dean and Voss 1999 Section 17.9)

$$Y_{ijk} = \mu + \tau_i + \beta_j + (\tau\beta)_{ij} + \varepsilon_{ijk},$$

(3.24)
$$i = 1, \ldots, t, \quad j = 1, \ldots, b, \quad k = 1, \ldots, r,$$

where

(1) The random variables $\varepsilon_{ijk} \sim \mathrm{N}(0,\sigma^2)$ for $i = 1,\ldots,t,,\ j = 1,\ldots,b$ and $k = 1,\ldots,r$ (normal errors with equal variances), with

$$\mathrm{Corr}(\varepsilon_{ijk}, \varepsilon_{i'jk'}) = \rho_\varepsilon$$

(2) The random variables $(\tau\beta)_{11}, \ldots, (\tau\beta)_{tb}$, are $\mathrm{N}(0, \sigma_{\tau\beta}^2)$ are independent of β_j and ε_{ij} for all i, j, k, and satisfy

$$\mathrm{Corr}((\tau\beta)_{ij}, (\tau\beta)_{i'j}) = \rho_{\tau\beta}.$$

(3) The random variables $\beta_1, \ldots, \beta_b$, are iid $\mathrm{N}(0, \sigma_\beta^2)$ and are independent of ε_{ijk} for all i, j.

The variation on the standard Model II is that we add the assumption of correlation between interaction effects within the same block, and between errors within the same cell. A consequence of the model is

$$\mathrm{Corr}(Y_{ijk}, Y_{i'j'k'}) = 0 \text{ unless } j = j',$$

$$(3.25) \qquad \mathrm{Corr}(Y_{ijk}, Y_{ijk'}) = \frac{\sigma_\beta^2 + \sigma_{\tau\beta}^2 + \rho_\varepsilon\sigma_\varepsilon^2}{\sigma_\beta^2 + \sigma_{\tau\beta}^2 + \sigma_\varepsilon^2} \qquad [\text{inside T} \times \text{B}],$$

$$\mathrm{Corr}(Y_{ijk}, Y_{i'jk'}) = \frac{\sigma_\beta^2 + \rho_{\tau\beta}\sigma_{\tau\beta}^2}{\sigma_\beta^2 + \sigma_{\tau\beta}^2 + \sigma_\varepsilon^2} \qquad [\text{inside B}].$$

The standard Model II has all $(\tau\beta)_{ij}$ and ε_{ijk} independent, which implies that $\rho_{\tau\beta} = \rho_\varepsilon = 0$. Except in special cases, this does not make good sense. That is, once we have chosen the random block, the interactions must take place within that block, so the interaction effects should, in fact, be correlated. (See Section 3.9.1 for a discussion of the difference between a block and a *random factor*, where the standard Model II makes sense for the latter.) Of course, $\mathrm{Cov}(\bar{Y}_{ij\cdot}, \bar{Y}_{i'j\cdot}) \neq 0$ even if we assume $\rho_{\tau\beta} = 0$, (Exercise 3.10) but this is a reflection of the intraclass correlation in a block design.

Note: The Y_{ijk}s are correlated even if ρ_ε and $\rho_{\tau\beta}$ are both zero. The question is whether we should model the error random variables to be correlated.

Now we take the expected values of the effects in (3.24) under these assumptions, using some of the details given in Exercise 3.30. As an illustration, consider

$$\mathrm{E}\,\mathrm{SS}(\mathrm{Blocks}) = \mathrm{E}\sum_j rt(\bar{Y}_j - \bar{Y})^2 = rt\mathrm{E}\sum_j \left[\beta_j - \bar{\beta} + (\bar{\tau\beta})_j - (\bar{\tau\beta}) + \bar{\varepsilon}_j - \bar{\varepsilon}\right]^2$$

$$= rt(b-1)\sigma_\beta^2 + r(b-1)\sigma_{\tau\beta}^2[1 + (t-1)\rho_{\tau\beta}] + (b-1)\sigma_\varepsilon^2[1 + (r-1)\rho_\varepsilon],$$

Table 3.7. Expected Mean Squares for RCB anova with random blocks and replication.

Source	df	EMS
Blocks	$b-1$	$\sigma_\varepsilon^2[1+(r-1)\rho_\varepsilon]+r\sigma_{\tau\beta}^2[1+(t-1)\rho_{\tau\beta}]+rt\sigma_\beta^2$
Treatments	$t-1$	$\sigma_\varepsilon^2[1+(r-1)\rho_\varepsilon]+r\sigma_{\tau\beta}^2[1-\rho_{\tau\beta}]+\frac{rt}{t-1}\sum_i(\tau_i-\bar\tau)^2$
T × B	$(t-1)(b-1)$	$\sigma_\varepsilon^2[1+(r-1)\rho_\varepsilon]+r\sigma_{\tau\beta}^2[1-\rho_{\tau\beta}]$
Within	$bt(r-1)$	$(1-\rho_\varepsilon)\sigma_\varepsilon^2$

where we have used the fact that the β_j, $(\tau\beta)_{ij}$ and ε_{ijk} are all independent of one another, but we have carefully kept track of the correlation in the $(\tau\beta)$ and ε. Similarly,

$$\mathrm{E\,SS(Trts)} = \mathrm{E}\sum_i rb(\bar Y_i - \bar Y)^2 = rb\mathrm{E}\sum_j \left[\tau_i - \bar\tau + (\tau\beta)_i - (\bar{\tau\beta}) + \bar\varepsilon_i - \bar\varepsilon\right]^2$$

$$= rt\sum_i(\tau_i - \bar\tau)^2 + r(t-1)\sigma_{\tau\beta}^2[1-\rho_{\tau\beta}] + (t-1)\sigma_\varepsilon^2[1+(r-1)\rho_\varepsilon],$$

and continuing in this fashion we can produce the anova in Table 3.7. By examining the expected mean squares, we see that Treatments are again to be tested against the T × B interaction. It turns out that formal justification of this is quite straightforward (Technical Note 3.8.3), as the same argument used in the no-interaction model works here.

_F_-test on treatments _____

The test on treatments is against MS(T × B) with no other assumptions needed. Formally, the null hypothesis of no treatment effect is

$$H_0 : \tau_i - \bar\tau = 0 \text{ for all } i$$

and is tested by

$$F_{t-1,(b-1)(t-1)} = \frac{\mathrm{MS(Trts)}}{\mathrm{MS(T \times B)}}$$

Again, treatments are *never* tested against MS(Within), in fact, here again SS(Within) is of no use unless we assume that $\rho_\varepsilon = 0$, that is, the replications are true replications.

Other _F_-tests _____

From Table 3.7, to get a valid F-test on interactions we need to assume that $\rho_\varepsilon = 0$, meaning we have true replication and not just technical subsampling. If $\rho_\varepsilon = 0$ we can test the hypothesis of no interaction

Table 3.8. Expected Mean Squares for RCB anova with random blocks and replication, assuming that $\rho_\varepsilon = \rho_{\tau\beta} = 0$.

Source	df	EMS
Blocks	$b-1$	$\sigma_\varepsilon^2 + r\sigma_{\tau\beta}^2 + rt\sigma_\beta^2$
Treatments	$t-1$	$\sigma_\varepsilon^2 + r\sigma_{\tau\beta}^2 + \frac{rt}{t-1}\sum_i(\tau_i - \bar{\tau})^2$
T × B	$(t-1)(b-1)$	$\sigma_\varepsilon^2 + r\sigma_{\tau\beta}^2$
Within	$bt(r-1)$	σ_ε^2

$$(3.26) \qquad H_0 : \sigma_{\tau\beta}^2 = 0 \text{ or } \rho_{\tau\beta} = 1$$

with

$$F_{(b-1)(t-1),r(b-1)(t-1)} = \frac{\text{MS(T × B)}}{\text{MS(Within)}}.$$

We also note that under the assumption $\rho_\varepsilon = 0$ there is a test on blocks, but there is usually little reason to care about this.

> **Note:** We again see that unless there is true replication, which implies that $\rho_\varepsilon = 0$, the within sum of squares in an RCB is of almost no use.

Finally, we note that the most commonly used RCB model assumes that $\rho_\varepsilon = \rho_{\tau\beta} = 0$, resulting in the anova given in Table 3.8 with all tests being straightforward. Thus, if enough assumptions can be made, things become simple.

3.6 Variations on a Theme

Here we discuss some strategies and designs that build from the RCB. There are fewer details here, as all of the substantive theory has already been covered in the previous sections.

3.6.1 Replicating the Experiment

In Section 3.5 we saw that if there is true replication of the experimental unit, then we get a valid test of the Treatment × Block interaction. Another way to obtain replication is to repeat the entire experiment, a strategy that we examine in this section.

Although it is somewhat of a luxury to be able to replicate an experiment, there are often good reasons to do so. For example, an agricultural experiment may be replicated in different years to allow for such variation when

Table 3.9. RCB replicated with blocks nested in replications.

	Replications						
	1		2	....		r	
	Block		Block			Block	
Trt	1 $\cdots$ b	Trt	1 $\cdots$ b		Trt	1 $\cdots$ b	
1	x $\cdots$ x	1	x $\cdots$ x	....	1	x $\cdots$ x	
$\vdots$	$\vdots$ $\vdots$ $\vdots$	$\vdots$	$\vdots$ $\vdots$ $\vdots$		$\vdots$	$\vdots$ $\vdots$ $\vdots$	
t	x $\cdots$ x	t	x $\cdots$ x		t	x $\cdots$ x	

looking at treatment differences; a microarray experiment may be repeated by different labs to allow for laboratory variation. We look at RCB models that are replicated, perhaps over years or over locations. We find, somewhat surprisingly (although maybe less so after reading Section 3.6.2) that we will need some further assumptions to get valid tests!

There are two kinds of models to look at, depending on whether blocks are nested in or crossed with replications. We treat the nested case here, and leave the crossed case for Exercise 3.15.

If an RCB is replicated in such a way that the blocks are nested in replications, we can model this as

$$(3.27) \qquad Y_{ijk} = \mu + R_k + \tau_i + \beta_{jk} + (\tau R)_{ik} + (\tau\beta)_{ijk} + \varepsilon_{ijk},$$
$$i = 1, \ldots, t, \quad j = 1, \ldots, b, \quad k = 1, \ldots, r,$$

where

$$
\begin{aligned}
Y_{ijk} &= \text{the response,} \\
R_k &= \text{replications, N}(0, \sigma_R^2), \\
\tau_i &= \text{treatments, crossed with replications,} \\
\beta_{jk} &= \text{blocks, nested in replications, N}(0, \sigma_\beta^2), \\
(\tau R)_{ik} &= \text{Trt} \times \text{Rep interaction, N}(0, \sigma_{\tau R}^2), \\
(\tau\beta)_{ijk} &= \text{Trt} \times \text{Block interaction, nested in Reps, N}(0, \sigma_{\tau\beta}^2), \\
\varepsilon_{ijk} &= \text{the experimental error, N}(0, \sigma_\varepsilon^2),
\end{aligned}
$$

and all of the error terms are independent. A schematic of this design is in Table 3.9, where the blocks are assumed to be different in each replication. Here is an example.

Example 3.11. REPLICATED RCBs, NESTED Five varieties of alfalfa were arranged in an RCB with five blocks, with the goal of finding the variety with maximum yield. The RCB was replicated at three different locations and, for each variety the yield, in tons of dry hay per acre, was recorded. The data

Table 3.10. Expected mean squares for a replicated RCB design, blocks nested in replications.

Source	df	EMS
Replications	r-1	$\sigma_\varepsilon^2 + \sigma_{\tau\beta}^2 + t\sigma_\beta^2 + b\sigma_{\tau R}^2 + bt\sigma_R^2$
Blocks (in Reps)	r(b-1)	$\sigma_\varepsilon^2 + \sigma_{\tau\beta}^2 + t\sigma_\beta^2$
Treatments	t-1	$\sigma_\varepsilon^2 + \sigma_{\tau\beta}^2 + b\sigma_{\tau R}^2 + \frac{rb}{t-1}\sum_i \tau_i^2$
Trt × Rep	(t-1)(r-1)	$\sigma_\varepsilon^2 + \sigma_{\tau\beta}^2 + b\sigma_{\tau R}^2$
Trt × Block (in Rep)	r(t-1)(b-1)	$\sigma_\varepsilon^2 + \sigma_{\tau\beta}^2$
Total	btr-1	

(in dataset `AlfalfaTrial-1`) were analyzed according to model (3.27), and resulted in the following anova:

Source	df	SS	MS
Location	2	3.119	1.559
Blocks(in Locations)	12	17.017	1.418
Variety	4	4.516	1.129
Variety × Location	8	1.702	0.213
Variety × Block (in Location)	48	5.843	0.122

Now it would be wonderful to use those 48 degrees of freedom for the treatment test, but things are not that straightforward.   ‖

The expected mean squares for model (3.27) are given in Table 3.10, where we see that the test on treatments does not come from the interaction with blocks, but from the interaction with reps. Thus, in Example 3.11, the F-test on treatments has 8 degrees of freedom in the denominator, rather than 48.

Here is the bad news

Note: To use the Trt × Block (in Rep) mean square to test treatments, we need to assume that $\sigma_{\tau R}^2 = 0$, that is, there is no interaction of the treatments with the replications.

If the replications truly represent a repeat of the experiment, then this may be a tenable assumption.

Example 3.12. REPLICATED RCBs, NESTED, CONTINUED Here, the locations are functioning as bigger blocks, and there is no basis for the assumption that $\sigma_{\tau R}^2 = 0$. In fact, the test of $H_0 : \sigma_{\tau R}^2 = 0$ yields

Table 3.11. Expected mean squares for a replicated RCB design, blocks crossed with replications.

Source	df	EMS
Replications	$r-1$	$\sigma_\varepsilon^2 + \sigma_{\tau\beta R}^2 + t\sigma_{\beta R}^2 + b\sigma_{\tau R}^2 + bt\sigma_R^2$
Blocks	$b-1$	$\sigma_\varepsilon^2 + \sigma_{\tau\beta R}^2 + t\sigma_{\beta R}^2 + r\sigma_{\tau\beta}^2 + rt\sigma_\beta^2$
Blocks × Reps	$(b-1)(r-1)$	$\sigma_\varepsilon^2 + \sigma_{\tau\beta R}^2 + t\sigma_{\beta R}^2$
Treatments	$t-1$	$\sigma_\varepsilon^2 + \sigma_{\tau\beta R}^2 + b\sigma_{\tau R}^2 + r\sigma_{\tau\beta}^2 + \frac{rb}{t-1}\sum_i \tau_i^2$
Trt × Rep	$(t-1)(r-1)$	$\sigma_\varepsilon^2 + \sigma_{\tau\beta R}^2 + b\sigma_{\tau R}^2$
Trt × Block	$(t-1)(b-1)$	$\sigma_\varepsilon^2 + \sigma_{\tau\beta R}^2 + r\sigma_{\tau\beta}^2$
Trt × Block × Rep	$(t-1)(b-1)(r-1)$	$\sigma_\varepsilon^2 + \sigma_{\tau\beta R}^2$
Total	$btr-1$	

$$F_{8,48} = \frac{0.213}{0.122} = 1.746 \text{ with } p\text{-value} = .112.$$

Although this is not significant at the magic levels, this is hardly strong evident of no interaction. Pooling the error terms, or using the Trt × Block (in Rep), to test treatments, may not be a good idea (see Exercise 3.16).   ‖

The situation does not improve (in fact, it worsens) if, instead of being nested in replications, the blocks are crossed with replications. A model for that situation is

$$(3.28) \quad Y_{ijk} = \mu + R_k + \tau_i + \beta_j + (\tau\beta)_{ij} + (\tau R)_{ik} + (\beta R)_{jk} + (\tau\beta R)_{ijk} + \varepsilon_{ijk},$$

and with R and β random, as well as the interactions with these terms, the EMS is given in Table 3.11.

Even more bad news ‖ In Table 3.11 we see that there is no straightforward test on treatments unless we make assumptions about the interactions (alternatively, we could use Satterthwaite's approximation; see Technical Note 5.8.1).

In particular, if either $\sigma_{\tau\beta}^2 = 0$ or $\sigma_{\tau R}^2 = 0$, or both, then we can test treatments. We of course can test these interactions, but then we are back in the situation of basing on test on the outcome of another, which is never pleasant (see Exercise 3.15).

3.6.2 Crossed Blocks

The crossed blocks model is a straightforward extension of the RCB, where we have two blocking factors that are crossed, and within this blocking structure treatments are randomized. Do not confuse this design with the *strip plot* design of Section 5.6.1.

Table 3.12. Crossed blocks design. Factors B and C are blocks. The treatment T is randomized on the intersection of B and C.

A schematic of this design is given in Table 3.12. This is another design that has its origins in agriculture, where it was sometimes noticed that a field plot has two gradients running in perpendicular directions - typically having to do with soil composition, moisture level, light levels, etc.

Example 3.13. CROSSED BLOCKS   A plant breeder conducted an experiment to compare the yields of 3 new varieties and a standard variety of peanuts. The varieties were assigned to the plots where it was known that land had a slight sloping from east to west and differences in available nitrogen from north to south.

Thus, to account for this variability it was decided to block in the east-west direction and in the north-south direction, and assign the four varieties randomly in each intersection of the blocks. ||

A model for the crossed blocks design is

$$(3.29) \quad Y_{ijk} = \mu + \tau_i + \beta_j + \gamma_k + (\beta\tau)_{ij} + (\beta\gamma)_{jk} + (\tau\gamma)_{ik} + (\beta\tau\gamma)_{ijk} + \varepsilon_{ijk},$$

where $i = 1,\ldots,t$, $j = 1,\ldots,r$, $k = 1,\ldots,g$, τ is the treatment, β and γ are the blocks, and

$$\beta_j \sim N(0,\sigma_\beta^2), \quad \gamma_k \sim N(0,\sigma_\gamma^2), \quad \varepsilon_{ij} \sim N(0,\sigma_\varepsilon^2),$$

all independent. The interactions with B and G are also random normal with the appropriate error variances, and all random effects are independent.

Although this looks like an innocuous extension of the RCB, it turns out that this is an awful design, as will be evident by looking at Table 3.13.

We see from the EMS in Table 3.13 that there is no test on treatments unless we make some assumptions about the interactions (or try a Satterthwaite approximation). In particular, if we assume that either $\sigma_{\tau\gamma}^2 = 0$ or $\sigma_{\tau\beta}^2 = 0$, we can then test treatments against a twoway interaction.

We do have tests on the twoway interactions, however, and it might be tempting to first do these tests and then, based on their outcomes, move on to a treatment test if we accept the hypothesis that one of $\sigma_{\tau\gamma}^2$ or $\sigma_{\tau\beta}^2$ is zero. But we are never happy to base a test on the outcome of another test, as error terms get compounded.

Notice that if we could do the design and analysis as a typical RCB with bg blocks having no structure, there would be no problem with any of the tests. The problem arises when there is some structure in the blocking that precludes the ordinary RCB approach.

| The trouble with crossed blocks | The point of this section is to show what complications arise when blocks are crossed, and that to use such a design the only reasonable course is to assume that the Treatment × Block interactions are all zero. Once this is done we then have a straightforward analysis. |

And what is typically done with crossed blocks is not to do the full factorial of model (3.29), but to only do a piece of the design in a clever way. Such a design is called a Latin square.

3.6.3 Latin Squares

A Latin square is a design in which two gradients are controlled with crossed blocks, but in each intersection there is only one treatment level. Moreover,

Table 3.13. Expected mean squares for a crossed block design.

Source	df	EMS
Blocks B	$b - 1$	$\sigma_\varepsilon^2 + \sigma_{\beta\tau\gamma}^2 + t\sigma_{\beta\gamma}^2 + g\sigma_{\tau\beta}^2 + tg\sigma_\beta^2$
Blocks C	$g - 1$	$\sigma_\varepsilon^2 + \sigma_{\beta\tau\gamma}^2 + t\sigma_{\beta\gamma}^2 + r\sigma_{\tau\gamma}^2 + tr\sigma_\gamma^2$
T	$t - 1$	$\sigma_\varepsilon^2 + \sigma_{\beta\tau\gamma}^2 + g\sigma_{\tau\beta}^2 + r\sigma_{\tau\gamma}^2 + \frac{rg}{t-1}\sum_i \tau_i^2$
B × T	$(b-1)(t-1)$	$\sigma_\varepsilon^2 + \sigma_{\beta\tau\gamma}^2 + g\sigma_{\tau\beta}^2$
C × T	$(g-1)(t-1)$	$\sigma_\varepsilon^2 + \sigma_{\beta\tau\gamma}^2 + r\sigma_{\tau\gamma}^2$
B × C	$(b-1)(g-1)$	$\sigma_\varepsilon^2 + \sigma_{\beta\tau\gamma}^2 + t\sigma_{\beta\gamma}^2$
B × C × T	$(b-1)(g-1)(t-1)$	$\sigma_\varepsilon^2 + \sigma_{\beta\tau\gamma}^2$
Total	bgt-1	

each row contains exactly one level of each treatment, and each column contains exactly one level of each treatment.

Example 3.14. PEANUT LATIN SQUARE   The experiment of Example 3.13 was actually run as a Latin square. If we denote the four treatments by T_1, T_2, T_3, T_4, the field layout of the experiment was

Blocks B East-West

		1	2	3	4
Blocks	1	T_3	T_1	T_2	T_4
C	2	T_1	T_2	T_4	T_3
North-South	3	T_2	T_4	T_3	T_1
	4	T_4	T_3	T_1	T_2

Note that necessarily, if there are t levels of the treatment, we must have t rows and t columns. So the design is a *square*. For obvious reasons, this design is sometimes referred to as a *Row-Column* design. The anova for this experiment (dataset `Peanut`) is

Source	df	SS	MS	F	p-value
Row	3	9.427	3.142		
Column	3	245.912	81.971		
Treatment	3	23.417	7.806	1.953	.223
Residuals	6	23.984	3.997		
Total	15	302.74			

where here we really mean *residuals*. Comparing this design to the full crossed blocks design in Section 3.6.2, it should be clear that the residual terms contains pieces of many different factors - all of the twoway and threeway interactions.

Note that if we did not have a categorization for treatments, then the anova would have been

Source	df	SS	MS
Row	3	9.427	3.142
Column	3	245.912	81.971
Row × Column	9	47.401	5.267
Total	15	302.74	

and so the SS(Treatments) gets pulled out of the Row × Column effect and we have

$$\text{SS(Residual)} = \text{SS(Row} \times \text{Column)} - \text{SS(Treatments)},$$

which reinforces why it is essential that there is no Row × Column effect, that is, that the residual is only measuring experimental error.   ‖

The F-test in the Latin square anova can only be formally justified if we assume that there are no twoway or threeway interactions. Then from Table 3.13 the residual MS is an estimate of σ_ε^2, and will provide the denominator for the test on treatments. But note that if the assumptions about no interactions are violated, this will lead to conservative tests, as the denominator will tend to be bigger than expected (Technical Note 2.8.4).

A model for the Latin square design is

$$(3.30) \qquad\qquad Y_{ijk} = \mu + \tau_i + \beta_j + \gamma_k + \varepsilon_{jk},$$

where τ is the treatment, β and γ are the blocks, $j = 1, \ldots, t$, $k = 1, \ldots, t$, and

$$ijk \in I_i = \{(j,k) : Y_{ijk} \text{ is from Treatment } i\}.$$

We need to define this more complex index set in order to keep exact track of the observations. If we just say that $i = 1, \ldots, t$, then we are indexing t^3 observations where there are only t^2 observations. Finally, we assume that

$$\beta_j \sim N(0, \sigma_\beta^2), \quad \gamma_k \sim N(0, \sigma_\gamma^2), \quad \varepsilon_{jk} \sim N(0, \sigma_\varepsilon^2),$$

all independent.

Compare this to the crossed blocks model (3.29). The models are quite similar; the differences being that here we assume all of the interaction variances are zero, and there is not a complete set of treatments within each intersection of rows and columns.

However, the Latin square is balanced in the following sense. Since each treatment is in exactly one row and one column, when we calculate treatment means the row/column effects are balanced and, in fact, under model (3.30) they cancel out. Consider the parameter arrangement corresponding to the Peanut Latin square:

Blocks B East-West

		1	2	3	4
Blocks	1	$\tau_3 + \beta_1 + \gamma_1$	$\tau_1 + \beta_2 + \gamma_1$	$\tau_2 + \beta_3 + \gamma_1$	$\tau_4 + \beta_4 + \gamma_1$
C	2	$\tau_1 + \beta_1 + \gamma_2$	$\tau_2 + \beta_2 + \gamma_2$	$\tau_4 + \beta_3 + \gamma_2$	$\tau_3 + \beta_4 + \gamma_2$
North-South	3	$\tau_2 + \beta_1 + \gamma_3$	$\tau_4 + \beta_2 + \gamma_3$	$\tau_3 + \beta_3 + \gamma_3$	$\tau_1 + \beta_4 + \gamma_3$
	4	$\tau_4 + \beta_1 + \gamma_4$	$\tau_3 + \beta_2 + \gamma_4$	$\tau_1 + \beta_3 + \gamma_4$	$\tau_2 + \beta_4 + \gamma_4$

If we sum over the cells corresponding to any particular τ_i, we see that we bring along all the βs and γs in a balanced way. For example,

$$\bar{Y}_1 = \frac{1}{t} \sum_{(j,k)\in I_1} Y_{1jk} = \mu + \tau_1 + \bar{\beta} + \bar{\gamma} + \bar{\varepsilon}_1,$$

$$\bar{\bar{Y}} = \frac{1}{t^2} \sum_i \sum_{(j,k)\in I_1} Y_{ijk} = \mu + \bar{\tau} + \bar{\beta} + \bar{\gamma} + \bar{\bar{\varepsilon}},$$

and thus

$$\bar{Y}_1 - \bar{\bar{Y}} = \tau_1 - \bar{\tau} + \bar{\varepsilon}_1 - \bar{\bar{\varepsilon}},$$

where $\bar{\varepsilon}_i = (1/t)\sum_{(j,k)\in I_i} \varepsilon_{jk}$. Thus the least squares estimates of the treatment effects are free of all row/column effects. If we restrict $\bar{\tau} = 0$ then

(3.31) $$\mathrm{E}(\bar{Y}_i - \bar{\bar{Y}}) = \tau_i \text{ and } \mathrm{Var}(\bar{Y}_i - \bar{\bar{Y}}) = \left(1 - \frac{1}{t}\right)\frac{\sigma_\varepsilon^2}{t},$$

free of all blocking effects. In fact, under model (3.30), $\bar{Y}_i - \bar{\bar{Y}}$ is the least squares estimate of τ_i (Exercise 3.18).

Finally, contrasts are also free of block effects since for any set of contrast coefficients $(a_1, \ldots, a_t)$,

$$\mathrm{E}\left(\sum_i a_i \bar{Y}_i\right) = \sum_i a_i \tau_i$$

(3.32) $$\mathrm{Var}\left(\sum_i a_i \bar{Y}_i\right) = \frac{\sigma_\varepsilon^2}{t}\sum_i a_i^2.$$

The Latin square is quite restrictive in that we need to have t rows, t columns, and t treatments and, therefore, there are only $(t-2)(t-1)$ degrees of freedom for the residual. One way to alleviate this problem, if possible, is to replicate the Latin squares.

Example 3.15. REPLICATING LATIN SQUARES   Consider an extension of Example 3.13 where the experiment is replicated three times:

Replications

	1					2					3			
	1	2	3	4		1	2	3	4		1	2	3	4
1	T_3	T_1	T_2	T_4	1	T_1	T_2	T_3	T_4	1	T_4	T_1	T_2	T_3
2	T_1	T_2	T_4	T_3	2	T_2	T_3	T_4	T_1	2	T_3	T_4	T_1	T_2
3	T_2	T_4	T_3	T_1	3	T_3	T_4	T_1	T_2	3	T_1	T_2	T_3	T_4
4	T_4	T_3	T_1	T_2	4	T_4	T_1	T_2	T_3	4	T_2	T_3	T_4	T_1

In general, if there are t treatments and r replications, the anova for the Latin Square is

Crossed			Nested		
Source	df				
Replications	$r-1$			Source	df
Rows	$t-1$			Replications	$r-1$
Columns	$t-1$			Rows (in Reps)	$r(t-1)$
Treatments	$t-1$			Columns (in Reps)	$r(t-1)$
Rows × Reps	$(r-1)(t-1)$			Treatments	$t-1$
Columns × Reps	$(r-1)(t-1)$			Residual	$(t-1)(rt-r-1)$
Residual	$(t-1)(rt-r-1)$			Total	rt^2-1
Total	rt^2-1				

and we have slightly different anovas depending on whether the Rows and Columns are crossed or nested with Replications (the Treatments are always crossed). Depending on the situation, either configuration is possible, but note that in either case the residual term is the same (Exercise 3.17), giving us an increase in residual degrees of freedom, and keeping the treatment effects balanced across the rows and columns.

For the crossed Latin squares we have implicitly assume that the rows and the columns are the same in each replication. A slight variation produces a *Latin Rectangle*. Suppose that the rows are crossed, but the columns are different in each replication, labeled $1, \ldots, rt$, so we essentially have Latin squares lined up next to each other. The analysis is almost the same as the crossed anova above, except that the degrees of freedom for Reps, Columns and their interaction get combined, that is $(r-1)+(t-1)+(r-1)(t-1) = rt-1$.
‖

We close this section with a number of observations about Latin squares

(1) If rows are ignored, the Latin square is an RCB, similarly it is an RCB if columns are ignored. With both blocking factors the Latin square is balanced in the sense that the treatment contrasts are free of confounding with block effects. However, it is also important to realize that the Latin square really has nothing to do with blocking – it is a treatment design that balances a treatment against rows and columns - which may be other treatments (see Example 6.20).

(2) It is called a Latin square because, typically, the treatments are denoted by the Latin letters A, B, C, D, ... The Latin Square whose first row is ABCDE and first column is ABCDE is called a *standard square*. Note that we can always write down the standard square with a cyclic plan; first row is ABCDE, second row is BCDEA, third row is CDEAB, etc. This will work for any size square.

(3) Formally, the Latin square that we choose in an experiment should be chosen at random from all possible Latin squares. For example, there are 12 3×3 Latin squares, 576 4×4 Latin squares and $161,280$ 5×5 Latin

Fig. 3.2. Relationship of full factorial crossed blocks design with a Latin square. For $t = 4$ the full factorial would need 64 observations, while the Latin square uses 16. This is, formally, a fractional replication; see Section 6.3.

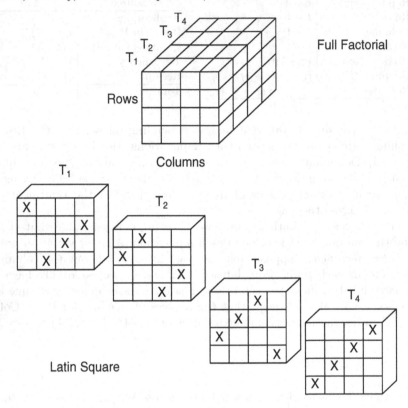

squares. However, choosing one or two at random is not a straightforward task. What can be easily done is to write down the standard square and then randomly permute rows and columns, which preserves the Latin square. In Example 3.15, the middle square is standard and the third square is a row-column permutation. (This is, of course, not the same as choosing a square at random from the entire set.)

(4) We do the Latin square to save observations (money) and (we hope) still get good information. Figure 3.2 shows the relationship of the Latin square to the full factorial with crossed blocks. The Latin square needs 1/4 the number of observations, and gives treatment effect estimates that are free of block effects.

Of course, we pay for this in that we need to make assumptions about the interactions in order to get good error estimates and valid tests. (We are also gambling that no observations will be missing or unusable. If there is missing data in the Latin square all the balance is lost.

> Remember –
> the assumption
> is NO
> interactions

(5) When using more than one Latin square, such as in Example 3.15, there is some thought that orthogonal squares should be used. Two Latin squares are *orthogonal* if, on superposition of one square on the other, each of the t^2 ordered pairs of treatment levels occurs exactly once. The mathematics of orthogonal Latin squares is very interesting, leading one into advanced group theory and there are advantages to using orthogonal Latin squares in designs (Morgan 1998). However, on this topic we tend to side more with Mead (1988, p. 181) who says, when discussing this topic, "At various points ... the mathematical ideas involved in the construction of useful designs offer a temptation to divert from the path of usefulness" (but see Section 6.4 for a specialized design that depends on orthogonal Latin squares).

3.7 Exercises

Essential

3.1 Referring to Example 3.3:
  (a) Recreate the anova tables and the t-tests
  (b) Break down the sums of squares of the RCB anova according to (3.3).
  (c) Show how to calculate the t-statistics using the component sums of squares in (3.3).
  (d) Show that under model (3.1), each component of SS(T $\times$ B) in (3.3) has the same expectation.
3.2 For the model (3.5):
  (a) Show that the solution to (3.13) gives the least squares estimates (3.14).
  (b) Show that the restriction $\bar{\tau} = \bar{\beta} = 0$ is equivalent to defining $\tau_i' = \tau_i - \bar{\tau}$, $\beta_j' = \beta_j - \bar{\beta}$, $\mu' = \mu + \bar{\tau} + \bar{\beta}$ and fitting the model $y_{ij} = \mu' + \tau_i' + \beta_j' + \varepsilon_{ij}$.
  (c) Restricting $\bar{\tau} = \bar{\beta} = 0$ is one of many possible restrictions to get a solution. Another common one is to set $\tau_t = \beta_b = 0$. Derive the least squares estimates under this constraint.
  (d) Show that the least squares estimates are unbiased and find the variance of the estimators.
3.3 Referring to Example 3.4 and using dataset **Alfalfa**:
  (a) Verify the anova table given in the example.
  (b) Show that the test on treatments remains the same if the three observations in each Variety $\times$ Block cell are replaced by their mean, creating a new dataset consisting of sixteen observations.
  (c) Suppose that 48 observations could be taken. Write out the anova table for experiments with 4 blocks, 8 blocks, and 12 blocks. Comment on situations when each design would be preferred.

3.4 In Section 3.1.2 we split the 2 degrees of freedom for treatments in the `Strawberry` dataset using two orthogonal contrasts $A - B$ and $\frac{1}{2}A + \frac{1}{2}B - C$, and saw that each of the two resulting RCB anovas was equivalent to the analogous paired t-test. Here we want to show that the individual anova sums of squares for treatment and residual add up to those in the full anova table given in Example 3.3. But we have to be careful about constants.
Suppose that we have b blocks, and take a treatment contrast $\sum_{i=1}^{k} a_i \bar{y}_i$ and split it into two pieces

$$\bar{w}_1 = \sum_{i=1}^{m} a_i \bar{y}_i \text{ and } \bar{w}_2 = -\sum_{i=m+1}^{k} a_i \bar{y}_i$$

(note the negative sign) where $m < k$ is arbitrary. If we define $\bar{w}_1$ and $\bar{w}_2$ to be the means of two treatment groups show that

(a) $\text{SS(Treatments)} = b[(\bar{w}_1 - \bar{w})^2 + (\bar{w}_2 - \bar{w})^2] = \frac{b}{2}(\sum_{i=1}^{k} a_i \bar{y}_i)^2$, (note that the choice of m is irrelevant) and hence
(b) $\text{SS(Treatments)}/\text{SS(Contrast)} = (1/2)\sum_i a_i^2$.
(c) Use the information in (b) to adjust the contrasts $A - B$ and $\frac{1}{2}A + \frac{1}{2}B - C$ to produce the two anova tables

Modified $A - B$			Modified $\frac{1}{2}A + \frac{1}{2}B - C$		
Source	df	Sum Sq	Source	df	Sum Sq
Block	3	1.724	Block	3	1.335
Trt	1	34.031	Trt	1	1.550
T × B	3	0.164	T × B	3	0.561

and show that the treatment and residual sums of squares for these table add to the full table given in Example 3.3.

3.5 To investigate the effect on potato yields of water loss due to transpiration, a horticulturist used shade covers on plots at various stages of their growth and development. Plots were shaded to reduce solar input (to the plants) by 0, 1/3, or 2/3. Each of the 3 shadings were applied to 4 plots for a one-month period during "early", "middle" and "late" stages of growth. The design was an RCB. Yields per plot (in lbs) were

		Growth Stage		
Block	Shading	Early	Middle	Late
	0	60	65	62
1	1/3	54	57	58
	2/3	41	53	56
	0	53	68	70
2	1/3	46	53	62
	2/3	42	58	54
	0	64	58	54
3	1/3	48	59	63
	2/3	36	50	60
	0	50	61	57
4	1/3	42	52	60
	2/3	50	49	51

(a) Construct the complete anova table. Be specific about your assumptions about blocks and include expected mean squares in your table.

(b) Construct and test an appropriate set of contrasts.

3.6 A variation of the RCB model (3.5) is used in the social sciences to measure the magnitude of a characteristic possessed by a person, when there are multiple (possibly subjective) measurements of the characteristic. Winer (1971) explores such models in detail; we give a small example. To fix the idea, look at the following data where 6 people were rated by 4 judges on a specific characteristic.

Person	Judge 1	Judge 2	Judge 3	Judge 4
1	1	4	3	3
2	5	7	5	6
3	1	3	1	2
4	7	9	9	8
5	2	4	6	1
6	6	8	8	4

Notice that the people are the blocks. It is now standard to assume that there is no treatment effect (judge effect), so that $\tau_i = 0$ in model (3.5) and we now have the model

$$Y_{ij} = \mu + P_j + \varepsilon_{ij}, \quad \varepsilon_{ij} \sim N(0, \sigma_\varepsilon^2), \quad P_j \sim N(0, \sigma_P^2), \text{ independent,}$$

with $1, \ldots, t$, $j = 1, \ldots, b$.

(a) Show that the EMS for this model is

Source	df	EMS
Between People	$b - 1$	$\sigma_\varepsilon^2 + t\sigma_P^2$
Within People	$b(t - 1)$	σ_ε^2

(b) The *reliability* of the mean of t measurements is defined as

$$\rho_t = \frac{\sigma_P^2}{\sigma_P^2 + (\sigma_\varepsilon^2/t)}.$$

Show that ρ_1 is the same as the *intraclass correlation*, and that we can write $\rho_t = t\rho_1/(1 + (t - 1)\rho_1)$. Show that ρ_t is an increasing function of t.

(c) Show how to estimate σ_P^2 and σ_ε^2 using mean square from an anova, and hence show how to estimate ρ_k.

(d) Estimate ρ_1 and ρ_k for the above data.

(The quantity ρ_t is known as the *Spearman-Brown prediction formula* in the psychometric literature. Winer (1971) also shows how to estimate reliability without the assumption that $\tau_i = 0$, which is know as adjusting for anchor points)

3.7 A small city in the Midwest was considering purchasing some outside sculpture pieces to complement its downtown. There were a total of 40 pieces of art that were under consideration, which were grouped into four categories (energetic, pastoral, spiritual, and representational), where each category contained 10 pieces of art. A total of 12 judges were available, and each judge rated each piece of art on a 7-point scale (from 1=dislike to 7=like very much). The data can be laid out in the following way:

Category

	1	2	3	4
	Art	Art	Art	Art
	1 2 $\cdots$ 10	11 12 $\cdots$ 20	21 22 $\cdots$ 30	31 32 $\cdots$ 40
1	$x\,x\,\cdots\,x$	$x\,x\,\cdots\,x$	$x\,x\,\cdots\,x$	$x\,x\,\cdots\,x$
Judges :	:	:	:	:
12	$x\,x\,\cdots\,x$	$x\,x\,\cdots\,x$	$x\,x\,\cdots\,x$	$x\,x\,\cdots\,x$

There are a total of $4 \times 10 \times 12 = 480$ observations.

(a) The original analysis was a fully nested design, as seems to be indicated by the data layout. Show the anova for this design (source and df) and indicate the tests. Also explain why this design is incorrect (other than the fact that with a 7-point ordinal scale the normality assumption is suspect).

(b) Explain why the correct design is an RCB. Show the anova for this design (source and df) and indicate the tests.

(c) Reconcile the designs in (a) and (b) in the following sense. Compare the error terms for the test on Art, and explain why the error term in (b) is the more appropriate one.

3.8 For the following experiment: (i) Specify the model equation (ii) Set up the anova table (source, df and EMS) (iii) Specify two hypotheses and how they would be tested.

An accounting firm, prior to introducing a wide-spread training in statistical sampling of auditing, tested three training methods: (1) study at home with programmed training methods, (2) training sessions at local offices conducted by local staff, and (3) training sessions in Chicago conducted by national staff. Thirty auditors were put into 10 groups of three, according to time elapsed since college graduation, and the auditors in each group were randomly assigned to the three training methods. At the end of the training each auditor was asked to analyze a complex case; a proficiency measure based on this analysis was obtained for each auditor.

3.9 Lake trout deposit their eggs on reefs in late fall and the young trout (fry) have developed enough to leave the reef by the following spring. Research in one large reef in Lake Ontario has clearly shown that almost all fry are in a small section of the reef in early spring, where past data indicate that most eggs are deposited.

The biologists are trying to understand the characteristics of this section of the reef that and important to the lake trout, and have listed several characteristics:

1. large-sized rocks   2. a deep rocky layer above the sand
3. location near the edge of a dropoff   4. location near the middle of the reef
5. strong currents

The experimenters constructed three "subreefs", each of which had large rocks in deep layers. One was near the edge of a dropoff, one was near the upper part of the reef, and the third was at the bottom of the dropoff. On each subreef five egg traps were placed, and were each checked four times throughout the fall (four trapping periods). When checked, the traps were reset. The biologists were interested in the total catch for the season.

After collecting data for one year, the biologists want to analyze the data:

(a) Describe the experiment; explain the factors, blocks (if any), what is nested and crossed.

(b) Construct and anova table (Source, df, and EMS) that corresponds to your answer in (a).

(c) According to your EMS, what is the variance for the mean number of eggs collected near the edge of the dropoff?

(d) If the variance due to cages is three times larger than the error variance, what would have a greater effect on reducing the variance in part (c) - doubling the number of cages or doubling the number of trapping periods?

3.10 For the RCB model with interaction (3.24):

(a) Show that

$$EY_{ijk} = \mu + \tau_i, \qquad \text{Var} Y_{ijk} = \sigma_\varepsilon^2 + \sigma_{\tau\beta}^2 + \sigma_\beta^2,$$

and the covariances are given by (3.25).

(b) The cell means $\bar{Y}_{ij}$ satisfy

$$E\bar{Y}_{ij} = \mu + \tau_i, \qquad \text{Var} \bar{Y}_{ij} = \frac{\sigma_\varepsilon^2}{r}(1 + (r-1)\rho_\varepsilon) + \sigma_{\tau\beta}^2 + \sigma_\beta^2,$$

and

$$\text{Cov}(Y_{ij}, Y_{i'j}) = \begin{cases} 0 & j \neq j' \\ \sigma_{\tau\beta}^2 + \sigma_\beta^2 & j = j'. \end{cases}$$

3.11 Starting with the RCB model (3.5):

$$Y_{ij} = \mu + \tau_i + \beta_j + \varepsilon_{ij}, \quad i = 1, \ldots, t, \quad j = 1, \ldots, b,$$

suppose that the treatments have a factorial structure, that is, the treatment effects $\tau_1, \ldots, \tau_t$ arise from the cells of two crossed factors C and D in a CRD, so that the model could be written

$$Y_{ijk} = \mu + \gamma_i + \delta_k + (\gamma\delta)_{ik}$$
$$+ \beta_j + (\gamma\beta)_{ij} + (\delta\beta)_{jk} + (\gamma\delta\beta)_{ijk} + \varepsilon_{ijk}.$$

(a) Write out the EMS for this model, and show that each treatment effect is tested against its corresponding interaction with blocks, that is, test C with $C \times B$, etc.

(b) What assumptions are needed to pool the three Trt $\times$ Block interactions into one error term? (Recall Section 1.6.)

3.12 A microarray experiment was planned to determine if human brain stem cells can be turned into neurons. The stem cells can be converted to neurons chemically, or by treating them with GFP(Green Florescence Protein) and then transplanting them into mice. There are four treatment conditions that the experimenters are interested in: Control, Chemical, GFP, and GFP+transplant. Brain stem cells from six subjects are available.

(a) It is suggested that for each subject, the stem cells should be divided into four groups and randomly assigned to the four treatments. A microarray will be run for each Subject $\times$ Treatment combination. What design is this? Write down the model, assumptions, and anova table.

(b) The experimenter thinks that two microarrays can be run for each subject/treatment condition. That is, the stem cells for each subject could be divided into eight groups, and two groups of cells would be given each treatment. Write down the model and anova for this design, and comment on whether this is a good use of resources.

(c) It turned out that there was another twist to this experiment, which was not mentioned at first. Apparently the transplanting into mice of the stem cells was thought to be quite variable, and the experimenter wanted to transplant the cells from each human subject into more than one mouse. Write out the model and anova for the part (a) experiment where the stems cells of each subject are transplanted into three mice. Show how the tests are done, and comment on any problems (variance assumptions?).

3.13 Another microarray project involving stem cells was done at the University of Florida. A hematologist was interested in measuring gene expression in subjects before and after the administration of the drug G-CSF (granulocyte-colony stimulating factor), a growth factor that stimulates the bone marrow to make more white blood cells. Before the treatment, RNA from each of five subjects was hybridized to an Affymetrix chip; after treatment with G-CSF RNA was again collected and hybridized. For each of the 54,000+ genes[1], the expression level, a measure of genomic activity, was measured. The dataset **StemCell** contains data for 250 genes, and looks like (for four genes)

Genes

Subject	Trt	AFFX-BioB-5-at	AFFX-BioB-M-at	AFFX-BioB-3-at	AFFX-BioC-5-at
1	Post	961	1734.3	825.7	2746.8
1	Pre	734.8	1239.7	607.3	2425
2	Post	1737.2	2926.7	1602.2	5256.6
2	Pre	755.5	1215.3	670.9	2306.3
3	Post	777.4	1597.8	750.3	2723.9
3	Pre	791.1	1349.7	711.2	2134.3
4	Post	1022.5	1761.7	871.8	2958.9
4	Pre	706.6	1145.8	596.1	2189
5	Post	754.9	1374.1	637.2	2334.4
5	Pre	809.8	1262.9	629.1	2100.7

(a) Write a model for this analysis. As mentioned in Miscellanea 1.9.1, when considering a microarray design we only need consider the response of one gene.

(b) Analyze the data using the response of the first gene. (It is best to log transform the response.)

(c) A complete analysis of the dataset would repeat part (b) for all the genes. Do this, and list the genes that are significant at a False Discovery Rate (FDR) of 5% (See Miscellanea 2.9.1). For another approach to the analysis of microarray data see Miscellanea 5.9.1.

3.14 There are three parental lines of *Persea americana*, or avocado, known as Bacon, Fuerte, and Zutano. Experimenters are interested in assessing genetic differences between these lines (treatments) and interactions with environment. These trees have been cloned, and four genetically identical trees were created. The trees are growing in California in two locations (environments) where two clones are randomly placed in each environment. Note that, since the same clone is used in both environments, clones are crossed with environments within parents. Four

---

[1] The chip actually does not have genes but "probe sets", which are segments of the genome that make up a gene.

genotypes from each parent are used in an experiment where the response is dry weight of fruit, and the anova table for the treatment design is

Source	df
Env	1
Parent	2
E × P	2
Genotype (in P)	9
G × E (in P)	9
Clones	24
Total	47

(a) Explain why this is an RCB, identify the blocks, and write down a model for the analysis.

(b) Calculate the EMS.

(c) Can we use Genotype (in Parents) to test the significance of Parents? Explain.

(d) Can G × E (in Parents) test Env and the Env × Parents interaction? Explain.

(e) Can we test the G × E (in P) interaction using the error term derived from the clone replicates? Explain.

3.15 Referring the Section 3.6.1, another variety trial was conducted with four varieties and 4 blocks, but here the replications were not locations but years (the same experiment was done in three successive years). In each year the same blocks were used, and thus replications (Years) are crossed with blocks. The data are in dataset AlfalfaTrial-2

(a) Construct an anova table based on model (3.28).

(b) Test the significance of the interactions with treatment. Based on those tests, and what you know about the experiment, are you willing to assume that one or both of the interactions is zero so that you can get a test of treatments?

(c) Discuss the consequences of an incorrect assumption about the interactions and the test on treatments with regard to Technical Note 2.8.4.

(d) As best as you can, estimate the treatment differences and their standard errors. What further conclusions can you make?

3.16 Referring to Example 3.11:

(a) Verify the anova table, and perform the F- tests as indicated by the EMS.

(b) If it is assumed that $\sigma^2_{\tau R} = 0$, and error terms for the treatment test are pooled based on this assumption, discuss the consequences if this assumption is incorrect (recall Technical Note 2.8.4).

(c) Estimate the treatment differences and their standard errors. What further conclusions can you make?

3.17 Referring to Examples 3.14 and 3.15:

(a) In Example 3.14:

   (*i*) Verify the anova table (dataset Peanut)

   (*ii*) Treatment 1 is the control treatment. Even though the anova F is not significant, do any of the other treatments have a significantly higher yield?

(*iii*) Using the current variance estimate, what power do you have to detect a .5 yield increase at $\alpha = .05$. How can you design another experiment, based on Latin squares, that would increase the power? Can you attain a power of .9 to detect a .5 yield increase at $\alpha = .05$?

(b) In Example 3.15, the peanut experiment was blocked at different locations in the field, but in each location the north-south/east-west gradients were identified. Analyze the blocked peanut data (dataset **Peanut2**) giving an anova table and seeing which pairwise differences are significant. Justify why this should be analyzed taking the Rows and Columns to be nested in replications.

(c) For the blocked case the residual is

$$\text{SS(Row} \times \text{Column)} + \text{SS(Rep} \times \text{Row} \times \text{Column)} - \text{SS(Treatment)}$$

and for the nested case it is

$$\text{SS(Row} \times \text{Column in Rep)} - \text{SS(Treatment)}.$$

Show that these must be the same.

3.18 Referring to Section 3.6.3:

(a) Verify the mean and variance given in (3.31).

(b) Show that $\bar{Y}_i - \bar{\bar{Y}}$ are the least squares estimates of τ_i and $\bar{\bar{Y}}$ is the least squares estimate of μ.

(c) Defining $\bar{Y}_j$ and $\bar{Y}_k$ in the obvious way, show that $\bar{Y}_j - \bar{\bar{Y}}$ and $\bar{Y}_k - \bar{\bar{Y}}$ are the least squares estimates of β_j and γ_k, respectively (although this is of lesser interest).

(d) Verify the contrast mean and variance in (3.32). In particular show that

$$\text{Var}(\bar{Y}_i) = \frac{1}{t}(\sigma_\varepsilon^2 + \sigma_\beta^2 + \sigma_\gamma^2), \quad \text{Cov}(\bar{Y}_i, \bar{Y}_{i'}) = \frac{1}{t}(\sigma_\beta^2 + \sigma_\gamma^2).$$

3.19 A poultry science professor used diets with low, medium, and high concentrations of protein to see if there were effects on the amount of food intake in leghorn chickens. Space limitations were such that the cages had to be stacked on top of one another and in front of each other (but with some space in between). This arrangement introduced two blockings; (*i*) the height of a cage was important because of a temperature differential (temperature affects food intake), and (*ii*) the depth of a cage was important because there were windows only on the front side of the cages (the amount of light affects food intake also). Therefore, cages were stacked 3 high and 3 deep (with spacers) and 10 chickens were randomly assigned to each cage. Treatments were then assigned to the cages according to the Latin square design below. After one week, total food intake of each cage was measured with the results given in the table (in ounces):

Depth	Height		
	Bottom Row	Middle Row	Top Row
Front Row	M(96)	H(81)	L(106)
Middle Row	H(94)	L(116)	M(114)
Top Row	L(100)	M(91)	H(89)

(a) Give the complete anova table.

(b) Test equality of mean food intakes for the concentrations of protein.

(c) Find two meaningful 1 df contrasts. Estimate the contrasts and give 90% confidence intervals.

3.20 (From Mead 1988, Chapter 10) A trial to compare four varieties of Brussels sprouts was designed as a 4×4 Latin square. Unfortunately, the experimental area was partially waterlogged during the course of the experiment, which produced at least one plot that was clearly not comparable with the other yields. Yields and scores of the extent of waterlogging (recorded for possible future use) are given below. Analyze the data, adjusting for the effect of waterlogging by covariance. Present the ancova table, with relevant tests, and the adjusted mean yields with standard errors.

Yields				Waterlogging Score			
98(B)	100(D)	127(A)	142(C)	0	0	0	0
141(C)	91(A)	110(D)	124(B)	0	0	0	0
98(D)	102(C)	103(B)	127(A)	.12	0	0	0
34(A)	71(B)	119(C)	118(D)	.45	.09	0	0

Accompaniment

3.21 Show details for the following calculations in Section 3.2.
(a) Verify (3.6) and (3.7).
(b) Verify (3.8), showing that the Y_{ij} are conditionally uncorrelated within a block.
(c) In (3.9), show that

$$\text{Cov}(\beta_j + \varepsilon_{ij}, \beta_j + \varepsilon_{i'j}) = \text{E}(\beta_j^2) - 2\text{E}(\beta_j + \varepsilon_{ij}) + \text{E}(\varepsilon_{ij}\varepsilon_{i'j}),$$

and hence that $\text{Cov}(Y_{ij}, Y_{i'j}) = \text{Var}(\beta_j) = \sigma_\beta^2$.
(d) Verify the EMS calculations in Table 3.6.

3.22 For the RCB model of (3.24),
(a) Show that the least squares estimates are the solutions to the equations

$$\bar{Y} = \mu + \bar{\tau} + \bar{\beta} + \overline{(\tau\beta)},$$
$$\bar{Y}_i = \mu + \tau_i + \bar{\beta} + \overline{(\tau\beta)}_i,$$
$$\bar{Y}_j = \mu + \bar{\tau} + \beta_j + \overline{(\tau\beta)}_j,$$
$$\bar{Y}_{ij} = \mu + \tau_i + \beta_j + (\tau\beta)_{ij}.$$

(b) Show how the estimates of the treatment effects follow from the least squares estimates.
(c) Comment on the desirability of having the parameters add to zero.

3.23 Refer to Section 3.4. To establish the expectation of (3.17), use Exercise 2.19, to show that

$$\sum_{ij}(\varepsilon_{ij} - \bar{\varepsilon}_{i\cdot} - \bar{\varepsilon}_{\cdot j} + \bar{\varepsilon})^2 = \sum_{ij}\varepsilon_{ij}^2 - b\sum_i \bar{\varepsilon}_{i\cdot}^2 - t\sum_j \bar{\varepsilon}_{\cdot j}^2 + bt\bar{\varepsilon}^2.$$

Then establish that

$$\text{E}\,\varepsilon_{ij}^2 = \sigma_\varepsilon^2, \quad \text{E}\,\bar{\varepsilon}_{i\cdot}^2 = \frac{\sigma_\varepsilon^2}{b}, \quad \text{E}\,\bar{\varepsilon}_{\cdot j}^2 = \frac{\sigma_\varepsilon^2}{t}, \quad \text{E}\,\bar{\varepsilon}^2 = \frac{\sigma_\varepsilon^2}{bt},$$

and combine everything to calculate the expectation of (3.17).

3.24 This exercise will establish some of the properties of the matrices needed in the application of Cochran's Theorem in the RCB anova, model (3.5), and in establishing Lemma 3.17.

(a) Establish the following properties of B_1, B_2, and B_3 (defined in Technical Note 3.8.2):

   (*i*) Show that B_1, B_2, and B_3 are all idempotent matrices.

   (*ii*) Show that $B_1 J = B_2 J = B_3 J = J$.

   (*iii*) Show that $B_1 B_2 = B_2 B_3 = B_1 B_3 = B_3$.

(b) Use the properties established in part (a) to prove Lemma 3.17.

3.25 Here we will prove Theorem 3.18.

(a) Show that under (3.5), the vector $\mathbf{Y} = \{Y_{ij}\}$ is multivariate normal with covariance matrix $\Sigma = \sigma_\varepsilon^2 I + \sigma_\beta^2 J$.

(b) If we order the vector Y as in (3.36), show that $\bar{Y}_{i\cdot} = B_1 Y$, $\bar{Y}_{\cdot j} = B_2 Y$, $\bar{\bar{Y}} = B_3 Y$, and that we can write the sums of squares as

$$\text{SS(Trts)} = Y'(B_1 - B_3)Y, \quad \text{SS(T} \times \text{B)} = Y'(I - B_1 - B_2 + B_3)Y,$$

with $B_1 - B_3$ and $I - B_1 - B_2 + B_3$ idempotent and $(B_1 - B_3)(I - B_1 - B_2 + B_3) = 0$.

(c) Referring to Exercise 3.24, show that the matrices

$$A_1^* = \frac{1}{\sigma_\varepsilon^2}(B_1 - B_3) \quad \text{and} \quad A_2^* = \frac{1}{\sigma_\varepsilon^2}(I - B_1 - B_2 + B_3)$$

satisfy the assumptions of Cochran's Theorem (Theorem 2.20) with $\Sigma = \sigma_\varepsilon^2 I + \sigma_\beta^2 J$.

(d) Show that, under $H_0 : \tau_i - \bar{\tau} = 0$, we have

$$Y'A_1^*Y \sim \chi_{t-1}^2 \quad \text{and} \quad Y'A_2^*Y \sim \chi_{(b-1)(t-1)}^2,$$

independently, and thus finish the proof of Theorem 3.18. Note that you must argue that, without loss of generality, for the test we can assume that $\mu = 0$ in (3.5).

3.26 Here we finish the proof of Theorem 3.18 with regard to the t-distribution of contrasts, and thus prove Theorem 3.10.

(a) Show that $B_1(I - B_1 - B_2 - B_3) = 0$, and thus argue that the vector $\bar{Y}_{i'\cdot}$ is uncorrelated with $(I - B_1 - B_2 - B_3)\mathbf{Y}$. Equivalently, establish $\text{Cov}(Y_{i'j'} - Y_{i'\cdot}Y_{i'\cdot} - Y_{\cdot j'} + \bar{\bar{Y}}, \bar{Y}_{i\cdot}) = 0$ for every i, i', j', that is, the covariance between treatment means and residuals is zero.

(b) Use the results of part (a) to show that, under normality, SS(T $\times$ B) is independent of $\sum_{i=1}^{k} a_i \bar{Y}_{i\cdot}$, and use the properties of the t-distribution (Section 2.8.2) to complete the proof.

3.27 Although Theorem 3.10 and Exercise 3.26 show that SS(T $\times$ B) has the proper distributional properties, they do not fully illustrate the quote of Fisher at the end of Section 3.1.2. This exercise will show the full treatment. (See also Exercise 3.4 for a numerical demonstration.)

From (3.17) we know that

$$\text{SS(T} \times \text{B)} = \sum_{i=1}^{t} \sum_{j=1}^{r} \hat{\varepsilon}_{ij}^2$$

and here we will show that the sum can be written as the sum of squares of $(r-1)(k-1)$ uncorrelated contrasts.

We construct our contrasts as a linear combination $\sum_{i=1}^{t}\sum_{j=1}^{r} a_{ij}\hat{\varepsilon}_{ij}$, where the a_{ij}s satisfy:

(1) $\sum_{i=1}^{t}\sum_{j=1}^{r} a_{ij} = 0$,

(2) $\sum_{j=1}^{r} a_{ij} = 0$ for each i,

(3) $\sum_{i=1}^{t} a_{ij} = 0$ for each j.

(4) Any two contrasts $\sum_{i=1}^{t}\sum_{j=1}^{r} a_{ij}\hat{\varepsilon}_{ij}$ and $\sum_{i=1}^{t}\sum_{j=1}^{r} a'_{ij}\hat{\varepsilon}_{ij}$ will be uncorrelated (hence independent under normality), and satisfy $\sum_{i=1}^{t}\sum_{j=1}^{r} a_{ij}a'_{ij} = 0$

We will construct $(r-1)(t-1)$ sets of constants whose corresponding contrasts are all independent, are each distributed as a $\sigma^2\chi_1^2$, and sum to $\sum_{i=1}^{t}\sum_{j=1}^{r} \hat{\varepsilon}_{ij}^2$. Note that conditions (2) and (3) are not required of interaction contrasts. They do make the argument easier, and are a natural result of the product construction used here. However, see the discussion in Section 5.2.3.

Refer to Table 3.14 for the construction. We start with the $t-1$ sets of contrasts defined by the rows. Each row sums to zero so each row can be used to define a contrast between treatments. Similarly, we also define the $r-1$ sets of contrasts by the columns. There, every column sums to zero and defines a contrast. We now construct our set of uncorrelated contrasts by taking products. For example, from row l and column m, we form a rectangular array of constants. Write row l across the top and column m down the left side. Now an element in the array is the product of the value on the top line and the value in the left column.

If we denote this set of constants by a_{ij}^{lm} (normalized so that $\sum_{i,j}(a_{ij}^{lm})^2 = 1$) and define $T_{lm} = \sum_{i=1}^{t}\sum_{j=1}^{r} a_{ij}^{lm}\hat{\varepsilon}_{ij}$, establish the following properties:

(a) For each l, m, $T_{lm} \sim n(0,\sigma^2)$.

(b) For each $l, m, l', m', l \neq l', m \neq m'$, $\text{Cov}(T_{lm}, T_{l'm'}) = 0$.

(c)
$$\frac{1}{\sigma^2}\sum_{l=1}^{t-1}\sum_{m=1}^{r-1}(T_{lm})^2 = \frac{1}{\sigma^2}\sum_{i=1}^{t}\sum_{j=1}^{r}\hat{\varepsilon}_{ij}^2.$$

Note that, by construction, there are only $(r-1)(t-1)$ quantities T_{lm}. (This requires a tedious amount of algebra.)

(d) Finally, show that (a)-(c) imply that the residual sum of squares is distributed $\sigma^2\chi_{(r-1)(t-1)}^2$.

3.28 Referring to Technical Note 3.8.3:

(a) Verify the expression for $\text{Cov}(\bar{Y}_{ij}, \bar{Y}_{i'j'})$.

(b) Prove Theorem 3.19.

3.29 Here we will establish properties of the matrices

$$A_1 = B_4'(I - B_1 - B_2 + B_3)B_4 \text{ and } A = I - \frac{1}{r}B_4'B_4,$$

in order to prove Lemma 3.20.

(a) Using the properties of B_1, B_2, and B_3, along with the facts that

$$B_4B_4' = rI, \quad B_4'B_4J = rJ, \quad B_4'B_4J^{(p)} = rJ^{(p)},$$

establish the following propertied of A_1 and A_2:

Table 3.14. Product construction of residual contrasts

	1	2	3	4	...	$t-1$	t
1	$t-1$	-1	-1	-1	...	-1	-1
2	0	$t-2$	-1	-1	...	-1	-1
3	0	0	$t-3$	-1	...	-1	-1
⋮	⋮	⋮	⋮	⋮	⋮	-1	-1
$t-1$	0	0	0	0	...	1	-1

	1	2	3	...	$r-1$
1	$r-1$	0	0	...	0
2	-1	$r-2$	0	...	0
3	-1	-1	$r-3$	...	0
4	-1	-1	-1	...	0
⋮	⋮	⋮	⋮	⋮	⋮
$r-1$	-1	-1	-1	...	1
r	-1	-1	-1	...	-1

	0 0 0 ... 0	$t-l$	-1	...	-1
0	0 0 0 ... 0	0	0	...	0
0	0 0 0 ... 0	0	0	...	0
·	 0	0	0	...	0
0	0 0 0 ... 0	0	0	...	0
$r-m$	0 0 0 ... 0	$(r-m)(t-l)$	$-(r-m)$	...	$-(r-m)$
-1	 0	$-(t-l)$	1	...	1
·	 0	.	1	...	1
-1	0 0 0 ... 0	$-(t-l)$	1	...	1

    (*i*) A_1 and A_2 are idempotent.

    (*ii*) $\mathbf{Y}'A_1\mathbf{Y} = \mathrm{SS}(\mathrm{T} \times \mathrm{B})$ and $\mathbf{Y}'A_2\mathbf{Y} = \mathrm{SS(Within)}$.

    (*iii*) $A_1A_2 = 0$.

    (*iv*) $A_1J = 0$ and $A_1J^{(p)} = rA_1$.

    (*v*) $A_2J = 0$ and $A_2J^{(p)} = 0$.

  (b) For Σ of (3.33), show that

$$A_1\Sigma = (\sigma_\varepsilon^2 + (1 - \rho_{\tau\beta})\sigma_{\tau\beta}^2)A_1, \quad A_2\Sigma = \sigma_\varepsilon^2 A_2.$$

  (c) We have now established enough properties of A_1 and A_2 to prove Lemma 3.20 and Theorem 3.21. Do so.

3.30 For the RCB model specified in (3.24):

  (a) Verify the variances given in Technical Note 3.8.1.

  (b) Verify the expected mean squares in Table 3.7.

3.31 (a) For the Scheffémodel (3.41), show that $\text{Corr}(\gamma_{ij}, \gamma_{i'j}) = -1/(t-1)$.

(b) For the RCB model (3.24), show that if $\rho_{\tau\beta} = -1/(t-1)$, then $\sum_i (\tau\beta)_{ij} = 0$ for every j. (Evaluate $\text{Var}(\sum_i (\tau\beta)_{ij})$.)

3.32 (a) In the usual version of model (3.24), the correlation $\rho_{\tau\beta}$ is taken to be zero. Produce an anova table (source, df, EMS) with this assumption, and compare it to the anova table without this assumption.

(b) Do the anova table (source, df, EMS) for the Scheffémodel (3.41), and compare it to the ones in part (a), in terms of what are the possible tests.

3.33 Referring to Technical Note 3.8.4:

(a) Verify that $V^* = \sigma_\beta^2 ZZ' + \sigma^2 I$, and has all row sums equal.

(b) Show that both the ordinary and generalized least squares estimates of θ are equal to the vector of cell means.

(c) Verify the expression for the prediction of β given in (3.40).

(d) For the joint distribution (3.39), verify the covariance of Y and β. (Note that $\text{Cov}(Y, \beta) = EY\beta'$, a $bt \times b$ matrix with entries of the form $EY_{ij}\beta_{j'}$.)

3.34 Referring to Section 3.6.1:

(a) Verify the EMS in Table 3.10.

(b) Verify the EMS in Table 3.11.

(c) For each of the two models in Section 3.6.1, calculate the variance of a treatment contrast and discuss how to estimate it.

3.35 Referring to Miscellanea 3.9.2:

(a) Show that $\hat{\sigma}_\varepsilon^2$ and $\hat{\sigma}_\beta^2$ are unbiased estimators of σ_ε^2 and σ_β^2.

The remainder of this problem refers to Example 3.4.

(b) Using model (3.5), which only has variance components σ_ε^2 and σ_β^2, derive the anova estimates for the variance components by equating mean squares with EMS. Use the anova table in Example 3.4 to get numerical values.

(c) Since the anova estimates in part (b) are nonnegative, they will agree with the REML estimates. Demonstrate this by calculating the REML estimates directly (see the code in Miscellanea 3.9.2).

(d) Since the data of Example 3.4 has replication, we can also use the model (3.24), which has three variance components, σ_ε^2, σ_β^2, and $\sigma_{\tau\beta}^2$. Using the EMS in Table 3.8, derive the anova estimators of the variance components and calculate the numerical values.

(e) Under the setup of part (d), do you expect the REML estimates to agree with the anova estimates. Calculate the REML estimates directly and discuss the outcome.

3.8 Technical Notes

3.8.1 Helpful Lemma II

When calculating variances of contrasts and effect estimates, and expected mean squares, we need the distribution of the errors, and sometimes also the differences of the errors. These follow from (3.24), and also get a little help from the following lemma, an extension of Lemma 2.16.

Lemma 3.16. *Suppose that $W_1, W_2, \ldots, W_n$ satisfy $EW = 0$, $\text{Var}W = \sigma^2$, and $\text{Cov}(W_i, W_{i'}) = \rho\sigma^2$. Then*

(1) $\text{Cov}(W_i - \bar{W}, W_{i'} - \bar{W}) = -\frac{1}{n}(1 - \rho)\sigma^2$.

(2) $\text{Var}(W_i - \bar{W}) = \text{E}(W_i - \bar{W})^2 = \left(1 - \frac{1}{n}\right)(1 - \rho)\sigma^2$.

If the W_i are normal, then

(3) $W_i - \bar{W} \sim \text{N}\left(0, \left(1 - \frac{1}{n}\right)(1 - \rho)\sigma^2\right)$.

3.8.2 Cochran's Theorem for RCBs – No Replication

Theorem 2.20 is quite general, applying to a multivariate normal distribution with covariance matrix Σ. Here we show how it applies to the sums of squares in an RCB. Define the matrices

$$B_1 = \frac{1}{b}\begin{pmatrix} I_t \\ \vdots \\ I_t \end{pmatrix}_{tb \times t} (I_t \cdots I_t)_{t \times tb},$$

$$B_2 = \frac{1}{t}\begin{pmatrix} 1_{t \times 1} & 0 & \cdots & 0 \\ 0 & 1_{t \times 1} & \cdots & 0 \\ \vdots & \vdots & \vdots & \vdots \\ 0 & 0 & \cdots & 1_{t \times 1} \end{pmatrix}_{tb \times b} \begin{pmatrix} 1'_{1 \times t} & 0 & \cdots & 0 \\ 0 & 1'_{1 \times t} & \cdots & 0 \\ \vdots & \vdots & \vdots & \vdots \\ 0 & 0 & \cdots & 1'_{1 \times t} \end{pmatrix}_{b \times tb}, \quad B_3 = \frac{1}{tb}J_{tb},$$

where I_m is the $m \times m$ identity matrix, J_m is the $m \times m$ matrix of ones, and $1_{m \times 1}$ is the $m \times 1$ vector of ones (note that $1_m 1'_m = J_m$).

Lemma 3.17. *Let* $\mathbf{Y} \sim \text{N}(0, \Sigma)$, *where* $\Sigma = \sigma_1^2 I + \sigma_2^2 J$. *Then*

$$A_1^* = \frac{1}{\sigma_1^2}(B_1 - B_3) \text{ and } A_2^* = \frac{1}{\sigma_1^2}(I - B_1 - B_2 + B_3)$$

satisfy the assumptions of Theorem 2.20 (Cochran's Theorem). That is

(1) $A^* \Sigma$ *is idempotent, where* $A^* = A_1^* + A_2^*$.
(2) $A_1^* \Sigma$ *and* $A_2^* \Sigma$ *are idempotent.*
(3) $A_1^* \Sigma A_2^* = 0$.

Using this lemma we can establish the following theorem.

Theorem 3.18 (Cochran's Theorem for RCBs – No Replication). *Under the RCB anova model (3.5),* $\mathbf{Y} = \{Y_{ij}\}$ *is multivariate normal with* $\Sigma = \sigma_\varepsilon^2 I + \sigma_\beta^2 J$. *Under* $H_0 : \tau_i - \bar{\tau} = 0$,
(1) $\mathbf{Y}' A_i^* \mathbf{Y} \sim \chi_{r_i}^2$ *for* $i = 1, 2$.
(2) $\mathbf{Y}' A_1^* \mathbf{Y}$ *and* $\mathbf{Y}' A_2^* \mathbf{Y}$ *are independent.*
(3) $\mathbf{Y}' A^* \mathbf{Y} \sim \chi_{r_1 + r_2}^2$, *where* $r_1 = \text{tr}(B_1 - B_3)$ *and* $r_2 = \text{tr}(I - B_1 - B_2 + B_3)$.
Thus,

$$\frac{\text{MS(Trts)}}{\text{MS}(T \times B)} = \frac{\sigma_\varepsilon^2 \mathbf{Y}' A_1^* \mathbf{Y}/r_1}{\sigma_\varepsilon^2 \mathbf{Y}' A_2^* \mathbf{Y}/r_2} \sim F_{r_1, r_2}.$$

Moreover, for any contrast $(a_1, \ldots, a_k)$,

$$\frac{\sum_{i=1}^k a_i \bar{Y}_i - \sum_{i=1}^k a_i \tau_i}{\sqrt{\frac{\text{MS(T$\times$B)}}{b} \sum_i a_i^2}} \sim t_{r_2}.$$

3.8.3 Cochran's Theorem for RCBs – With Replication

We now consider Cochran's Theorem for the RCB with replication, where we can separate the interaction from the error, given by model (3.24). We first need a bit more notation, and here we define the data vector $\mathbf{Y}$ by

$$\mathbf{Y}' = \{Yijk\}'$$
$$= (Y_{111}, \cdots, Y_{11r}, \cdots, Y_{t11}, \cdots, Y_{t1r}, \cdots, Y_{1b1}, \cdots, Y_{1br}, \cdots, Y_{tb1}, \cdots, Y_{tbr}),$$

the matrix B_4 by

$$B_4 = \begin{pmatrix} \mathbf{1}_{1 \times r} & 0 & \cdots & 0 \\ 0 & \mathbf{1}_{1 \times r} & \cdots & 0 \\ \vdots & \vdots & \vdots & \vdots \\ 0 & 0 & \cdots & \mathbf{1}_{1 \times r} \end{pmatrix}_{tb \times tbr},$$

and note that $\frac{1}{r} B_4 \mathbf{Y} = \{\bar{Y}_{ij}\}$, the vector of cell means.

From Table 3.3 we see that the test on treatments is only a function of $\{\bar{Y}_{ij}\}$, so to apply Cochran's Theorem we only have to deal with the cell means. Under model (3.24) is it straightforward to verify

$$\text{Cov}(\bar{Y}_{ij}, \bar{Y}_{i'j'}) = \begin{cases} 0 & \text{if } j \neq j' \\ \sigma_\beta^2 + \rho_{\tau\beta}\sigma_{\tau\beta}^2 & \text{if } j = j', \quad i \neq i' \\ \frac{\sigma_\varepsilon^2}{r}(1 + (r-1)\rho_\varepsilon) + \sigma_\beta^2 + \rho_{\tau\beta}\sigma_{\tau\beta}^2 & \text{if } j = j', \quad i = i'. \end{cases}$$

Therefore, if we identify

$$\sigma_1^2 = \frac{\sigma_\varepsilon^2}{r}(1 + (r-1)\rho_\varepsilon), \qquad \sigma_2^2 = \sigma_\beta^2 + \rho_{\tau\beta}\sigma_{\tau\beta}^2,$$

we can apply Lemma 3.17 to establish the following result.

Theorem 3.19 (Cochran's Theorem for RCB Treatment Test – With Replication). *For model (3.24), under $H_0 : \sum_i (\tau_i - \bar{\tau})^2 = 0$,*

$$\frac{\text{MS(Trts)}}{\text{MS}(T \times B)} \sim F_{t-1,(b-1)(t-1)}.$$

We next derive the test of the interaction in model (3.24). From the expected mean squares in Table 3.7, we see that unless we assume that $\rho_\varepsilon = 0$ there will be no test on the interaction. Thus, we make this assumption, that we have true replication.

Since the blocks are independent, $\text{Cov}(Y_{ijk}, Y_{i'j'k'}) = 0$ unless $j = j'$. We have $\text{Var}(Y_{ijk}) = \sigma_\beta^2 + \sigma_{\tau\beta}^2 + \sigma_\varepsilon^2$ and, within a block,

$$\text{Cov}(Y_{ijk}, Y_{i'jk'}) = \begin{cases} \sigma_\beta^2 + \sigma_{\tau\beta}^2 & \text{if } i = i' \\ \sigma_\beta^2 + \rho_{\tau\beta}\sigma_{\tau\beta}^2 & \text{if } i \neq i', \end{cases}$$

and thus $\{Y_{ijk}\}$ has covariance matrix

(3.33) $$\Sigma = \sigma_\varepsilon^2 I + (\sigma_\beta^2 + \rho_{\tau\beta}\sigma_{\tau\beta}^2)J + (1 - \rho_{\tau\beta})\sigma_{\tau\beta}^2 J^{(p)},$$

where

$$J^{(p)} = \begin{pmatrix} J_{r\times r} & 0 & \cdots & 0 \\ 0 & J_{r\times r} & \cdots & 0 \\ \vdots & \vdots & \vdots & \vdots \\ 0 & 0 & \cdots & J_{r\times r} \end{pmatrix}_{tbr \times tbr}.$$

We then have the following results.

Lemma 3.20. *Let* $\mathbf{Y} \sim N(0, \Sigma)$, *where* Σ *is given by (3.33). Then*

$$A_1^* = \frac{1}{\sigma_\varepsilon^2 + (1 - \rho_{\tau\beta})\sigma_{\tau\beta}^2}(B_4'(I - B_1 - B_2 + B_3)B_4) \ and \ A_2^* = \frac{1}{\sigma_\varepsilon^2}\left(I - \frac{1}{r}B_4'B_4\right),$$

satisfy the assumptions of Theorem 2.20 (Cochran's Theorem). That is
(1) $A^*\Sigma$ *is idempotent, where* $A^* = A_1^* + A_2^*$.
(2) $A_1^*\Sigma$ *and* $A_2^*\Sigma$ *are idempotent.*
(3) $A_1^*\Sigma A_2^* = 0$.

As before, the lemma lets us establish the following theorem.

Theorem 3.21 (Cochran's Theorem for RCB Interaction Test). *Under the RCB anova model (3.24) with* $\rho_\varepsilon = 0$, $\mathbf{Y} = \{Y_{ijk}\}$ *is multivariate normal with* Σ *given by (3.33). Under* $H_0 : \sigma_{\tau\beta}^2 = 0$:
(1) $\mathbf{Y}'A_i^*\mathbf{Y} \sim \chi_{r_i}^2$ *for* $i = 1, 2$.
(2) $\mathbf{Y}'A_1^*\mathbf{Y}$ *and* $\mathbf{Y}'A_2^*\mathbf{Y}$ *are independent.*
(3) $\mathbf{Y}'A^*\mathbf{Y} \sim \chi_{r_1+r_2}^2$, *where* $r_1 = tr(I - B_1 - B_2 + B_3)$ *and* $r_2 = tr(I - \frac{1}{r}B_4'B_4)$.
    Thus,

$$\frac{\mathrm{MS}(T \times B)}{\mathrm{MS}(\mathrm{Within})} = \frac{\sigma_\varepsilon^2\mathbf{Y}'A_1^*\mathbf{Y}/r_1}{\sigma_\varepsilon^2\mathbf{Y}'A_2^*\mathbf{Y}/r_2} \sim F_{r_1,r_2}.$$

3.8.4 Estimating Fixed and Random Effects I

The distinction between fixed and random effects is, in some cases, a convenient fiction. Whether "all levels of interest" are in the experiment, or the levels in the experiment are "a random sample from a larger population" is of lesser statistical consequence. What *does* matter is that the covariance structure of the experiment is adequately modeled. In fact, the assumption of a random effect is sometimes made just to get the covariance structure correct, not because the effects are necessarily random.

Of course, there are legitimate experiments in which certain effects should be modeled as random, as in Example 3.6. The consequences of modeling an effect as random are:
(1) There is an implied correlation structure, as once a random block (subject) is obtained, all observations are then taken within that subject.

(2) The parameter of interest is a variance, as there are no mean parameters to estimate. In a model like (3.34), if the β_j are random, we are only interested in estimating σ_β^2.

Perhaps one of the most important consequences is:

(3) Since the β_j are random variables, there is no sense in having them in a least squares equation. The proper estimation strategy is to used *mixed models* and estimate the other fixed effects while taking account of the variance in the random effects.

It seems to be common, in designing experiments, to ignore this implication and use least squares throughout. Although we will ultimately use least squares estimation, we want to understand the implications of our action. We look at the RCB model

$$(3.34) \quad y_{ij} = \mu + \tau_i + \beta_j + \varepsilon_{ij} = \theta_i + \beta_j + \varepsilon_{ij}, \quad i = 1,\ldots,t, \quad j = 1,\ldots,b,$$

where, for convenience, we define $\theta_i = \mu + \tau_i$, producing the *cell means model*. This will make the algebra a bit easier (but not much!). Also, we are not going to directly work with the interaction model, but the techniques here can be extended to that case. For more details see Searle *et al.* (1992, Section 4.9 and Chapter 7).

In matrix form (3.34) is

$$(3.35) \quad\quad\quad\quad\quad Y = X\theta + Z\beta + \varepsilon,$$

where X and Z are the incidence matrices associated with the model. In more general regression problems X and Z contain covariate estimation - most of what we do here extends to those cases.

Note that we write the matrices in a particular order, where the treatment index moves first, giving

$$(3.36) \quad Y = \begin{pmatrix} y_{11} \\ \vdots \\ y_{t1} \\ \vdots \\ y_{1b} \\ \vdots \\ y_{tb} \end{pmatrix}, \quad\quad X = \left.\begin{pmatrix} I_t \\ I_t \\ \vdots \\ I_t \end{pmatrix}\right\} b \text{ times},$$

$$Z = \left.\begin{pmatrix} \mathbf{1}_t & 0 & 0 & \cdots & 0 \\ 0 & \mathbf{1}_t & 0 & \cdots & 0 \\ & & \vdots & & \\ 0 & 0 & \cdots & 0 & \mathbf{1}_t \end{pmatrix}\right\} b \text{ times},$$

where I_t is the $t \times t$ identity matrix, $\mathbf{1}_t$ is a $t \times 1$ vector of ones, Y is $bt \times 1$, X is $bt \times t$ and Z is $bt \times b$.

Now to our RCB models. Here we consider the case of random blocks, leaving the (easier!) fixed blocks case to Technical Note 4.8.2. The assumption that the blocks are random puts us in the realm of the *mixed model*, a model that has both fixed and random effects. The estimation here is a bit more complex. We work with model (3.34), and assume as in (3.5),

$$(3.37) \quad\quad \beta_j \sim N(0, \sigma_\beta^2), \quad\quad \varepsilon_{ij} \sim N(0, \sigma_\varepsilon^2), \quad\quad \text{independent.}$$

Under (3.37), model (3.35) is a linear model with a different error structure:

(3.38) $$Y = X\theta + (Z\beta + \varepsilon) = Y = X\theta + \varepsilon^*,$$

where
$$V^* = \text{Cov}(\varepsilon^*) = \text{Cov}(Z\beta + \varepsilon) = \sigma_\beta^2 ZZ' + \sigma_\varepsilon^2 I.$$

The generalized least squares estimator of θ is then

$$\hat{\theta} = (X'V^{*-1}X)^{-1}X'V^{*-1}Y.$$

Now, to "estimate" β, which is a random variable under this formulation, we can *predict* it from the joint distribution of Y and β (see Searle *et al.* 1992, Chapter 7, for other options). The joint distribution is

(3.39) $$\begin{pmatrix} Y \\ \beta \end{pmatrix} \sim \text{N} \left(\begin{pmatrix} X\theta \\ 0 \end{pmatrix}, \begin{pmatrix} V^* & \sigma_\beta^2 Z \\ \sigma_\beta^2 Z' & \sigma_\beta^2 I \end{pmatrix} \right),$$

and the conditional distribution of $\beta|Y$ is

$$\beta|Y \sim \text{N} \left(\sigma_\beta^2 Z'V^{*-1}(Y - X\theta), [\sigma_\beta^2 I - \sigma_\beta^2 Z'V^{*-1}\sigma_\beta^2 Z]^{-1} \right),$$

showing that the estimate (or prediction) of the random β is the mean

$$\hat{\beta} = \sigma_\beta^2 Z'V^{*-1}(Y - X\hat{\theta}).$$

Evaluating this expression shows that

(3.40) $$\hat{\beta} = \frac{t\sigma_\beta^2}{t\sigma_\beta^2 + \sigma_\varepsilon^2} \begin{pmatrix} \bar{y}_{\cdot 1} - \bar{\bar{y}} \\ \bar{y}_{\cdot 2} - \bar{\bar{y}} \\ \vdots \\ \bar{y}_{\cdot t} - \bar{\bar{y}} \end{pmatrix},$$

where we can also write

$$\frac{t\sigma_\beta^2}{t\sigma_\beta^2 + \sigma_\varepsilon^2} = \frac{\rho t}{1 - \rho + \rho t},$$

where ρ is the intraclass correlation. Thus, to use the mixed model prediction of β requires knowing σ_β^2 or ρ, or having estimates. On this latter point, if ρ needs to be estimated then the optimal variance properties of the estimates may be lost. (See Harville 1976 and Kacker and Harville 1984.)

3.9 Miscellanea

3.9.1 Models for Random Blocking
The literature contains many variations on models for the RCB with random blocking, mostly dealing with how to model the correlations and the interactions. In this section we will illustrate some of the concerns, without going into great detail. The material in this section draws from Hocking(1973, 1985), Scheffé(1959, Section 8.1) and Samuels *et al.* (1991, 1993). Hocking(1973) defines Models I, II, and III as variations on the RCB model, and Samuels *et al.*

(1991) are concerned with the modeling of the interaction term, and differentiating between a "block" and a "random factor". The real differences lie between Model I, which is the model developed by Scheffé(1959), and Model II, which is essentially (3.24).

The Scheffé model is constructed as follows. We start with

$$Y_{ikj} = m(i, j) + \varepsilon_{ijk},$$

where $m(i, j)$ represents the true response of treatment i in block j. We think of i as a finite index, $i = 1, \ldots, t$, and j as an infinite index, going over the infinite population of the random blocks. From the $m(i, j)$ we define the treatment and block effects as follows. The overall mean is calculated by summing over all treatments and taking the expectation over all levels of j, that is,

$$\mu = m(\cdot, \cdot) = \frac{1}{t} \sum_{i=1}^{t} E_j m(i, j).$$

Similarly, we define all of the effects as follows:

$$\text{Treatment Effect } \tau_i = m(i, \cdot) - m(\cdot, \cdot),$$
$$\text{Block Effect } \beta_i = m(\cdot, j) - m(\cdot, \cdot),$$
$$\text{Interaction Effect } \gamma_{ij} = m(i, j) - m(i, \cdot) - m(\cdot, j) + m(\cdot, \cdot)$$
$$= m(i, j) - \tau_i - \beta_j + \mu.$$

So the Scheffé model is

(3.41)
$$Y_{ikj} = \mu + \tau_i + \beta_j + \gamma_{ij} + \varepsilon_{ijk},$$

where, by construction,

$$\frac{1}{t} \sum_{i=1}^{t} \tau_i = 0, \quad E_j \beta_j = 0, \text{ and } \frac{1}{t} \sum_{i=1}^{t} \gamma_{ij} = 0, \quad E_j \gamma_{ij} = 0.$$

Here, the interaction terms γ_{ij} are defined as residual parameters, so the above constraints immediately follow. Note that the randomness in γ_{ij} comes directly from the randomness in the blocks, making this model substantially different from (3.24).

Finally, if we define

$$\text{Var } m(i, j) = \sigma_m^2,$$
$$\text{Cov}(m(i, j), m(i', j)) = \rho_B \sigma_m^2,$$

then direct calculation shows

(3.42)
$$\text{Var } \beta_j = \frac{\sigma_m^2}{t}[1 + (t - 1)\rho_B], \quad \text{Cov}(\gamma_{ij}, \gamma_{i'j}) = \frac{\sigma_m^2}{t}(1 - \rho_B),$$
$$\text{Var } \gamma_{ij} = \frac{(t-1)\sigma_m^2}{t}(1 - \rho_B), \quad \text{Cov}(\gamma_{ij}, \beta_j) = 0.$$

Scheffé motivates this model by supposing that $m(i, j)$ is the effect of machine i, one of a finite set of machines, when run by a randomly chosen worker j,

selected from an infinite population of workers. The multivariate random
variable that is the basis of the model is

$$m(j) = (m(1,j), m(2,j), \ldots, m(t,j)).$$

This is an interesting and useful model. Although it does put strong constraints on the parameters, it should be considered as an alternative to (3.24).
One of the major differences between (3.24) and the Scheffémodel is the fact
that in the latter model the interaction terms sum to zero, while in the former they do not. Of course, the construction of the Scheffémodel forces this
to happen, but we can also ask if it can happen in (3.24). The answer is
quite interesting. If the interaction terms in (3.24) sum to zero, that is, if
$\sum_i (\tau\beta)_{ij} = 0$, then

$$\sum_i (\tau\beta)_{ij} = 0 \rightarrow \mathrm{Var}\left(\sum_i (\tau\beta)_{ij}\right) = 0 \rightarrow \rho_{\tau\beta} = -\frac{1}{t-1},$$

which is exactly the correlation in (3.42). Thus, the Schefféconstruction imposes a certain value for this correlation, and assuming that value in model
(3.24) results in the Schefféinteraction constraint (Exercise 3.31).

3.9.2 Variance Component Estimation

Although a book about statistical design is necessarily concerned about variances, we have not directly addressed the problem of estimation of variance
components. For example, in the RCB of Example 3.4 there are two vari-
ance components, σ_ε^2 and σ_β^2 (see Table 3.6). In the split plot model (5.14) of
Chapter 5 there are four variance components: σ_β^2, σ_ε^2, $\sigma_{\beta\gamma}^2$, and σ_δ^2.

From the EMS in Table 3.6 we see that the expected mean squares can provide
unbiased estimates of some combinations of these variances, but this does not
directly get to the estimation of the individual components.

For most of our purposes, we mainly care about producing valid F-tests and
variances of treatment contrasts, leading to confidence intervals and inferences
about the contrasts. For this reason we have not discussed the estimation of
the individual variance components that, to do the topic justice, requires a
book-length treatment (see Searle *et al.* 1992 or Cox and Solomon 2003).
However, we will give a small illustration of variance component estimation.
Although we will only discuss RCB designs, the principles here will apply all
of the designs considered in this book, and beyond.

A first method of estimating variance components was to equate the observed
mean squares with the expected mean squares, and solve for the components,
a type of method of moments approach (sometimes called the *anova method*).
This can produce unbiased estimators of the variance components, but can
also get us into trouble. For example, from Table 3.6 we equate

$$\mathrm{MS(Blocks)} = \sigma_\varepsilon^2 + t\sigma_\beta^2 \text{ and } \mathrm{MS(T \times B)} = \sigma_\varepsilon^2,$$

leading to the estimators

$$\hat{\sigma}_\varepsilon^2 = \mathrm{MS(T \times B)} \text{ and } \hat{\sigma}_\beta^2 = \frac{1}{t}\left[\mathrm{MS(Blocks)} - \mathrm{MS(T \times B)}\right],$$

which are unbiased estimators of the variance components. Applying this to
Example 3.3 yields

(3.43) $\hat{\sigma}_\varepsilon^2 = .121$ and $\hat{\sigma}_\beta^2 = \dfrac{1}{3}[.574 - .121] = .151,$

which are both reasonable estimates, and show that the variability due to blocks is about at the same level as the residual variability.

However, unbiased estimates of variance components can have problems and, as a general rule, we can expect problems with unbiased estimators of positive quantities that can, possibly, take on negative values. Here, it can be the case that MS(Blocks) < MS(T × B), which would result in a negative estimate of σ_β^2. For example, if the data of Table 3.1 were

		Blocks			
		1	2	3	4
	A	10.1	10.8	9.8	10.5
Variety of Strawberry	B	8.4	6.9	5.3	6.2
	C	6.3	9.4	9.0	9.2

where only the values 8.4 and 6.3 were exchanged (which will increase the interaction), the anova table is

Source	df	SS	MS	F	p-value
Block	3	1.722	0.574		
Trt	2	25.921	12.960	7.488	0.023
T × B	6	10.385	1.731		

This leads to the estimates

(3.44) $\hat{\sigma}_\varepsilon^2 = 1.731$ and $\hat{\sigma}_\beta^2 = -0.386,$

> Yikes! Negative variance estimates!

which is clearly ridiculous. This is a fault of the estimation procedure, not the model – the requirement of unbiasedness here can lead to negative estimates. What is the solution? Some might say that this is indication that $\sigma_\beta^2 = 0$, but there is really no compelling argument for this (unless a formal hypothesis test can be derived). Rather, we should use a better estimation procedure, one that will only give nonnegative estimates. The references mentioned above detail many methods, but perhaps the most popular is REML (REstricted Maximum Likelihood), a variation of the maximum likelihood method. REML can be implemented in R, but with a bit of effort. To get the REML variance estimates for the Strawberry data, we can use the commands

```
library(nlme)
strawmodel<-lme(yield ~ 1+trt,strawdata,random= 1|block)
summary(strawmodel)
VarCorr(strawmodel)
```

which says that the intercept and treatment are fixed effects, and the blocks are random. If we run this with the original data, which produced the estimates (3.43), we see that the REML estimates are identical to the anova estimates. This is a property of REML: When the anova estimates are nonnegative, they agree with the REML estimates. However, when we run REML on the second data set, which produced the anova estimates (3.44), we obtain

$$\hat{\sigma}_{\varepsilon}^2 = 1.345 \text{ and } \hat{\sigma}_{\beta}^2 = 0.00027,$$

which are much more sensible. This tells us that the block variance is small (essentially negligible), but not negative!!

As mentioned above, in this book we are mainly concerned with estimation of treatment contrasts and their variances, which we can accomplish here with anova methods (giving nonnegative answers), sometimes enhanced with Satterthwaithe's approximation (see Miscellanea 5.8.1). Exercise 3.35 looks at REML estimation in other RCB models.

4

Interlude: Assessing the Effects of Blocking

The first principle is that you must not fool yourself...

Richard P. Feynman
Surely You're Joking, Mr. Feynman

4.1 Introduction

In Chapter 3 we modeled blocks as a random factor, one in which the levels that actually appear in the experiment are considered a random sample from all levels. However, the concept of "random factor" can sometimes be puzzling, as most of the time we do not actually take a random sample of blocks. Rather, we *choose* blocks to represent a wide variety of situations. In a sense the concept of a random factor is a fallacy (see Section 3.8.4). That is, the important implication is that blocking induces a *correlation* in the design. This only makes sense, as experimental units within a particular block should behave similarly, and hence will be correlated. This correlation can be modeled directly, or can arise as a byproduct of assuming that blocks are a random effect. In either case we end up with the similar analyses.

Whether blocks are random or are fixed, the important point is that a correlation structure is induced. As far as the model calculations go – variances, covariances, etc, they are quite similar.

It is the correlation structure, either induced by the assumption of randomness or modeled directly as a block correlation, that is the important thing. Thus, we should make the appropriate assumption about randomness of blocks, but always take care to model the correlation structure.

> Random
> Factor
> =
> Correlation

It is thus our preference to model a factor according to what it really is, as the following example shows.

Example 4.1. FIXED OR RANDOM BLOCKS   Consider the following three experiments:

(a) Five varieties of tomato plant are subjected to different levels of light. The experiment is done in a greenhouse, which is divided into four areas for the experiment. The response variable is time to flowering.

(b) Eighteen subjects are recruited for an experiment that will measure their ability to memorize as a function of distraction. They will be asked to memorize a 10-digit number when subjected to various levels of distraction (conversations, music, etc.). The response variable is the amount of time needed to correctly recite the number.

(c) Within a school district, five elementary schools are used in an experiment to assess three different methods of teaching. In each school three third-grade classes are chosen, and the three teaching methods are used. The response is the students' scores on a pre-test and post-test.

In each of these experiments what are the blocks, and should they be considered fixed or random?

(a) Here, blocks = areas of the greenhouse. Unless we want to assume that the greenhouse is a random sample from all greenhouses, and the areas are selected at random, both of which are somewhat far-fetched assumptions, these blocks are fixed. They represent four specific areas in a particular greenhouse.

(b) Here, blocks = subjects. If the experiment is carefully done, and the subjects are actually selected at random from some population, then blocks are random.

(c) Here, blocks = schools. As in (a), the blocks are fixed. Unless we want to make the somewhat unrealistic assumption that these five schools are a representative sample of some larger population of schools, the only sensible assumption is that they are a fixed factor.

‖

Fortunately, as we will see, whether the blocks are fixed or random does not affect the calculation of treatment variances, nor the formation of tests and confidence intervals. The *scope* of the inference will be affected, however, as the breath of the treatment inference can only go to the blocks that are assumed to be represented.

4.2 Model and Distribution Assumptions _____

The calculations of expectations and variances done in Section 3.2 remain essentially the same if the blocks are assumed to be fixed, as long as we make the assumption that the ε_{ij} within a block are correlated. So, for the model with fixed blocks, we observe random variables Y_{ij} according to a model just like (3.5), where the β_j are fixed values. That is,

(4.1) $\qquad Y_{ij} = \mu + \tau_i + \beta_j + \varepsilon_{ij}, \quad i = 1, \ldots, t, \quad j = 1, \ldots, b,$

where

(1) The random variables $\varepsilon_{ij} \sim N(0, \sigma^2)$ for $i = 1, \ldots, t$, and $j = 1, \ldots, b$ (normal errors with equal variances).

(2)

$$\text{Corr}(\varepsilon_{ij}, \varepsilon_{i'j'}) = \begin{cases} \rho & \text{if } j = j', i \neq i' \\ 0 & \text{otherwise} \end{cases} \quad \left\| \begin{array}{l} \text{CORRELATION within blocks} \\ \text{but not between blocks} \end{array} \right.$$

The mean and variance of Y_{ij}s are

(4.2) $\qquad E(Y_{ij}) = \mu + \tau_i + \beta_j, \quad \text{Var}(Y_{ij}) = \sigma^2,$

and the correlation is $\text{Corr}(Y_{ij}, Y_{i'j'}) = \rho$ by the model assumption. Note that these calculations are identical to those in Section 3.2 (see (3.10)) if we identify

(4.3) $\qquad \rho\sigma^2 = \sigma_\beta^2$ and $\sigma^2 = \sigma_\beta^2 + \sigma_\varepsilon^2.$

Here, by specifying the covariance directly, the correlation in the blocks is not necessarily restricted to be a positive correlation, as it must be when it arises from random blocking.

4.3 Expected Squares and F-tests _____

Here we first look at EMS calculations for the case of one observation per treatment-block combination, assuming the blocks are fixed. Although the overall conclusions and inferences that we get here are virtually the same as those in Section 3.3, there are important differences in the details.

Using the RCB model (4.1) with blocks fixed, the ordinary least squares estimates (Exercise 3.22) are given by

(4.4) $\qquad \bar{Y}_i - \bar{Y} = \tau_i - \bar{\tau}$ and $\bar{Y}_j - \bar{Y} = \beta_j - \bar{\beta},$

where we set $\bar{\tau} = \bar{\beta} = 0$. The residuals are

$$\hat{\varepsilon}_{ij} = Y_{ij} - \bar{Y}_i - \bar{Y}_j + \bar{Y}.$$

We now apply those results in the EMS calculations.

$$\text{ESS(Blocks)} = E \sum_j t(\bar{Y}_j - \bar{Y})^2 = t \sum_j \beta_j^2 + t E \sum_j (\bar{\varepsilon}_j - \bar{\varepsilon})^2$$

(4.5) $$= t \sum_j \beta_j^2 + (b-1)\sigma^2 \left(1 + (t-1)\rho\right),$$

Table 4.1. Expected Mean Squares for RCB anova with fixed blocks and no subsampling.

Source	df	EMS
Blocks	$b-1$	$\sigma^2\left(1+(t-1)\rho\right)+\frac{t}{b-1}\sum_j \beta_j^2$
Treatments	$t-1$	$\sigma^2(1-\rho)+\frac{b}{t-1}\sum_i \tau_i^2$
T × B	$(t-1)(b-1)$	$\sigma^2(1-\rho)$

$$\mathrm{ESS(Trts)} = \mathrm{E}\sum_i b(\bar{Y}_i - \bar{Y})^2 = b\sum_i \tau_i^2 + b\mathrm{E}\sum_i (\bar{\varepsilon}_i - \bar{\varepsilon})^2$$

(4.6)
$$= b\sum_i \tau_i^2 + (b-1)\sigma^2(1-\rho),$$

$$\mathrm{ESS(T \times B)} = \mathrm{E}\sum_{ij}(Y_{ij} - \bar{Y}_i - \bar{Y}_j + \bar{Y})^2 = \mathrm{E}\sum_{ij}(\varepsilon_{ij} - \bar{\varepsilon}_i - \bar{\varepsilon}_j + \bar{\varepsilon})^2$$

(4.7)
$$= bt\left[\mathrm{Var}\varepsilon_{ij} - \mathrm{Var}\bar{\varepsilon}_i - \mathrm{Var}\bar{\varepsilon}_j + \mathrm{Var}\bar{\varepsilon}\right]$$
$$= \sigma^2(1-\rho),$$

where the simplification in the interaction term is a direct result of algebra (see Exercise 2.19). Dividing the expected sums of squares by their degrees of freedom yields the expected mean squares given in Table 4.1.

The null hypothesis of no treatment effect is

$$H_0 : \tau_i = 0 \text{ for all } i$$

and, from Cochran's Theorem (Exercise 4.13), H_0 is tested by

(4.8)
$$F_{t-1,(b-1)(t-1)} = \frac{\mathrm{MS(Trts)}}{\mathrm{MS(T \times B)}}.$$

Note that, unless $\rho = 0$, there is no test on blocks, that is, there is no test of the null hypothesis $H_0 : \beta_j = 0$ for all j. However, there is no cause for concern about this, as the blocks are merely in the experiment to represent a wide variety of experimental conditions, and are not of direct interest. If we can make the assumption that $\rho = 0$, then we are, in effect, saying that the design is not really a blocked design, and the analysis (and sampling) would be the same as a twoway CRD.

Table 4.1 may appear more complicated than Table 3.6, but it is actually quite similar. In fact, recall the intraclass correlation (3.10) and, using the notation of Table 3.6, write

(4.9)
$$\sigma_\varepsilon^2 = \left(\frac{\sigma_\varepsilon^2}{\sigma^2 + \sigma_\beta^2}\right)(\sigma_\varepsilon^2 + \sigma_\beta^2) = (1-\rho)(\sigma_\varepsilon^2 + \sigma_\beta^2),$$

Table 4.2. A comparison of the assumptions of the two RCB models.

	Blocks Fixed	Blocks Random
ε_{ij}	Correlated within block	Independent throughout design
Y_{ij}	Correlated within block	Correlated within block
ρ, intraclass correlation	$-1 \leq \rho \leq 1$	$0 \leq \rho \leq 1$

where $\rho = \sigma_\beta^2/(\sigma_\varepsilon^2 + \sigma_\beta^2)$ is the intraclass correlation. Thus, if we identify σ^2 of Table 4.1 with $\sigma_\varepsilon^2 + \sigma_\beta^2$ of Table 3.6, the EMS calculations for treatments and T × B are the same. Although the test for treatments is the same as (4.8), the inference about blocks is different. In Table 3.6, the EMS tells us that there is a test on blocks, but here we can only test the null hypothesis (Exercise 4.13)

$$(4.10) \qquad\qquad H_0 : \sum_j \beta_j^2 = 0 \text{ and } \rho = 0.$$

In the random blocks case, the correlation becomes part of the variance σ_β^2, and the EMS tells us there is a test. Here, things are both more clear and a bit more complex, as the correlation is more apparent. If the null hypothesis in (4.10) is true, the model would be equivalent to that of a a twoway CRD, but the experiment design is different.

> **Note:** Although we have been looking at similarities, it is important to realize that the two tables, Table 3.6 and Table 4.1, come from models with very different assumptions.

In any particular problem it is important to use the model that best describes the process. Table 4.2 lists some of the differences.

4.4 Estimating Contrasts ⎯⎯⎯⎯⎯⎯⎯⎯⎯⎯⎯

As compared to the calculations in Section 3.4, here we find that point estimation is a bit more straightforward, while variance calculation is somewhat more involved. The point estimation is simpler because here all effects are fixed effects, while the explicit covariance in (4.1) makes the variance calculations more involved.

Point Estimates ⎯⎯⎯⎯⎯⎯⎯⎯⎯⎯⎯⎯⎯⎯⎯⎯⎯

We write the RCB model with no subsampling as

$$(4.11) \qquad y_{ij} = \mu + \tau_i + \beta_j + \varepsilon_{ij}, \quad i = 1, \dots, t, \quad j = 1, \dots, b.$$

The least squares estimates of μ, τ_i and β_j are just as in Section 3.4, given by

$$\begin{aligned}
\bar{y}_{i\cdot} - \bar{\bar{y}} &= \tau_i - \bar{\tau}, \\
\bar{y}_{\cdot j} - \bar{\bar{y}} &= \beta_j - \bar{\beta}, \\
\bar{\bar{y}} &= \mu + \bar{\tau} + \bar{\beta}.
\end{aligned}$$

(4.12)

Here, requiring $\bar{\tau} = \bar{\beta} = 0$, or redefining the parameters as $\tau_i' = \tau_i - \bar{\tau}$ and $\beta_j' = \beta_j - \bar{\beta}$ is straightforward because all of the effects are fixed effects.

The least squares estimators are unbiased under model (4.11), and we have that

(4.13) $\mathrm{E}(\bar{Y}_{i\cdot} - \bar{Y}) = \hat{\tau}_i, \quad \mathrm{E}(\bar{Y}_{\cdot j} - \bar{Y}) = \beta_j,$

and a contrast estimate $\sum_i a_i \hat{\tau}_i$ is an unbiased estimate of the contrast $\sum_i a_i \hat{\tau}_i$ with variance

$$\begin{aligned}
\mathrm{Var}\left(\sum_i a_i \hat{\tau}_i\right) &= \mathrm{Var}\left(\sum_i a_i (\bar{Y}_{i\cdot} - \bar{Y})\right) \\
&= \mathrm{Var}\left(\sum_i a_i (\bar{\varepsilon}_i - \bar{\varepsilon})\right) \\
&= \mathrm{Var}\left(\sum_i a_i \bar{\varepsilon}_i\right) \quad [\textstyle\sum_i a_i \bar{\varepsilon} = 0] \\
&= \sum_i a_i^2 \mathrm{Var}(\bar{\varepsilon}_i) + 2 \sum_{i>i'} a_i a_{i'} \mathrm{Cov}(\bar{\varepsilon}_i, \bar{\varepsilon}_{i'})
\end{aligned}$$

From the model, $\mathrm{Var}(\bar{\varepsilon}_i) = \sigma^2/b$, and we can get $\mathrm{Cov}(\bar{\varepsilon}_i, \bar{\varepsilon}_{i'})$ from Technical Note 4.8.1. However, for completeness we calculate

$$\mathrm{Cov}(\bar{\varepsilon}_i, \bar{\varepsilon}_{i'}) = \mathrm{E}\left(\bar{\varepsilon}_i \bar{\varepsilon}_{i'}\right) = \mathrm{E}\left(\frac{1}{b^2} \sum_j \varepsilon_{ij} \sum_{j'} \varepsilon_{i'j'}\right) = \mathrm{E}\left(\frac{1}{b^2} \sum_j \varepsilon_{ij} \varepsilon_{i'j}\right),$$

since, if $j \neq j'$ the observations are in different blocks and the covariance is zero. Recalling that $\mathrm{Cov}(\varepsilon_{ij}\varepsilon_{i'j}) = \rho\sigma^2$, we have $\mathrm{Cov}(\bar{\varepsilon}_i, \bar{\varepsilon}_{i'}) = \rho\sigma^2/b$ and

(4.14) $\mathrm{Var}\left(\sum_i a_i \hat{\tau}_i\right) = \dfrac{\sigma^2}{b} \sum_i a_i^2 + 2\dfrac{\rho\sigma^2}{b} \sum_{i>i'} a_i a_{i'} = \dfrac{(1-\rho)\sigma^2}{b} \sum_i a_i^2$

where the last equality follows after a bit of algebra (Exercise 4.10).

This is very important!

Thus we see that the variance of a contrast depends quite strongly on the intrablock correlation, with a large positive correlation yielding a decrease in the variance. However, the only quantity that is directly under the experimenters control is b, the number of blocks, and we see that increasing the number of blocks is the surest way of decreasing the variance of a contrast estimate.

This message, about the effect of correlation, was also present in Section 3.4, but perhaps it was a bit more subtle. Recall the variance of a contrast in the random blocks case (3.16),

$$(4.15) \qquad \text{Var}\left(\sum_i a_i \hat{\tau}_i\right) = \frac{\sigma_\varepsilon^2}{b} \sum_i a_i^2.$$

Now, from (4.9) we can write $\sigma_\varepsilon^2 = (1-\rho)(\sigma_\beta^2 + \sigma_\varepsilon^2) = (1-\rho)\text{Var}(Y)$. Comparing this last expression to (4.14) shows that the intraclass correlation impacts the variance whether or not the blocks are modeled as random.

Variance Estimates

From (4.11) and (4.12) the residuals from the model are

$$y_{ij} - \hat{\mu} - \hat{\tau}_i - \hat{\beta}_j = y_{ij} - \bar{y}_{i\cdot} - \bar{y}_{\cdot j} + \bar{\bar{y}} = \varepsilon_{ij} - \bar{\varepsilon}_{i\cdot} - \bar{\varepsilon}_{\cdot j} + \bar{\bar{\varepsilon}},$$

and we would typically base our variance estimate on the residual mean square,

$$(4.16) \qquad \text{SS(Res)} = \sum_{ij} (\varepsilon_{ij} - \bar{\varepsilon}_{i\cdot} - \bar{\varepsilon}_{\cdot j} + \bar{\bar{\varepsilon}})^2.$$

This should be the basis of our variance estimate, and we now look at its expectation. A relatively straightforward, but somewhat lengthy calculation (see Exercise 4.14) will show that

$$\text{ESS(Res)} = (b-1)(t-1)(1-\rho)\sigma^2.$$

Of course, we realize that SS(Res) is exactly SS(T × B). Thus, the mean squared $\hat{\sigma}^2 = \text{MS(T} \times \text{B})$ is an unbiased estimator of the treatment variance, and for any contrast we can estimate its variance with

$$(4.17) \qquad \widehat{\text{Var}}\left(\sum_i a_i \hat{\tau}_i\right) = \frac{\text{MS(T} \times \text{B})}{b} \sum_i a_i^2.$$

Example 4.2. GREENHOUSE EXPERIMENT   A researcher is planning an experiment to determine the effectiveness of four house plant fertilizers. The researcher has arranged to use three benches (blocks) in different areas of a greenhouse. There are four pots on each bench, and the fertilizers will be randomly assigned to the pots. At the end of the experiment plant heights will be recorded (in inches). The data are in dataset Greenhouse1, and result in the following analysis:

Source	df	SS	MS	F	p-value
Bench	2	106.08	53.04	1.652	
Fertilizer	3	915.69	305.23	9.508	0.011
Bench × Fertilizer	6	192.62	32.10		

This is a case where there is reason to model the blocks as fixed, as they are locations within a specific greenhouse. The variance of any contrast will come from MS(Bench × Fertilizer) and (4.17). See Exercise 4.1. ||

Inference

The most important inference from a RCB anova concerns the treatments and, in particular, the estimation of contrasts between the treatments. We do all calculations under the RCB model (4.1), where the β_j are fixed values and there is an intrablock correlation. We will be concerned with inference on specific treatment contrasts.

Working under the RCB anova assumptions we have that,

$$Y_{ij} \sim N(\mu + \tau_i + \beta_j, \sigma^2), \quad i = 1, \ldots, t, \quad j = 1, \ldots, b$$

(4.18) and

$$\text{Cov}(Y_{ij}, Y_{i'j}) = \rho \sigma^2.$$

The parameter of interest is the treatment contrast $\sum_{i=1}^{k} a_i \tau_i$, whose estimator $\sum_{i=1}^{k} a_i \overline{Y}_{i\cdot}$ satisfies

$$(4.19) \quad E\left(\sum_{i=1}^{t} a_i \overline{Y}_{i\cdot}\right) = \sum_{i=1}^{k} a_i \tau_i \text{ and } \text{Var}\left(\sum_{i=1}^{t} a_i \overline{Y}_{i\cdot}\right) = \frac{(1-\rho)\sigma^2}{b} \sum_i a_i^2.$$

Since the Y_{ij}s are normal, from (4.18) and (4.19) we have

$$\sum_{i=1}^{k} a_i \overline{Y}_{i\cdot} \sim N\left(\sum_{i=1}^{k} a_i \tau_i, \frac{(1-\rho)\sigma^2}{b} \sum_i a_i^2\right)$$

and, therefore,

$$(4.20) \quad \frac{\sum_{i=1}^{k} a_i \overline{Y}_{i\cdot} - \sum_{i=1}^{k} a_i \tau_i}{\sqrt{\frac{(1-\rho)\sigma^2}{b} \sum_i a_i^2}} \sim N(0,1).$$

From (4.17) we can estimate $(1-\rho)\sigma^2$ with MS(T × B), and it remains to establish the distribution of

$$(4.21) \quad \frac{\sum_{i=1}^{k} a_i \overline{Y}_{i\cdot} - \sum_{i=1}^{k} a_i \tau_i}{\sqrt{\frac{\text{MS(TxB)}}{b} \sum_i a_i^2}}.$$

That this, in fact, has Student's t-distribution with $(t-1)(b-1)$ degrees of freedom follows from arguments similar to those establishing Theorem 3.18 in Technical Note 3.8.2 (Exercise 4.13).

Thus, to test

$$H_0: \sum_{i=1}^{t} a_i \tau_i = 0 \qquad \text{versus} \qquad H_1: \sum_{i=1}^{t} a_i \tau_i \neq 0$$

at level α, we

$$(4.22) \qquad \text{reject } H_0 \text{ if } \left| \frac{\sum_{i=1}^{t} a_i \bar{Y}_{i\cdot}}{\sqrt{\left(\frac{\text{MS(T} \times \text{B)}}{b}\right) \sum_{i=1}^{t} a_i^2}} \right| > t_{(b-1)(t-1), \alpha/2}.$$

More importantly, we get an interval estimator of $\sum a_i \tau_i$. With probability $1-\alpha$,

$$(4.23) \qquad \sum_{i=1}^{t} a_i \bar{Y}_{i\cdot} - t_{(b-1)(t-1), \alpha/2} \sqrt{\frac{\text{MS(T} \times \text{B)}}{b} \sum_{i=1}^{t} a_i^2}$$

$$\leq \sum_{i=1}^{t} a_i \tau_i \leq \sum_{i=1}^{t} a_i \bar{Y}_{i\cdot} + t_{(b-1)(t-1), \alpha/2} \sqrt{\frac{\text{MS(T} \times \text{B)}}{b} \sum_{i=1}^{t} a_i^2}.$$

4.5 Modeling the Interaction

For fixed blocks with interaction, a model is

$$(4.24) \qquad \begin{aligned} Y_{ijk} &= \mu + \tau_i + \beta_j + (\tau\beta)_{ij} + \varepsilon_{ijk}, \\ \varepsilon_{ijk} &\sim N(0, \sigma^2), \\ \text{Cov}(\varepsilon_{ijk}, \varepsilon_{ijk'}) &= \rho_\varepsilon \sigma^2 \text{ for } k \neq k', \\ \text{Cov}(\varepsilon_{ijk}, \varepsilon_{i'jk'}) &= \rho_B \sigma^2 \text{ for } i \neq i', \end{aligned}$$

where $i = 1, \ldots, a$, $j = 1, \ldots, b$, $k = 1, \ldots, r$, and

$$\bar{\tau} = \bar{\beta} = (\overline{\tau\beta})_i = (\overline{\tau\beta})_j = 0.$$

We distinguish between

(1) ρ_ε, the correlation between two observations within the same treatment-block combination, induced by technical replication,

(2) ρ_B, the correlation between two true replicates in the same block.

Table 4.3. Expected Mean Squares for RCB anova with fixed blocks and replication.

Source	df	EMS
Blocks	$b-1$	$\sigma^2\left[1+(r-1)\rho_\varepsilon+r(t-1)\rho_B\right]+\frac{rt}{b-1}\sum_j \beta_j^2$
Treatments	$t-1$	$\sigma^2[1-r\rho_B+(r-1)\rho_\varepsilon]+\frac{br}{t-1}\sum_i \tau_i^2$
TxB	$(t-1)(b-1)$	$\sigma^2[1-r\rho_B+(r-1)\rho_\varepsilon]+\frac{r}{(b-1)(t-1)}\sum_{ij}(\tau\beta)_{ij}^2$
Within	$bt(r-1)$	$(1-\rho_\varepsilon)\sigma^2$

If we have true within replication of the experimental units, then $\rho_\varepsilon = 0$.

We will see that if $\rho_\varepsilon \neq 0$, then the within replication plays no essential role in any of the analysis. This underscores, once again, the fact that subsamples in an RCB do not contribute much to the information in the design.

As the treatment design is a twoway crossed design, the ordinary least squares estimates are given by

(4.25)
$$\bar{Y}_i - \bar{Y} = \hat{\tau}_i,$$
$$\bar{Y}_j - \bar{Y} = \hat{\beta}_j,$$
$$\bar{Y}_{ij} - \bar{Y}_i - \bar{Y}_j + \bar{Y} = (\hat{\tau\beta})_{ij}.$$

Calculation of the EMS is straightforward, for example,

$$\text{ESS(Within)} = \text{E}\sum_{ijk}(Y_{ijk}-\bar{Y}_{ij})^2 = \text{E}\sum_{ijk}(\varepsilon_{ijk}-\bar{\varepsilon}_{ij})^2$$
$$= rbt(\text{Var}\varepsilon_{ijk}-\text{Var}\bar{\varepsilon}_{ij}) \quad \text{[Lemma 3.16]}$$
(4.26)
$$= rbt\left(\sigma^2-\frac{\sigma^2}{r}(1+(r-1)\rho_\varepsilon)\right) \quad \text{[(4.1)]}$$
$$= bt(r-1)(1-\rho_\varepsilon)\sigma^2$$

$$\text{ESS(Blocks)} = \text{E}\sum_j rt(\bar{Y}_j-\bar{Y})^2 = rt\sum_j \beta_j^2 + rt\text{E}\sum_j(\bar{\varepsilon}_j-\bar{\varepsilon})^2$$
$$= rt\sum_j \beta_j^2 + (b-1)\sigma^2\left(1+(r-1)\rho_\varepsilon+r(t-1)\rho_B\right),$$

where we note that the cross terms all have zero expectation. We can continue on like this, and calculate the expectation of all of the sums of squares for the RCB anova (see Exercise 4.15). Dividing the expected sums of squares by their degrees of freedom yields the expected mean squares given in Table 4.3.

If $r = 1$ this reduces to Table 4.1. Also, from looking at the EMS there is an implicit restriction in this model on the size of the correlations, notably that

$$1 - r\rho_B + (r-1)\rho_\varepsilon \geq 0 \text{ and } 1 + (r-1)\rho_\varepsilon + r(t-1)\rho_B \geq 0.$$

This is examined in Exercise 4.11.

To justify F-tests and contrast distributions, the theorems of Technical Note 3.8.3 can be adapted. Sparing everyone the details, we have the following observations about F-tests in the RCB anova with blocks fixed and replication.

F-test on treatments

The test on treatments is against MS(T × B) under the assumption that $(\tau\beta)_{ij} = 0$. Formally, the null hypothesis of no treatment effect is

$$H_0 : \tau_i = 0 \text{ for all } i$$

and is tested by

$$F_{t-1,(b-1)(t-1)} = \frac{\text{MS(Trts)}}{\text{MS(T × B)}}.$$

under the assumption that $(\tau\beta)_{ij} = 0$ for all i and j. This means that we must assume that the effect of the jth block is the same for every treatment. If this assumption is violated, the F-test can still be done, but it becomes a conservative test, since the denominator will tend to be larger, making the statistic smaller and rejection more difficult. (See Technical Note 2.8.4.)

Treatments are *never* tested against MS(Within), in fact, SS(Within) is of no use unless we assume either that $\rho_\varepsilon = 0$ (and we have true replication) or $\rho_\varepsilon = \rho_B$, that the correlation within each treatment-block combination is the same as the correlation between different treatments within the same block. In general, this latter assumption does not seem reasonable, as we would expect $\rho_\varepsilon > \rho_B$ if the replication is technical. (Note that $\rho_\varepsilon = \rho_B$ is trivially true if $r = 1$.)

Although the assumptions needed to get a valid test on treatments may seem both stringent and startling, they turn out to be similar to the assumptions needed if the blocks are modeled as random effects, as we will see in Section 4.6.

Other F-tests

It is interesting that with this covariance structure, and with blocks fixed, there is no straightforward way to test interactions. This is similar to the case for blocks random, as we saw in Section 3.5.

From Table 4.3 to get a valid F-test we need to make more assumptions, some of which see untenable. The within error can be used to test treatments or interaction if we assume that $\rho_\varepsilon = \rho_B$, which is an unfounded assumption as, again, we would expect $\rho_\varepsilon > \rho_B$ if the replication is technical.

If there is true replication then $\rho_\varepsilon = 0$ and we can test the hypothesis

(4.27) $$H_0 : (\tau\beta)_{ij} = 0 \text{ for all } i, j \text{ and } \rho_B = 0$$

with

$$F_{(b-1)(t-1), r(b-1)(t-1)} = \frac{\text{MS(T} \times \text{B)}}{\text{MS(Within)}},$$

which seems to be a rather strange interaction hypothesis. However, this is not very different from what we saw in Section 3.5 and display (3.26). There the correlation term becomes part of the variance to be tested, so we do not need to explicitly deal with it.

However, we should realize that everything we do is an artifact of the model that we assume, and we must decide which model, and hence which set of assumptions, are appropriate for our design and analysis. $\|$ Its all about the model

We also note that under the assumption $\rho_\varepsilon = 0$ there is a test on blocks,

$$H_0 : \beta_j - \bar{\beta} + (\bar{\tau\beta})_j - (\bar{\tau\beta}) = 0 \text{ for all } j \text{ and } \rho_B = 0,$$

but, again, there is usually little reason to care about this.

> **Note:** One main message from this analysis is that unless there is true replication, which implies that $\rho_\varepsilon = 0$, the within sum of squares in any RCB is almost of no use.

Example 4.3. FIXED BLOCKS WITH INTERACTION   A chemist wishes to test the effect of four levels of a chemical agent on the strength of cloth. There are five particular types of cloth that the agent will be used on, and the performance is to be evaluated against those five types. For each of the types of cloth, a roll is selected, and the chemist applies all four levels of the chemical, in a random order, to each roll. Two observations are taken at each cloth-chemical combination. The resulting tensile strengths are:

Cloth

Chemical	1	2	3	4	5
1	73, 74	68, 67	74, 80	71, 73	67, 70
2	73, 75	67, 70	75, 74	72, 71	70, 70
3	75, 75	68, 69	78, 72	73, 72	68, 68
4	73, 75	71, 69	75, 73	75, 72	69, 68

The cloth is acting as a block here, but there are only five types so there is no sampling from a larger population. The experimental unit is the roll of cloth, and the two observations in each cell are subsamples. Referring to model (4.24), we cannot assume that $\rho_\varepsilon = 0$ and thus, from the EMS in Table 4.3 we can test the main effect of Chemical but not the interaction. (See Exercise 4.12.)   ‖

4.6 Reconciliation

In general, attempting to reconcile two models, or two designs is usually futile and often pointless, as the design, and hence the analysis, should be dictated by the physical reality of the experiment under consideration. However, it sometimes is useful to examine the effect of an assumption, in the hopes that it will bring greater understanding and, if there is uncertainty in choosing the appropriate model, we may better understand the consequences of the assumption. Thus, we take a deeper look at the effect of modeling the interaction when the blocks are fixed or random.

Although Table 3.7 appears simple compared to Table 4.3, it is based on some additional assumptions. In particular, as a consequence of the interaction being a random effect, two effects have a common expectation and end up simplifying the EMS. In Table 4.4 the EMS for both the fixed blocks and random blocks models are lined up, so we can see the relevant mean squares for the test on treatments.

In the random blocks model the test is clear: Under $H_0 : \sum_i \tau_i^2 = 0$ the two expected mean squares are the same, so we have a central F-distribution under H_0.

Table 4.4. Comparison of EMS from RCB anovas with fixed and random blocks and replication.

Source	Blocks Fixed
Treatment	$\sigma^2[1 - r\rho_B + (r-1)\rho_\varepsilon] + \frac{br}{t-1}\sum_i \tau_i^2$
Interaction	$\sigma^2[1 - r\rho_B + (r-1)\rho_\varepsilon] + \frac{r}{(b-1)(t-1)}\sum_{ij}(\tau\beta)_{ij}^2$

Source	Blocks Random
Treatment	$\sigma_\varepsilon^2[1 + (r-1)\rho_\varepsilon] + r\sigma_{\tau\beta}^2[1 - \rho_{\tau\beta}] + \frac{br}{t-1}\sum_i \tau_i^2$
Interaction	$\sigma_\varepsilon^2[1 + (r-1)\rho_\varepsilon] + r\sigma_{\tau\beta}^2[1 - \rho_{\tau\beta}]$

For the fixed blocks model this does not happen, and the EMS are not equal unless we also assume that $\sum_{ij}(\tau\beta)_{ij}^2 = 0$. However, in a sense that will be explained, this assumption is implicit in the random blocks test.

We examine the EMS for treatments and interactions in an extra careful way, leaving some of the details to Exercise 4.18. Start with the model

$$Y_{ijk} = \mu + \tau_i + \beta_j + (\tau\beta)_{ij} + \varepsilon_{ijk},$$

but, for now, we do not make any assumptions about whether $(\tau\beta)_{ij}$ is fixed or random. For the treatment EMS we have

$$\text{EMS(Treatments)} = \frac{rb}{t-1}\text{E}\left(\sum_i (\bar{Y}_i - \bar{\bar{Y}})^2\right)$$

$$= \frac{rb}{t-1}\text{E}\sum_i \left(\tau_i + [(\bar{\tau\beta})_i - (\bar{\bar{\tau\beta}})] + [\bar{\varepsilon}_i - \bar{\bar{\varepsilon}}]\right)^2,$$

and for the interaction EMS we have

$\text{EMS(T} \times \text{B)}$

$$= \frac{r}{(t-1)(b-1)}\text{E}\left(\sum_{ij}(\bar{Y}_{ij} - \bar{Y}_i - \bar{Y}_j + \bar{\bar{Y}})^2\right)$$

$$= \frac{r}{(t-1)(b-1)}\text{E}\sum_{ij}\left([(\tau\beta)_{ij} - (\bar{\tau\beta})_i - (\bar{\tau\beta})_j + (\bar{\bar{\tau\beta}})] + [\bar{\varepsilon}_{ij} - \bar{\varepsilon}_i - \bar{\varepsilon}_j + \bar{\bar{\varepsilon}}]\right)^2.$$

Now for either EMS, the terms involving ε_{ijk} have zero cross terms with the other factors. Moreover, under either the random blocks model (3.24) or the fixed blocks model (4.24) we have

$$\frac{rb}{t-1}\text{E}\sum_i(\bar{\varepsilon}_i - \bar{\bar{\varepsilon}})^2 = \frac{r}{(t-1)(b-1)}\text{E}\sum_{ij}(\bar{\varepsilon}_{ij} - \bar{\varepsilon}_i - \bar{\varepsilon}_j + \bar{\varepsilon})^2$$

(4.28) $$= r\text{Var}\bar{\varepsilon}_{ij},$$

so these terms do not contribute to any discrepancy in EMS.

We next look at the remaining term in the interaction MS, and find

$$\frac{r}{(t-1)(b-1)}\text{E}\sum_{ij}[(\tau\beta)_{ij} - (\bar{\tau\beta})_i - (\bar{\tau\beta})_j + (\bar{\bar{\tau\beta}})]^2$$

(4.29) $$= \begin{cases} \frac{r}{(t-1)(b-1)}\sum_{ij}(\tau\beta)_{ij}^2 & \text{blocks fixed} \\[2mm] r\text{Var}(\tau\beta)_{ij} & \text{blocks random.} \end{cases}$$

These are almost the same terms, in their fixed and random incarnations, each calibrating the variation in $(\tau\beta)_{ij}$ (in the fixed blocks model we set $(\bar{\bar{\tau\beta}}) = 0$

for identifiability). So, thus far, there is no real discrepancy in the calculations. This will change when we calculate the remaining term for the treatment mean squares. We have

$$\frac{rb}{t-1}\text{E}\sum_i[\tau_i + (\bar{\tau\beta})_i - (\bar{\tau\beta})]^2$$

(4.30)
$$= \begin{cases} \frac{rb}{t-1}\sum_i \tau_i^2 & \text{blocks fixed} \\ \frac{rb}{t-1}\sum_i \tau_i^2 + r\text{Var}(\tau\beta)_{ij} & \text{blocks random,} \end{cases},$$

and now we see the difference, and the implicit inequality that makes the random block model work.

In the random blocks model we have

$$\frac{rb}{t-1}\text{E}\sum_i[(\tau\beta)_i - (\bar{\tau\beta})]^2 =$$

(4.31)
$$\frac{r}{(t-1)(b-1)}\text{E}\sum_{ij}[(\tau\beta)_{ij} - (\bar{\tau\beta})_i - (\bar{\tau\beta})_j + (\bar{\tau\beta})]^2,$$

which cannot happen in the fixed blocks model because the identifiability constraint forces us to restrict the values of $(\tau\beta)_{ij}$, with the most common restriction being $(\bar{\tau\beta})_i = (\bar{\tau\beta})_j = (\bar{\tau\beta}) = 0$. Thus, in Section 4.5, when we made the assumption that $(\tau\beta)_{ij} = 0$ in order to get a treatment test, we were invoking the fixed blocks analog of (4.31).

The point of this section is not to decide which design is "better", because the comparison of designs is an academic, but not a practical, question. The type of design is almost totally dictated by the actual experimental situation; we do not have a *choice* of making blocks fixed or random, they *are* fixed or random. Rather, the point of this section is to understand the differences, the compromises, and hence the inferences. Here, we have seen that the major difference in the models lies in the covariance assumption, and any inferences must be made in light of this assumption.

4.7 Exercises

Essential

4.1 Referring to Example 4.2:
   (a) Explain why, regardless of how many plants are in each pot, we only use one number per pot.
   (b) Verify the anova table
   (c) Calculate the variance of the difference of two treatment means.
   (d) Estimate the Helmert contrasts for these data, taking Fertilizer 1 to be the control.
   (e) Test the significance of the Helmert contrasts. What can you conclude?

(f) Compute the variance for a difference in treatment means.

(g) The researcher wants to be able to detect a difference in means as small as $\delta = 1.5$cm at $\alpha = .05$. What power does she have to do this?

4.2 Referring to Example 4.2, suppose instead that the researcher was able to use two independent pots in each Bench $\times$ Fertilizer combination. These data are in dataset **Greenhouse2**.

(a) Redo parts (b)-(e) of Exercise 4.1. Has anything changed?

(b) For her next experiment, the researcher would like to first do a power calculation, and be able to detect a difference in means of 5in at $\alpha = .05$. What combination of blocks and plants per pot would guarantee a power of at least $(i)0.60$ $(ii)0.80$?

4.3 This question refers to Example 3.2, with data in **BrownieData**.

(a) If "Brand" is considered a block, then the experiment is an RCB. Write out the anova (source, df, and EMS) for this RCB. Analyze the data as an RCB.

(b) A naive approach to this experiment would say that, since there are only three brands of interest, and they are in the experiment, "brand" is therefore a fixed factor and the experiment is a CRD. Reanalyze the data as a CRD. Clearly identify the denominator for testing all effects. Are the conclusions from the RCB analysis different from the conclusions of the CRD analysis?

(c) Another approach, which also denies the RCB structure, would be to have a term representing "sample nested within brands" with 15 degrees of freedom.

   (i) How would you calculate the sum of squares for "sample nested within brands"?

   (ii) Construct the anova table for this experiment? In particular, where do the degrees of freedom come from for "sample nested within brands"? How can you test power and baking time?

4.4 For model (4.11), with $\text{Var}(\varepsilon_{ij}) = \sigma^2$ and $\text{Cov}(\varepsilon_{ij}, \varepsilon_{i'j'}) = \rho$:

(a) Show that the least squares estimates are given by (3.14)

(b) Show that the restriction $\bar{\tau} = \bar{\beta} = 0$ is equivalent to defining $\tau_i' = \tau_i - \bar{\tau}$, $\beta_j' = \beta_j - \bar{\beta}$, $\mu' = \mu + \bar{\tau} + \bar{\beta}$ and fitting the model $y_{ij} = \mu' + \tau_i' + \beta_j' + \varepsilon_{ij}$.

(c) Restricting $\bar{\tau} = \bar{\beta} = 0$ is one of many possible restrictions to get a solution. Another common one is to set $\tau_t = \beta_b = 0$. Derive the least squares estimates under this constraint.

4.5 Continuing from Exercise 4.4:

(a) Show that the least squares estimates are unbiased.

(b) Find the variance of the estimators (see (4.14)).

4.6 This problem is similar to Exercise 3.11, but here we will start with the fixed blocks model (4.1). Suppose that the treatments have a factorial structure, that is, the treatment effects $\tau_1, \ldots, \tau_t$ arise from the cells of two crossed factors C and D in a CRD, so that the model could be written

$$Y_{ijk} = \mu + \gamma_i + \delta_k + (\gamma\delta)_{ik}$$
$$+ \beta_j + (\gamma\beta)_{ij} + (\delta\beta)_{jk} + (\gamma\delta\beta)_{ijk} + \varepsilon_{ijk}.$$

(a) Write out the EMS for this model, and indicate how each treatment effect is to be tested.

(b) Compare the tests here to those found in Exercise 3.11.

4.7 Consider an experiment that is intended to test the effect of ozone on plants. The researcher assigned four environmental chambers to four different levels of added ozone. Twelve plants were placed in each chamber. Fourteen days later the plants were removed and analyzed. Data were recorded for each plant.

(a) Draw a schematic of the layout of the experiment

(b) What is the experimental unit? Explain why this should be considered an RCB with fixed blocks. Write the model equation and state any assumptions.

(c) After the experiment began, the experimenter realized that he could not analyze all 48 plants in one day. He could analyze 25 or 30 plants in one day, so he has to break the plants up into groups for analysis. Should he analyze all of the plants from two chambers on the first day, half the plants from each chamber on the first day, or some other allocation? Discuss the consequences.

4.8 The following data are a portion of the responses collected during an interlaboratory study involving several laboratories. Each laboratory was sent samples from four different materials that would have a range of values on the characteristic of interest. The laboratories were each sent three samples from each type of material. The four materials were labeled A, B, C and D for the laboratories. Assume that there is no laboratory by material interaction. The data are in data set **Laboratory**.

Laboratory	Material	Measurements
1	A	12.20 12.28 12.16
	B	15.51 15.02 15.29
	C	18.14 18.08 18.21
	D	18.54 18.36 18.45
2	A	12.59 12.30 12.67
	B	14.98 15.46 15.22
	C	18.54 18.31 18.60
	D	19.21 18.77 18.69
3	A	12.72 12.78 12.66
	B	15.33 15.19 15.24
	C	18.00 18.15 17.93
	D	18.88 18.12 18.03

(a) From the description above is the interest in this experiment on the comparison among laboratories or on the comparison among materials?

(b) Write the model equation and give assumptions needed for the analysis. Explain why the laboratories should be considered fixed blocks.

(c) Construct the appropriate anova table, consistent with your answer in (a). Include EMS.

4.9 Use the data of Exercise 4.8 here. The researchers were interested in the protein content of the materials that they sent to the laboratories The materials are actually different samples of grain of two species using two cultivation methods. The A samples are from three fields of winter wheat, all from fields that have been producing wheat for at least the two previous years. The B sample is from three fields of winter wheat, all from fields that were planted to soybeans in the previous season. The C samples are from fields of triticale that have been planted to wheat or triticale for at least the two previous years. The D samples are from fields of triticale that were planted to soybeans in the previous season.

(a) What set of orthogonal contrasts would be sensible for comparisons among the four materials (answer in terms of population parameters)?

(b) Modify the anova table of Exercise 4.8 to include the tests on contrasts.

(c) Estimate the contrasts and their standard errors and summarize the results of your analysis.

4.10 In (4.14) we derived the variance of a contrast as a function of the covariance, but some gaps need to be filled.

(a) Show that $\mathrm{Var}\left(\sum_i a_i \hat{\tau}_i\right) = \frac{\sigma^2}{b}\left(\sum_i a_i^2 + 2\rho \sum_{i>i'} a_i a_{i'}\right)$.

(b) Using the fact that $\sum_i a_i = 0$, show that

$$\sum_i a_i^2 + 2\rho \sum_{i>i'} a_i a_{i'} = (1-\rho)\sum_i a_i^2,$$

and hence derive (4.14).

(c) These calculations assumed that blocks are fixed. Show that if we instead assume that blocks are random, as in Section 3.4, we get the same formula for the variance of a contrast.

4.11 Referring to Table 4.3:

(a) Show that the inequality $1 - r\rho_B + (r-1)\rho_\varepsilon \geq 0$ must be satisfied by considering the correlation between $\bar{\varepsilon}_{ij}$ and $\bar{\varepsilon}_{i'j}$.

(b) Show that the inequality $1 + (r-1)\rho_\varepsilon + r(t-1)\rho_B \geq 0$ is always satisfied if $\rho_B \geq 0$, and find the smallest negative value of ρ_B that can be accommodated.

4.12 Referring to Example 4.3:

(a) Produce the anova table corresponding to the model in the example. Explain why we lose no information by replacing the two observations per cell with their average.

(b) Suppose that instead of subsampling the rolls, the two observations per cell came from two different rolls. Explain how this would change you assessment of the model (4.24) and the EMS of Table 4.3. Produce an anova table and run all the tests that can now be done.

(c) Suppose that each observation had been taken from a different, randomly selected roll. What is this design? Produce an anova table and run all of the tests that can now be done.

Note: The data are in dataset `Cloth`, where replications are identified. For part (a) that column in the dataset should be ignored.

Accompaniment

4.13 (a) Verify the EMS calculations in Table 4.1.

(b) Show that the ratio of mean squares in (4.8) has the stated F distribution by applying Cochran's Theorem. (The argument is almost exactly the same as the one used to establish Theorem 3.18, as the covariance matrix Σ of that theorem has the same form as the covariance matrix here (which is given in (4.32)).

(c) Show that $\text{MS(Blocks)}/\text{MS(T} \times \text{B)}$ can test the null hypothesis in (4.10).

(d) Show that the statistic in (4.21) has the indicated t-distribution by adapting Theorem 3.18.

4.14 Referring to Section 4.4, we will establish the expectation of (4.16) in two separate ways.

(a) Using Exercise 2.19, show that

$$\sum_{ij}(\varepsilon_{ij} - \bar{\varepsilon}_{i\cdot} - \bar{\varepsilon}_{\cdot j} + \bar{\varepsilon})^2 = \sum_{ij}\varepsilon_{ij}^2 - b\sum_i\bar{\varepsilon}_{i\cdot}^2 - t\sum_j\bar{\varepsilon}_{\cdot j}^2 + bt\bar{\varepsilon}^2,$$

$$\text{E}\,\varepsilon_{ij}^2 = \sigma^2, \quad \text{E}\,\bar{\varepsilon}_{i\cdot}^2 = \frac{\sigma^2}{b}, \quad \text{E}\,\bar{\varepsilon}_{\cdot j}^2 = \frac{\sigma^2}{t}[1+(t-1)\rho], \quad \text{E}\,\bar{\varepsilon}^2 = \frac{\sigma^2}{bt}[1+(t-1)\rho],$$

and combine everything to calculate the expectation of (4.16).

(b) As an alternative derivation, expand SS(Res) as

$$\text{SS(Res)} = \sum_{ij}(\varepsilon_{ij} - \bar{\varepsilon}_{i\cdot})^2 + \sum_{ij}(\bar{\varepsilon}_{\cdot j} - \bar{\varepsilon})^2 - 2 \times \sum_{ij}(\varepsilon_{ij} - \bar{\varepsilon}_{i\cdot})(\bar{\varepsilon}_{\cdot j} - \bar{\varepsilon}),$$

and use the results of Section 3.8 to calculate

$$\text{E}\left(\sum_{ij}(\varepsilon_{ij} - \bar{\varepsilon}_{i\cdot})^2\right) = t(b-1)\sigma^2,$$

$$\text{E}\left(\sum_{ij}(\bar{\varepsilon}_{\cdot j} - \bar{\varepsilon})^2\right) = (b-1)\sigma^2(1 + (t-1)\rho),$$

$$\text{E}\left(\sum_{ij}(\varepsilon_{ij} - \bar{\varepsilon}_{i\cdot})(\bar{\varepsilon}_{\cdot j} - \bar{\varepsilon})\right) = (b-1)\sigma^2(1 + (t-1)\rho),$$

and combine everything to calculate the expectation of (4.16).

4.15 (a) Verify the variances and covariances in Technical Note 4.8.1. Although the calculations are tedious, there is a pattern and after doing the first two, you should see the pattern.

(b) Verify the expected mean squares in Table 4.3.

4.16 (For the strong-willed.) In Technical Note 4.8.1 we calculated variances and covariances of the error means. Sometimes in calculating variances of treatment effects we are also interested in the variances of error deviations. Explain that all of the error deviations are again normal, and show that their variances and covariances are given by

$$\text{Var}(\varepsilon_{ijk} - \bar{\varepsilon}_{ij\cdot}) = \left(1 - \tfrac{1}{r}\right)(1 - \rho_\varepsilon)\sigma_\varepsilon^2, \quad \text{Cov}(\varepsilon_{ijk} - \bar{\varepsilon}_{ij\cdot}, \varepsilon_{ijk'} - \bar{\varepsilon}_{ij\cdot}) = \frac{1-\rho_\varepsilon}{r}\sigma_\varepsilon^2$$

$$\text{Var}(\bar{\varepsilon}_{ij} - \bar{\varepsilon}_i) = \left(1 - \tfrac{1}{b}\right)\sigma_B^2, \quad \text{Cov}(\bar{\varepsilon}_{ij} - \bar{\varepsilon}_i, \varepsilon_{ij'} - \bar{\varepsilon}_i) = -\frac{\sigma_B^2}{b}$$

$$\text{Var}(\bar{\varepsilon}_{ij} - \bar{\varepsilon}_j) = \left(1 - \tfrac{1}{t}\right)(1 - \rho_B)\sigma_B^2, \quad \text{Cov}(\bar{\varepsilon}_{ij} - \bar{\varepsilon}_j, \bar{\varepsilon}_{i'j} - \bar{\varepsilon}_j) = \frac{1-\rho_B}{t}\sigma_B^2$$

$$\text{Var}(\bar{\varepsilon}_i - \bar{\varepsilon}) = \left(1 - \tfrac{1}{t}\right)(1 - \rho_B)\frac{\sigma_B^2}{b}, \quad \text{Cov}(\bar{\varepsilon}_i - \bar{\varepsilon}, \bar{\varepsilon}_{i'} - \bar{\varepsilon}) = -(1 - \rho_B)\frac{\sigma_B^2}{bt}$$

$$\text{Var}(\bar{\varepsilon}_j - \bar{\varepsilon}) = \left(1 - \tfrac{1}{b}\right)\frac{\sigma_B^2(1+(t-1)\rho_B)}{t}, \quad \text{Cov}(\bar{\varepsilon}_j - \bar{\varepsilon}, \bar{\varepsilon}_{j'} - \bar{\varepsilon}) = -\frac{\sigma_B^2(1+(t-1)\rho_B)}{bt}$$

Note that the first line above refers to the within cell variance, and is the only place that the parameters ρ_ε and σ_ε^2 appear. Also, the subsample size r only goes to reduce the subsampling variance, and hence the impact on the design is almost nothing.

4.17 Referring to Technical Note 4.8.2,

    (a) Show that the V matrix in (4.32) has all row sums equal.

    (b) Show, by direct calculation that the two estimates of θ, ordinary and generalized least squares, yield the same estimates, the vector of cell means.

    (c) Show, by direct calculation, that the two estimates of β, ordinary and generalized least squares, yield the same estimates, the vector of mean deviations.

    (d) Verify the equality of V and V^* under (4.9), and the estimate of β given in (4.35).

4.18 Referring to Section 4.6

    (a) Verify (4.28), which works because there is independence between blocks in both models.

    (b) Verify (4.31), which also works because there is independence between blocks.

4.19 Referring to Miscellanea 4.9.1:

    (a) For the oneway CRD:

        (i) Reproduce the anova table and perform the randomization test of H_0: no treatment effect. Do your results agree with Figure 4.1?

        (ii) For a oneway CRD with t treatments and r observations per cell, the randomization process can be summarized as follows. The observations (experimental units) can be written $y_1, y_2, \ldots, y_{rt}$, and we have a permutation random variable

$$\delta_{ijk} = \begin{cases} 1 \text{ if } y_k \text{ is given subscript } ij \\ 0 \text{ otherwise.} \end{cases}$$

        Show that $P(\delta_{ijk} = 1) = 1/rt$.

    (b) Repeat part (a) for the RCB of Example 4.4 of Miscellanea 4.9.1. Remember that here the permutation must respect the blocks, and with t treatments and b blocks we will have $P(\delta_{ijk} = 1) = 1/t$.

    (c) Referring to Example 4.2, run the usual anova and the randomization test. Comment on the agreement of the results.

4.8 Technical Notes

4.8.1 Error Distributions – Fixed Blocks

Using Lemma 3.16, it can be shown that all of the error terms arising from model (4.24) are normal, as are their sums and differences. Taking account of their covariances, it is straightforward (Exercise 4.15) to verify that

$$\begin{aligned}
\text{Var}\bar{\varepsilon}_{ij} &= \tfrac{\sigma^2}{r}(1 + (r-1)\rho_\varepsilon), & \text{Cov}(\bar{\varepsilon}_{ij}, \varepsilon_{i'j}) &= \rho_B \sigma_\varepsilon^2, \\
\text{Var}\bar{\varepsilon}_i &= \tfrac{\sigma^2}{rb}(1 + (r-1)\rho_\varepsilon), & \text{Cov}(\bar{\varepsilon}_i, \bar{\varepsilon}_{i'}) &= \tfrac{1}{b}\rho_B \sigma_\varepsilon^2, \\
\text{Var}\bar{\varepsilon}_j &= \tfrac{\sigma^2}{rt}(1 + (r-1)\rho_\varepsilon + r(t-1)\rho_B), & \text{Cov}(\bar{\varepsilon}_j, \bar{\varepsilon}_{j'}) &= 0, \\
\text{Var}\bar{\varepsilon} &= \tfrac{\sigma^2}{rbt}(1 + (r-1)\rho_\varepsilon + r(t-1)\rho_B),
\end{aligned}$$

where it also follows from models (3.24) and (4.24) that

$$\sigma_B^2 = \text{Var}(\bar{Y}_{ij\cdot}) = \frac{\sigma_\varepsilon^2}{r}(1 + (r-1)\rho_\varepsilon).$$

4.8.2 Estimating Fixed and Random Effects II

We continue our development of effect estimates, started in Technical Note 3.8.4, and develop the estimates in the case of fixed blocks. With blocks fixed, there is no mixed model, and (3.35) is a standard linear model, with estimation done through least squares. The ordinary least squares solution to (3.35) is

$$\begin{pmatrix} \hat\theta \\ \hat\beta \end{pmatrix} = \left[\begin{pmatrix} X' \\ Z' \end{pmatrix} (X Z) \right]^{-1} \begin{pmatrix} X' \\ Z' \end{pmatrix} Y = \begin{pmatrix} (X'X)^{-1}X'Y \\ (Z'Z)^{-1}Z'(Y - X\hat\theta) \end{pmatrix}.$$

If the error term ε has covariance matrix V, we might prefer the generalized least squares solutions

$$\begin{pmatrix} \hat\theta \\ \hat\beta \end{pmatrix} = \left[\begin{pmatrix} X' \\ Z' \end{pmatrix} V^{-1} (X Z) \right]^{-1} \begin{pmatrix} X' \\ Z' \end{pmatrix} V^{-1} Y$$

$$= \begin{pmatrix} (X'V^{-1}X)^{-1}X'V^{-1}Y \\ (Z'V^{-1}Z)^{-1}Z'V^{-1}(Y - X\hat\theta) \end{pmatrix},$$

which will have variance no larger than that of the least squares estimates. Note that both the least squares and generalized least squares estimates are unbiased.

With the correlation structure of (4.1), the error variance V has the form

$V = \text{Var } \varepsilon =$

$$(4.32) \qquad \sigma^2 \begin{pmatrix} (1-\rho)I_t + \rho J_t & 0 & 0 \cdots & 0 \\ 0 & (1-\rho)I_t + \rho J_t & 0 \cdots & 0 \\ \vdots & \vdots & \vdots & \vdots \\ 0 & 0 & 0 \cdots & (1-\rho)I_t + \rho J_t \end{pmatrix},$$

where J_t is a $t \times t$ matrix of all ones. Substituting above, we can get the least squares estimate for θ as

$$(4.33) \qquad \hat\theta = (X'X)^{-1}X'Y = (X'V^{-1}X)^{-1}X'V^{-1}Y = \begin{pmatrix} \bar{y}_{1\cdot} \\ \bar{y}_{2\cdot} \\ \vdots \\ \bar{y}_{t\cdot} \end{pmatrix},$$

the vector of treatment means. Note that in this case the least squares and generalized least squares estimates agree; this is a special case of a result first due to Zyskind (see Zyskind 1967, or Puntanen and Styan 1989 for a survey of these types of results). For certain types of V matrices the two estimates agree, the sufficient condition satisfied here is that the row sums of V are all the same. The same occurs for the estimates of the block effects, where we get (Exercise 4.17)

$$(4.34) \quad \hat\beta = (Z'Z)^{-1}Z'(Y - \hat\theta) = (Z'V^{-1}Z)^{-1}Z'V^{-1}(Y - \hat\theta) = \begin{pmatrix} \bar{y}_{\cdot 1} - \bar{\bar{y}} \\ \bar{y}_{\cdot 2} - \bar{\bar{y}} \\ \vdots \\ \bar{y}_{\cdot t} - \bar{\bar{y}} \end{pmatrix}.$$

These are the estimates that are typically used in an RCB anova, regardless of the assumption about blocks being fixed or random. For the fixed blocks model they are unbiased, have the smallest variance among linear estimators and, under normality are the maximum likelihood estimators and have minimum variance among all unbiased estimators.

Comparing V^* to V of (4.32), we see that these two matrices are identical if we use the relationship in (4.9), showing that the estimate of θ is the same whether blocks are fixed or random (as is its variance - see Exercise 3.33). However, it is not the same for the estimator of β. Evaluating $\hat{\beta}$ with the relationship in (4.9) yields

$$(4.35) \qquad \hat{\beta} = \sigma_\beta^2 Z' V^{*-1}(Y - X\hat{\theta}) = \frac{\rho t}{1 - \rho + \rho t} \begin{pmatrix} \bar{y}_{\cdot 1} - \bar{\bar{y}} \\ \bar{y}_{\cdot 2} - \bar{\bar{y}} \\ \vdots \\ \bar{y}_{\cdot t} - \bar{\bar{y}} \end{pmatrix}.$$

4.9 Miscellanea

4.9.1 Randomization Tests

Throughout this book we have used a type of inference known as *model-based* inference. That is, for each experiment we have written down a model for the population, such as (3.24) or (4.1), made assumptions about error distributions, and then derived tests and confidence intervals. While this remains a most popular approach, there is another approach to inference, based on *randomization* or *permutation* of the data, that is an alternate strategy.

Randomization tests can be traced back, at least, to the famous example of Fisher about "The Lady Tasting Tea", explained in Chapter II of his *Design of Experiments*. A lady claims that she can distinguish whether the milk or the tea was first put in the teacup. Fisher designs an experiment, based on randomization principles[1], to test this declaration. (see also Hinkelmann and Kempthorne 1994, Section 5.2)

An appeal of randomization tests is that the randomization plan of the experiment is the driving force behind the inference. That is, the null distribution, sometimes called the *reference distribution*, which is typically an F-distribution in model-based inference, is derived directly from the form of the randomization used to do the experiment. We illustrate this with the following numerical example.

Example 4.4. ONEWAY CRD AND RCB TESTS THROUGH RANDOMIZATION
Suppose that data were collected as a oneway CRD. The principle of randomization uses the fact that, under the null hypothesis of no treatment effect, any random arrangement of the observations that are possible according to the design, is equally likely. So we can have

---

[1] Much of the original underlying theory of randomization tests was developed by Kempthorne, and can be found in his 1952 book, and in the revised and updated Hinkelmann and Kempthorne 1994.

Observed Data		One Possible Randomization	

Observed Data

Treatment
1 2 3

14	18	27
11	26	31
18	9	24
18	22	34

One Possible Randomization

Treatment
1 2 3

11	22	14
18	34	26
9	27	24
18	31	18

Notice that the randomization is carried out throughout the table – reflecting the completely randomized design. So we are assuming that under H_0: no treatment effect, we could have observed any permutation of the data that respects the randomization scheme.

Compare this to what would happen if the experiment were carried out as an RCB. For illustration, we keep the same numbers and suppose we have

Observed Data

Treatment
1 2 3

	1	14	18	27
Block	2	11	26	31
	3	18	9	24
	4	18	22	34

One Possible Randomization

Treatment
1 2 3

	1	27	18	14
Block	2	26	31	11
	3	18	24	9
	4	22	18	34

Notice that here the randomization must respect the block structure – and is only carried out within the blocks. The randomization of the CRD is not an allowable randomization for the RCB as it violates the block structure.   ∥

When randomization tests were first developed, much effort was put into showing that the randomization distribution could be approximated by the usual F-distribution. (See, for example, Kempthorne 1952, Section 8.2 or Hinkelmann and Kempthorne 1994, Section 9.2.) However, now we can directly calculate the randomization distribution and test the null hypothesis of no treatment effect in the following way.

Example 4.5. RANDOMIZATION TESTS CONTINUED   Here we look at testing for the treatment effect in the CRD of Example 4.4. To calculate the null distribution we do the following:
(1) For each of $i = 1, 2, \ldots, m$ permutations
    (a) Obtain a random permutation of the data
    (b) Calculate the F-statistic for the data
(2) Order the m F-statistics and obtain the upper α cutoff point, say F_{cut}
(3) Reject H_0 if the calculated F statistic for the observed data, F_{obs}, satisfies
    $F_{obs} > F_{cut}$.
This is illustrated in Figure 4.1, which shows a histogram of $10,000$ permutations of the oneway CRD data of Example 4.4. The observed F-ratio is in the tail of this histogram, which is the null distribution of the permutation test, above the 5% cutoff but not above the 1% cutoff. Of the $10,000$ F-ratios calculated from the permutations, the observed F-ratio is greater than 9808 of them, giving a p-value of 0.0192. Compare this analysis to the usual anova table:

Fig. 4.1. Reference distribution for the oneway CRD randomization test of Example 4.4. The observed F-ratio is 7.31. The randomization distribution has the 5% cutoff at 4.77 and the 1% cutoff at 8.73.

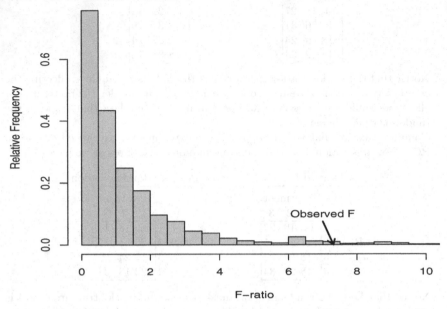

Source	df	SS	MS	F	p-value
Treatment	2	408.50	204.25	7.31	.013
Residuals	9	251.50	27.94		

A similar analysis can be done for the RCB of Example 4.4. This is left to Exercise 4.19.   ‖

We end this discussion with a few observations about randomization tests.

(1) Since the distribution of the test statistic is based on the actual randomization, this clearly ties the resulting inference to the experiment design. In model-based inference this is done through the model, and this connection is not as transparent (although it is there).

(2) Conversely, with the randomization test the actual scope of inference is not as clear as in model-based inference, as there is no formal model tying the observations to the population. It seems that if the blocks are considered to be fixed, then the randomization inference is a bit clearer: The randomization distribution is one observation on the true reference distribution, based on the observed data. If we replicated the experiment we could obtain more observations of the reference distribution, and improve our estimate of it.

(3) From the construction of the randomization distribution we see that it is *distribution free*, that is, we have made no parametric assumptions about

the error structure. However, we still need some assumptions about the permuted random variables to ensure the validity of the reference distribution. If they are *iid* that is enough, but this can be relaxed to *exchangeable*.

Definition 4.6. The random variables $Y_1, \ldots, Y_n$ are *exchangeable* if any permutation of any subset of them of size k $(k \leq n)$ has the same distribution.

For example, equicorrelated random variables are exchangeable.

(4) The randomization approach encounters some difficulty in testing interactions, for example, the Treatment $\times$ Block interaction in an RCB. This results from the violation of exchangeability (see, for example, Good 2005, Chapter 7 or Hinkelmann and Kempthorne 1994, Section 9.6).

(5) One attraction of randomization tests is that they are always available (if we have exchangeability) so they have found use in complex situations where the models can be difficult to deal with. For example, the permutation tests of Churchill and Doerge (1994) were a major breakthrough in inference in QTL problems.

(6) An alternative to permutation based inference is the *bootstrap* (Efron and Tibshirani 1993), which also provides a distribution-free approach to inference.

Book length treatments of randomization tests are Edgington and Onghena (2007) and Good (2005).

5

Split Plot Designs

"How absurdly simple!", I cried.

"Quite so!", said he, a little nettled. "Every problem becomes very childish when once it is explained to you."

Dr. Watson and Sherlock Holmes
The Adventure of the Dancing Men

5.1 Introduction

Split plot experiments are the workhorse of statistical design. There is a saying that if the only tool you own is a hammer, then everything in the world looks like a nail. It might be fair to say that, from now on, almost every design that you see will be some sort of split plot.

A *split plot design (or split unit design)* is one in which there is more than one type of experimental unit. Although split unit is probably the more accurate term, this design also grew out of agriculture, and the historical term seems to be the more popular one.

5.1.1 A Split Plot Model

Example 5.1. DIETARY SPLIT PLOT   In a study of dietary composition on health, four diets were randomly assigned to 12 subjects, all of similar health status. Baseline blood pressure was established, and one measure of health was blood pressure change after two weeks. Blood pressure was measured in the morning and the evening. The data layout is

Diet

	1	2	3	4
	Subject	Subject	Subject	Subject
	1 2 3	4 5 6	7 8 9	10 11 12
Morning	$x\ x\ x$	$x\ x\ x$	$x\ x\ x$	$x\ x\ x$
Evening	$x\ x\ x$	$x\ x\ x$	$x\ x\ x$	$x\ x\ x$

Notice that although there are 12 subjects, which are the experimental units, there are 24 numbers in the data. This is because the experimental unit was split and two observations were taken on one experimental unit. The anova for this design looks like

Source	df
Diets	3
Subjects (in Diets)	8
Time	1
Time × Diet	3
Time × Subjects (in Diets)	8
Total	23

||

There are a number of things to note about the split plot design that are illustrated in this example.

(1) It is important to note that the split plot design is an experiment design, not a treatment design. In Example 5.1 the treatment design is a twoway crossed design; Diet and Morning/Evening are crossed (or in a factorial arrangement). If we did not know the experiment design, we might just see a data layout like

	Diet			
	1	2	3	4
Morning	x x x	x x x	x x x	x x x
Evening	x x x	x x x	x x x	x x x

and (mistakenly) might presume that this is a twoway CRD.

(2) There is an implied correlation structure, as the multiple observations taken on one unit are correlated. In the above data layout the two observations in each column (Morning-Evening) are correlated.

(3) The whole plots, the experimental units at the whole plot level (the Subjects in Example 5.1) act as blocks for the split plot treatment (Morning-Evening in Example 5.1).

(4) There are essentially two parts to the anova table - "above the line" for the whole plots, or true experimental units, and "below the line" for the split plots, or split units. The above anova table usually has the names:

Source	df
Diets	3
Whole Plot Error	8
Time	1
Time × Diet	3
Split Plot Error	8
Total	23

which shows how to do the tests. Note that

Subjects (in Diets) = Whole Plot Error

Time × Subjects (in Diets) = Split Plot Error

(5) Comparisons "below the line", done with the split units, tend to have greater precision. These comparisons are within a subject, so each subject is his own control, and there are more degrees of freedom. A general design rule is, if possible, to put the treatment of greatest interest at the split level (below the line).

The split plot advantage

(6) Just as with blocks, the split plot is more about restricting randomization, and the modeling is more about correlation. This is most evident when we look at an example of the classical agricultural split plot experiment (see Example 5.3).

A model for this split plot design is

$$Y_{ijk} = \mu + \tau_i + S_{ij} + \gamma_k + (\tau\gamma)_{ik} + \varepsilon_{ijk},$$

where

$$Y_{ijk} = \text{the response to diet } i \text{ of subject } j \text{ at time } k,$$
$$\tau_i = \text{the effect of diet } i,$$
$$S_{ij} = \text{the effect of subject } j \text{ in diet, } i \text{ (whole plot error)}$$
$$(\tau\gamma)_{ik} = \text{the interaction between diet } i \text{ and time } j,$$
$$\varepsilon_{ijk} = \text{the experimental error, N}(0, \sigma^2). \text{ (split plot error)}$$

As we would clearly want subjects to be random, say $S_{ij} \sim \text{N}(0, \sigma_S^2)$, this would be a classical model for a split plot analysis. The subjects provide the whole plot error, and the residuals, estimated by Time × Subjects (in Diets), estimate σ_ε^2.

5.1.2 Dissecting the Split Plot

There are, in effect, two separate analyses in a split plot design but they are intertwined, which allows for a better estimate of error. The data for the diet experiment are

	Diet			
	1	2	3	4
	Subject	Subject	Subject	Subject
	1 2 3	4 5 6	7 8 9	10 11 12
Morning	123 120 122	117 125 122	114 109 115	140 141 138
Evening	135 136 129	139 136 142	123 132 132	150 147 154

and the full anova table is

Source	df	SS	MS	F
Diets	3	1873.46	624.49	85.16
Subjects (in Diets)	8	58.67	7.33	
Time	1	1190.04	1190.04	73.60
Time × Diet	3	53.13	17.71	1.10
Time × Subjects (in Diets)	8	129.33	16.17	

where the correct tests are shown. The above the line tests are done against the Subjects (in Diets) term, and the below the line tests are against the Time × Subjects (in Diets) mean square.

The "above the line" analysis on the whole plots is equivalent to a oneway CRD on the blood pressure averages:

Diet

Average	129 128 125.5	128 130.5 132	118.5 120.5 123.5	145 144 146

with anova table

Source	df	SS	MS	F
Diets	3	936.73	312.24	85.16
Within	8	29.33	3.67	

Note that the sums of squares are half of those in the split plot analysis (Exercise 5.2), but the F-statistic is the same.

In the "below the line" analysis, each subject acts as a block for the treatment Time, so we could analyze this as an RCB(see Exercise 5.2). However, since there are only two levels to Time, we can actually do the below the line analysis on the differences:

Diet

	1	2	3	4
Difference	−12 − 16 − 7	−22 − 11 − 20	−9 − 23 − 17	−10 − 6 − 16

with anova table

Source	df	SS	MS	F
Diets	3	106.25	35.42	1.10
Within	8	258.667	32.33	

Note here that the ordinary oneway layout calculations give us this table. But since we are working with differences, the sum of squares attributed to diets (106.25) is actually the interaction between Diets and Time, and is exactly the sum of squares for Time × Diet in the complete analysis (note that the F-statistics match). Moreover, the "within" based on the differences is exactly Time × Subjects (in Diets).

Finally, where is the sum of squares due to Time? It is simply the mean of the differences squared (with an adjustment, see Exercise 5.2).

This highlights how the split plot works. Below the line we can analyze treatment differences, so each subject is his/her own control. This gives us greater precision in the below the line analysis.

5.2 CRD on the Whole Plots

In the split plot design, the whole plot treatments effectively act as blocks for the split plot treatments, so we would expect that the analysis "below the line" would be largely unaffected by the design on the whole plot treatments. However, this is not quite the case, and there are enough differences that it makes sense to treat two separate cases for the whole plots: CRD and RCB. We will see that with the whole plot treatments in a CRD the analysis is quite straightforward, but if they are in an RCB we get an extra error term, and need some extra assumptions to get the "classical" split plot analysis.

5.2.1 Model and Distribution Assumptions

The classical split plot model, with whole plot treatments in a CRD, is

$$(5.1) \qquad Y_{ijk} = \mu + \tau_i + \varepsilon_{ij} + \gamma_k + (\tau\gamma)_{ik} + \delta_{ijk},$$

where $i = 1, \ldots, t$, $j = 1, \ldots, r$, $k = 1, \ldots, g$, and

$$
\begin{aligned}
Y_{ijk} &= \text{response,} \\
\mu &= \text{overall mean effect,} \\
\tau_i &= \text{whole plot treatment,} \\
\varepsilon_{ij} &= \text{whole plot error, } \varepsilon_{ij} \overset{\text{iid}}{\sim} N(0, \sigma_\varepsilon^2), \\
\gamma_k &= \text{split plot treatments,} \\
(\tau\gamma)_{ik} &= \text{interaction,} \\
\delta_{ijk} &= \text{split plot error, } \overset{\text{iid}}{\sim} N(0, \sigma_\delta^2), \\
&\quad \text{independent of } \varepsilon_{ij}.
\end{aligned}
$$

Identifiability considerations lead us to make the parameter restriction

$$(5.2) \qquad \bar{\tau} = \bar{\gamma} = (\bar{\tau\gamma})_i = (\bar{\tau\gamma})_k = 0,$$

where we recall again that this is merely a renaming of the effects, and does not signify any change in the real parameter space. (These are the same restrictions needed for identifiability in any twoway crossed treatment design.)

The full anova for this model, identifying all terms, is given in Table 5.1, where we see that the whole plot error comes from the replication of the whole plot treatments, just as in any CRD. The split plot error, in this design, comes from the interaction of the split plot treatment with the replications – just like in a regular RCB. However, here this interaction is nested in the whole plot treatments, which we must take into account. Look carefully at Table 5.2, and it should be clear that the design on the split plot treatment is an RCB *within* each level of the whole plot treatment.

Table 5.1. Anova for split plot design with CRD on whole plot treatments. T = Whole Plot Treatment, G = Split Plot Treatment, R = Replication

Source	df	SS
Whole Plot (WP)Trt	$t-1$	$rg\sum_i(\bar{y}_i - \bar{\bar{y}})^2$
Replication (in WP)	$t(r-1)$	$g\sum_{ij}(\bar{y}_{ij} - \bar{y}_i)^2$
(Whole Plot Error)		
Split Plot (SP) Trt	$g-1$	$rt\sum_k(\bar{y}_k - \bar{\bar{y}})^2$
SP Trt $\times$ WP Trt	$(g-1)(t-1)$	$r\sum_{ik}(\bar{y}_{ik} - \bar{y}_i - \bar{y}_k + \bar{\bar{y}})^2$
SP Trt $\times$ Rep (in WP)	$t(g-1)(r-1)$	$\sum_i\left[\sum_{jk}(y_{ijk} - \bar{y}_{ij} - \bar{y}_{ik} + \bar{y}_i)^2\right]$
(Split Plot Error)		
Total	grt-1	

Realize that, without any trouble, we can have another treatment design on either the whole plot or split plot treatments, for example, the whole plot treatment many actually be a factorial arrangement of treatments, as in Example 5.3 (see also Example 6.20). This extension does not change the basic analysis.

Before we further analyze the model, there are some things to note:

(1) The whole plot analysis is based only on the $\bar{y}_{ij}$, and thus can be done in ignorance of what goes on below the line. Moreover, it should be clear that the $\bar{y}_{ij}$ are independent.
(2) There is correlation below the line, which can be viewed as a consequence of the random effects in model (5.1). However, we can see this directly, as the split plot treatments are randomized within the levels of the whole plot treatments. Thus, we should have

Table 5.2. Data Layout for split plot design with CRD on whole plot treatments. T = Whole Plot Treatment, G = Split Plot Treatment, R = Replication

(5.3) $$\mathrm{Corr}(Y_{ijk}, Y_{ijk'}) = \rho \neq 0.$$

This correlation will appear as a consequence of the model.
(3) The split plot error is a pooled interaction term; within each whole plot treatment level we calculate the Split Plot $\times$ Rep interaction and sum.

The covariance structure of the split plot design is actually quite simple. From (5.1) we see that the observations in different levels of the whole plot treatment are independent, as are observations in different replications inside the whole plot treatment. The only nonzero correlation is between observations within a replication inside a whole plot (where there is the RCB on the split plot treatment). Thus, from (5.1)

$$\mathrm{Cov}(Y_{ijk}, Y_{ijk'}) = \mathrm{E}(\varepsilon_{ij} + \delta_{ijk})(\varepsilon_{ij} + \delta_{ijk'}) = \sigma_{\varepsilon}^2,$$

and

$$\mathrm{Corr}(Y_{ijk}, Y_{ijk'}) = \frac{\sigma_{\varepsilon}^2}{\sigma_{\varepsilon}^2 + \sigma_{\delta}^2}.$$

Two things about this correlation structure are worthy of note.

(1) As a consequence of the random effects model, this correlation is nonnegative.
(2) The correlation between any two observations is the same – so we have the same equicorrelation structure as in the RCB. The reason why this is important is that the split plot design is often used for *repeated measures* analysis, but doing so can result in difficulties. (See Section 5.6.3.)

5.2.2 Expected Squares and F-tests

Even though the split plot design is a somewhat complex design, it turns out that estimation and testing is, for the most part, straightforward. We do run into some trouble when the whole plots are in an RCB, but the trouble is avoided when the whole plots are in a CRD. Since calculation of the expected mean squares is straightforward in the split plot design, here we indicate the calculations and leave some of the details to the exercises.

Under model (5.1), together with (5.2), the expected value of the whole plot treatment sum of squares is

$$\mathrm{ESS(WP\ Trts)} = rg \sum_i \mathrm{E}\left(\tau_i + \bar{\varepsilon}_{ij} + \bar{\delta}_i - \bar{\bar{\varepsilon}\bar{\delta}}\right)^2.$$

Now use the fact that all of the εs and δs are independent, which means that when we expand the square there are no cross terms. We thus have

$$\mathrm{ESS(WP\ Trts)} = rg \left(\sum_i \tau_i^2 + \sum_i \mathrm{E}(\bar{\varepsilon}_{ij} - \bar{\varepsilon})^2 + \sum_i \mathrm{E}(\bar{\delta}_i - \bar{\delta})^2\right).$$

Finally, using Lemma 3.16, $\mathrm{E}(\bar{\varepsilon}_{ij} - \bar{\bar{\varepsilon}})^2 = (1 - 1/t)(\sigma_\varepsilon^2/r)$ and $\mathrm{E}(\bar{\delta}_i - \bar{\bar{\delta}})^2 = (1 - 1/t)(\sigma_\delta^2/rg)$ and we can write

$$\mathrm{ESS(WP\ Trts)} = rg \sum_i \tau_i^2 + g(t-1)\sigma_\varepsilon^2 + (t-1)\sigma_\delta^2.$$

The calculation for the other expected sums of squares are similar, and somewhat routine (see Exercise 5.29), but we want to highlight one below the line calculation. So, for example, consider the expected value of the split plot treatment sum of squares. Again from (5.1), with a bit more detail, we have

ESS(SP Trts)

$$= rt \sum_k \mathrm{E} \left[(\mu + \bar{\tau} + \bar{\bar{\varepsilon}} + \gamma_k + (\bar{\tau\gamma})_k + \bar{\delta}_k) - (\mu + \bar{\tau} + \bar{\bar{\varepsilon}} + \bar{\gamma} + (\bar{\tau\gamma}) + \bar{\bar{\delta}}) \right]^2.$$

As expected, the fixed effects, other than γ cancel out. But the interesting thing is that the random effect due to ε also cancels out.

Note: The random variables cancel each other out! This is not just in expectation.

Because the split plot treatment is balanced across the whole plots in the replications, under the model assumptions the average ε error is the same in each split plot treatment level, and equal to the overall whole plot error. Thus, the whole plot error "disappears" in the split plot part of the analysis.

Continuing the calculation gives

$$\mathrm{ESS(SP\ Trts)} = rt \sum_k \mathrm{E} \left[\gamma_k + (\bar{\delta}_k - \bar{\bar{\delta}}) \right]^2$$

$$= rt \sum_k \gamma_k^2 + rt \sum_k (\bar{\delta}_k - \bar{\bar{\delta}})^2$$

$$= rt \sum_k \gamma_k^2 + (s-1)\sigma_\delta^2,$$

as $\mathrm{E}(\bar{\delta}_k - \bar{\bar{\delta}})^2 = (1 - 1/g)(\sigma_\delta^2/rt)$. Similarly, for all of the other split plot factors, the whole plot effects and errors cancel out, resulting in the EMS in Table 5.3.

The EMS in Table 5.3 are pleasingly simple, giving straightforward tests from the anova. The null hypothesis of no whole plot treatment effect is

$$H_0 : \tau_i = 0 \text{ for all } i$$

and is tested by

$$F_{t-1, t(ri1)} = \frac{\mathrm{MS(WP\ Trts)}}{\mathrm{MS(Replication\ in\ WP)}}.$$

Table 5.3. Expected mean squares for a split plot design with the whole plots in a CRD, with data layout as in Table 5.2

Source	df	EMS
Whole Plot (WP) Trt	$t - 1$	$\sigma_\delta^2 + g\sigma_\varepsilon^2 + \frac{rg}{t-1}\sum_i \tau_i^2$
Replication (in WP)	$t(r - 1)$	$\sigma_\delta^2 + g\sigma_\varepsilon^2$
Split Plot (SP) Trt	$g - 1$	$\sigma_\delta^2 + \frac{rt}{g-1}\sum_k \gamma_k^2$
SP Trt $\times$ WP Trt	$(g - 1)(t - 1)$	$\sigma_\delta^2 + \frac{r}{(g-1)(t-1)}\sum_{ik}(\tau\gamma)_{ik}^2$
SP Trt $\times$ Replication (in WP)	$t(g - 1)(r - 1)$	σ_δ^2
Total	grt-1	

At the split plot level, both the split plot main effect $H_0 : \gamma_k = 0$ for all k and the interaction $H_0 : (\tau\gamma)_{ik} = 0$ for all i, k are tested against the split plot error.

Note that the split plot error here is very interesting. It is very reminiscent of an RCB error in that it is an interaction, but there is a difference. The split plot treatment is not fully crossed with replications, but rather is crossed with the replications in a particular level of the whole plot. The fact that the split plot treatment is not fully crossed with the replication is a major reason why the tests are so straightforward here, as opposed to what we will find when the whole plots are in an RCB.

Other things to note

(1) The design on the whole plots is a CRD with each experimental unit being the sum over the split plot observations. Thus, any correlation structure at the split plot level has no effect on the whole plot analysis.
(2) The split plot level will have more degrees of freedom that the whole plot level, as there are more experimental units.
(3) The comparisons at the split plot level are more tightly controlled and hence more precise; so the rule is to always put the most important factor at the split plot level (if you can).
(4) Justification of the F-tests using Cochran's Theorem is straightforward, and somewhat similar to what was done for the RCB (although the algebra becomes a bit more involved). See Technical Note 5.8.2.

5.2.3 Estimating Contrasts

Estimates of contrasts are also straightforward in the split plot, as unbiased estimates are provided by the analogous contrasts in the least squares estimates. However, we do run into some difficulties when estimating the variances.

There are four types of contrasts to consider:

(1) Whole Plot Means: $\sum_i a_i\tau_i$, where $\sum_i a_i = 0$.

(2) Split Plot Means: $\sum_k a_k \gamma_k$, where $\sum_k a_k = 0$.

(3) Interaction Means, Same Level of Whole Plot: $\sum_k a_k (\tau\gamma)_{ik}$, where $\sum_k a_k = 0$.

(4) Interaction Means, Different Whole Plot Level: $\sum_{ik} a_{ik} (\tau\gamma)_{ik}$, where $\sum_{ik} a_{ik} = 0$.

In each case, we estimate the contrast with the analogous least squares estimate, giving an unbiased estimate.

As we will see, the first three cases turn out to be relatively simple, but the fourth case will give us problems. These problems will be caused by the fact that although $\sum_{ik} a_{ik} = 0$, it need not be the case that $\sum_k a_{ik} = 0$ or $\sum_i a_{ik} = 0$.

> **Note:** Using either model (5.1), or the RCB model (5.14), it is easy to obtain the estimates of the effects.

This is because, as we have seen, if we use least squares, the estimation procedure only depends on the treatment design, not on the experiment design. This is an important distinction.

If we look at (5.1) and ignore the error terms, we see that the least squares estimates satisfy

$$(5.4) \qquad \min_{\mu, \tau_i, \gamma_k, (\tau\gamma)_{ik}} \sum_{ijk} [y_{ijk} - (\mu + \tau_i + \gamma_k + (\tau\gamma)_{ik})]^2,$$

which will lead to the same estimates as we got from the twoway CRD least squares fit in (2.7) and (2.8)! There, the treatment design is a twoway crossed design, as is the split plot (5.1), so using (2.8) we can immediately see that the least squares estimates are given by

$$(5.5) \qquad \begin{aligned} \hat{\mu} &= \bar{\bar{y}}, \\ \hat{\tau}_i &= \bar{y}_i - \bar{\bar{y}}, \\ \hat{\gamma}_k &= \bar{y}_k - \bar{\bar{y}}, \\ \hat{(\tau\gamma)}_{ik} &= \bar{y}_{ik} - \bar{y}_i - \bar{y}_k + \bar{\bar{y}}. \end{aligned}$$

Thus, the least squares estimates are based on the cell means and, of course, are unbiased (Exercise 5.30).

Note that because these are contrasts, when we use $\hat{\tau}$, etc., the $\bar{\bar{y}}$ piece drops out, so we are only left with cell means.

Case (1): Whole Plot Comparisons _____

When we calculate $\text{Var}(\sum_i a_i \bar{Y}_i)$ we use the fact that $\bar{Y}_i$ and $\bar{Y}_{i'}$ are independent, so there is no covariance term in the variance of the sum. From (5.1),

$$\text{Var}(\bar{Y}_i) = \text{Var}(\bar{\varepsilon}_i + \bar{\delta}_i) = \text{Var}(\bar{\varepsilon}_i) + \text{Var}(\bar{\delta}_i) = \frac{\sigma_\varepsilon^2}{r} + \frac{\sigma_\delta^2}{rg},$$

and thus

$$\text{Var}\left(\sum_i a_i \bar{Y}_i\right) = \left(\frac{\sigma_\varepsilon^2}{r} + \frac{\sigma_\delta^2}{rg}\right) \sum_i a_i^2.$$

Note that replication at the split plot level has less of an effect here, in that it only cuts down a piece of the variance. Also note that the split plot variance appears in the variance of a treatment contrast at the whole plot level.

From the anova EMS we see that $E(\text{MS(WP Error)}) = g\sigma_\varepsilon^2 + \sigma_\delta^2$, and thus we estimate the variance of a whole plot contrast with

$$\widehat{\text{Var}}\left(\sum_i a_i \bar{Y}_i\right) = \frac{\text{MS(WP Error)}}{rg} \sum_i a_i^2.$$

Realize that at the whole plot level we are essentially dealing with a CRD, so calculation of contrast variances are relatively easy. Things get a little trickier at the split plot level.

Case (2): Split Plot Comparisons _____

Next we look at the variance of $\sum_k a_k \bar{Y}_k$, a contrast in the split plot means, first looking at the variance and covariances of the split plot means. We have

$$\text{Var}(\bar{Y}_k) = \text{Var}\left(\frac{1}{t} \sum_{i=1}^t \bar{Y}_{ik}\right)$$

$$= \frac{1}{t^2} \sum_{i=1}^t \text{Var}(\bar{Y}_{ik}) \quad \text{[whole plot independence]}$$

$$= \frac{1}{t^2 r^2} \sum_{i=1}^t \sum_{j=1}^r \text{Var}(Y_{ijk}) \quad \text{[independence of reps in plots]}$$

$$= \frac{\sigma_\varepsilon^2 + \sigma_\delta^2}{tr}.$$

Next look at the covariance between split plot means to find

$$\text{Cov}(\bar{Y}_k, \bar{Y}_{k'}) = \text{Cov}\left(\frac{1}{t} \sum_{i=1}^t \bar{Y}_{ik}, \frac{1}{t} \sum_{i=1}^t \bar{Y}_{ik'}\right)$$

$$= \frac{1}{t^2} \sum_{i=1}^t \text{Cov}\left(\bar{Y}_{ik}, \bar{Y}_{ik'}\right) \quad \text{[covariance} = 0 \text{ if } i \neq i']$$

and then

$$\mathrm{Cov}\left(\bar{Y}_{ik}, \bar{Y}_{ik'}\right) = \mathrm{Cov}\left(\frac{1}{r}\sum_{j=1}^{r}\bar{Y}_{ijk}, \frac{1}{r}\sum_{j=1}^{r}t\bar{Y}_{ij'k'}\right)$$

(5.6)
$$= \frac{1}{r^2}\sum_{j=1}^{r}\mathrm{Cov}\left(\bar{Y}_{ijk}, \bar{Y}_{ijk'}\right) = \frac{1}{r}\sigma_{\varepsilon}^2, \quad \begin{bmatrix} \text{since covariance} = 0 \\ \text{if } j \neq j' \end{bmatrix}$$

and putting it all together gives $\mathrm{Cov}(\bar{Y}_k, \bar{Y}_{k'}) = \frac{1}{rt}\sigma_{\varepsilon}^2$ and, hence,

(5.7)
$$\mathrm{Var}\left(\sum_k a_k\bar{Y}_k\right) = \frac{\sigma_{\varepsilon}^2 + \sigma_{\delta}^2}{tr}\sum_k a_k^2 + 2\frac{\sigma_{\varepsilon}^2}{tr}\sum_{k>k'}a_k a_{k'}.$$

We can finally put everything together, rearranging (5.7) to get

$$\mathrm{Var}\left(\sum_k a_k\bar{Y}_k\right) = \frac{\sigma_{\delta}^2}{tr}\sum_k a_k^2 + \frac{\sigma_{\varepsilon}^2}{tr}\left(\sum_k a_k^2 + 2\sum_{k>k'}a_k a_{k'}\right)$$

(5.8)
$$= \frac{\sigma_{\delta}^2}{tr}\sum_k a_k^2,$$

where we use the fact that

(5.9)
$$\sum_k a_k^2 + 2\sum_{k>k'}a_k a_{k'} = \left(\sum_k a_k\right)^2 = 0$$

since this is a contrast.

> **Note:** So the use of the contrast eliminated the presence of σ_{ε}^2, the whole plot variance, from the comparison of the split plot means.

This is the advantage of the split plot design, and the reason why the treatment of greatest interest should be at the split plot level if possible. Finally, since $E[\mathrm{MS}(G \times R \ (\text{in } T))] = \sigma_{\delta}^2$, we have an estimate of the variance.

The most interesting, and typically the most important, case for variances in the split plot designs is for the interaction mean comparisons. Above we noted two cases (3) and (4), depending on whether the comparisons were in the same level of the whole plots. This is a common way of presenting these variances because, as we will see, variance estimation in Case (4) is more difficult than in Case (3).

Case (3): Interaction Mean Comparisons: Same Whole Plot _____

Calculation of the variance of $\sum_k a_k\bar{y}_{ik}$, within the same level of the whole plot treatment, is quite similar to that of Case (2), and is left to Exercise 5.34. The result is

$$(5.10) \quad \text{Var}\left(\sum_k a_k \bar{Y}_{ik}\right) = \frac{\sigma_\varepsilon^2 + \sigma_\delta^2}{r} \sum_k a_k^2 + 2\frac{\sigma_\varepsilon^2}{r} \sum_{k>k'} a_k a_{k'} = \frac{\sigma_\delta^2}{r} \sum_k a_k^2,$$

where again we see that the whole plot variance σ_ε^2 does not appear in the interaction contrasts *within* a whole plot.

Case (4): Interaction Mean Comparisons: Different Whole Plot ___

Finally we turn to Case (4), and calculate the variance of a contrast between interaction means at different levels of a whole plot. The result here is very interesting, and shows what comparisons can be made with greater accuracy. Note that these are not the least squares estimates of $(\tau\gamma)_{ik}$; see Section 5.4.

A general contrast among the interaction means is given by Case (4): $\sum_i \sum_k a_{ik} \bar{y}_{ik}$, which is merely a sum of the type of contrasts considered in Case (3). Using this fact greatly reduces the calculation load, and we have

$$\text{Var}\left(\sum_i \sum_k a_{ik} \bar{Y}_{ik}\right) = \sum_i \text{Var}\left(\sum_k a_{ik} \bar{Y}_{ik}\right) \qquad \text{[whole plot independence]}$$

$$(5.11) \qquad = \sum_i \left(\frac{\sigma_\varepsilon^2 + \sigma_\delta^2}{r} \sum_k a_{ik}^2 + 2\frac{\sigma_\varepsilon^2}{r} \sum_{k>k'} a_{ik} a_{ik'}\right) \quad \begin{bmatrix} \text{using} \\ (5.10) \end{bmatrix}$$

$$= \frac{\sigma_\delta^2}{r} \sum_{ik} a_{ik}^2 + \frac{\sigma_\varepsilon^2}{r} \left(\sum_i \left[\sum_k a_{ik}^2 + 2\sum_{k>k'} a_{ik} a_{ik'}\right]\right),$$

where we rearrange as before, hoping to use something like (5.9) to get rid of the whole plot variance. When we add the terms in the square brackets we get

$$(5.12) \qquad \text{Var}\left(\sum_{ik} a_{ik} \bar{Y}_{ik}\right) = \frac{\sigma_\delta^2}{r} \sum_{ik} a_{ik}^2 + \frac{\sigma_\varepsilon^2}{r} \sum_i \left(\sum_k a_{ik}\right)^2,$$

but this is as far as we can go because, in general, for the interaction contrast we have $\sum_{ik} a_{ik} = 0$, and we cannot assume that $\sum_k a_{ik} = 0$ for every i.

Note the implications of these contrast variances:

(i) In Cases (2) and (3), we will typically have greater precision of comparisons, as the contrast variance only depends on the parameter σ_δ^2, which is estimated with the (presumably) smaller MS(Split Plot Error).

(ii) For Case (1), there is less precision in the comparison, as we must estimate the variance parameter $\sigma_\delta^2 + g\sigma_\varepsilon^2$ with the (presumably) larger MS(Whole Plot Error).

(iii) Case (4) is the most interesting, as depending on the particular contrast, we sometimes can eliminate σ_ε^2 from the variance. If we cannot, we must use both σ_δ^2 and σ_ε^2 in the variance estimate. This entails the use of Satterthwaite's approximation (Technical Note 5.8.1) and is best used with the variant of (5.12), whose derivation we leave to Exercise 5.33,

$$(5.13)\ \mathrm{Var}\left(\sum_{ik} a_{ik}\bar{Y}_{ik}\right) = \frac{\sigma_\delta^2}{r}\sum_{ik}(a_{ik}-\bar{a}_i)^2 + \frac{\sigma_\delta^2 + g\sigma_\varepsilon^2}{rg}\sum_i\left(\sum_k a_{ik}\right)^2.$$

We continue with Example 5.1 to illustrate the contrast inferences

Example 5.2. DIET SPLIT PLOT CONTINUED
The anova for the data of Example 5.1 (dataset Diet) is

Source	df	SS	MS	F	p-value
Diet	3	1873.46	624.49	85.16	< .0001
Subject(in Diet)	8	58.667	7.333		
Time	1	1190.04	1190.04	73.611	< .0001
Diet × Time	3	53.13	17.71	1.095	0.405
Split Plot Error	8	129.33	16.17		

where we see that both Diet and Time are significant, but there is no Diet × Time interaction. A closer examination of the interactions means may show us more, however. The interaction means are

		Time	
		AM	PM
	1	121.67	133.33
Diet	2	121.33	139.00
	3	112.67	129.00
	4	139.67	150.33

and we consider the following four interaction mean contrasts

	Same WP		Within WP		Between WP		Interaction	
	AM	PM	AM	PM	AM	PM	AM	PM
1	1	−1	1	−1	1	0	1	−1
2	0	0	1	−1	−1	0	−1	1
3	0	0	0	0	0	0	0	0
4	0	0	0	0	0	0	0	0

For the first two contrasts, where the comparisons are on the split plot treatments within the levels of the whole plot, we have $\sum_k a_{ik} = 0$, so we are in Case (3), and the error term for the contrast is given by (5.10), where we estimate σ_δ^2 with 16.17 on 8 degrees of freedom. So, for example, for the first contrast we have under H_0: no contrast difference,

$$t_8 = \frac{121.67 - 133.33}{\sqrt{\frac{2}{3} \times 16.17}} = -3.55,$$

which results in a two-sided p-value of .0075, suggesting a strong effect.

The third contrast is between cell means from different whole plots, and we have $\sum_k a_{ik} \neq 0$, so we are in Case (4). Using (5.13), we have the variance of this contrast given by

$$\frac{\text{MS(SP error)}}{r} \sum_{ik} (a_{ik} - \bar{a}_i)^2 + \frac{\text{MS(WP error)}}{rg} \sum_i \left(\sum_k a_{ik} \right)^2 = \frac{16.17 + 7.33}{3},$$

and the Satterthwaite approximation gives $\hat{\nu} = 14.01$, so we test H_0 with

$$t_{14} = \frac{121.67 - 121.33}{\sqrt{(16.17 + 7.33)/3}} = 0.121,$$

which is not significant.

Finally, the fourth contrast is an interaction of cell means where we have $\sum_k a_{ik} = 0$, so we are back again in Case (3). Thus, we must always use a bit of care when calculating the variance of the split plot interaction means, and we know that some interactions can be contrasted using only the split plot error. ‖

5.3 RCB on the Whole Plots

Thus far, we have concentrated on split plot designs with a CRD on the whole plots treatments. However, as we have mentioned earlier, there is no inherent restriction to this whole plot treatment design. In fact, a somewhat more popular setup, which we now address, is to have the whole plot treatments in an RCB. (Recall that, by construction, the split plot treatments are already in an RCB with the whole plots as blocks.) This variation from Section 5.2 does not change computations and inference too much, but it does have an interesting effect on the split plot error terms.

Example 5.3. VARIETY SPLIT PLOT   A classical split plot design was run at the Cornell Experiment Station to compare alfalfa varieties and their response to fertilizer treatments. There were six varieties (Narragannsut, Ontario, Ranger, Grimm, K. Command and Atlantic) and four fertilizer treatments, high and low levels of Potassium (k and K) and Phosphorus (p and P) in a 2×2 factorial. The experiment was replicated twice, with field layout given in Table 5.4 ‖

5.3.1 Model and Distribution Assumptions

A model for the split plot design with whole plots in blocks is

$$(5.14) \qquad Y_{ijk} = \mu + \tau_i + \beta_j + \varepsilon_{ij} + \gamma_k + (\tau\gamma)_{ik} + (\beta\gamma)_{jk} + \delta_{ijk},$$

Table 5.4. Field layout for split plot experiment of Example 5.3. Whole plots are in an RCB; split plot treatments (varieties) are completely randomized in whole plot treatments.

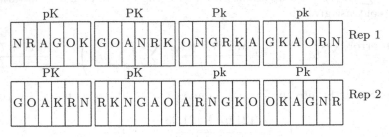

where $i = 1, \ldots, t$, $j = 1, \ldots, r$, $k = 1, \ldots, g$, and

$$
\begin{aligned}
Y_{ijk} &= \text{response,} \\
\mu &= \text{overall mean effect,} \\
\tau_i &= \text{whole plot treatment,} \\
\beta_j &= \text{whole plot block} \overset{\text{iid}}{\sim} N(0, \sigma_\beta^2), \\
\varepsilon_{ij} &= \text{whole plot error, } \varepsilon_{ij} \overset{\text{iid}}{\sim} N(0, \sigma_\varepsilon^2), \\
\gamma_k &= \text{split plot treatments,} \\
(\tau\gamma)_{ik} &= \text{treatment interaction,} \\
(\beta\gamma)_{ik} &= \text{block-treatment interaction} \overset{\text{iid}}{\sim} N(0, \sigma_{\beta\gamma}^2), \\
\delta_{ijk} &= \text{split plot error, } \overset{\text{iid}}{\sim} N(0, \sigma_\delta^2),
\end{aligned}
$$

where we assume that all error terms are independent. The big difference between this model and the CRD split plot (5.1) is that with the addition of the block structure we have two new random effects. This variation will result in a more complicated split plot error term, which we will look at below. Finally, note that the ε_{ij} is the Block $\times$ Treatment interaction (which we could have also designated as $(\tau\beta)_{ij}$).

Identifiability considerations again lead us to impose the parameter restrictions

(5.15) $$\bar{\tau} = \bar{\gamma} = (\bar{\tau\gamma})_i = (\bar{\tau\gamma})_k = 0,$$

where we again recall that this is merely a renaming of the effects, and does not signify any change in the real parameter space.

The full anova table for this model, identifying all terms, is given in Table 5.5. Comparing this anova table to the CRD split plot anova in Table 5.1, we see two big differences. One, all of the factors are crossed, which should also be evident from Table 5.6. Two, there are now two error terms below the line, representing the Block-Treatment interactions at the split plot level.

The correlation structure induced by the RCB split plot design is a bit more complicated than that of the CRD split plot. Note that if $j \neq j'$, then since the blocks are independent $\text{Corr}(Y_{ijk}, Y_{i'j'k'}) = 0$. However, within a block, we can calculate from (5.14)

$$\text{Cov}(Y_{ijk}, Y_{i'jk}) = \sigma_\beta^2 + \sigma_{\beta\gamma}^2,$$

the correlation between two observations in different WP treatments but the same SP treatment (within a block). See Exercise 5.7 for the remainder of the covariances.

We see from the anova table that "above the line", at the whole plot level, we have a simple RCB, with the WP error being the usual treatment by block interaction. Below the line is a bit more complicated, as now we formally have two split plot errors. When we calculate expected mean squares below, we will see that the two split plot error terms provide the correct denominator for their respective tests, keeping the rule that in an RCB, treatment effects are tested against their interaction with blocks. Often, these two error terms are pooled into one " split plot error", and we will look at the assumptions needed for, and consequences of, this strategy.

Example 5.4. VARIETY SPLIT PLOT CONTINUED   The anova for the data of Example 5.3 (in dataset `VarietySP`), considering the WP treatments to be the fertilizer and the SP treatment the variety, is given by

Table 5.5.   Anova for split plot design with RCB on whole plot treatments. T=Whole Plot Treatment, G = Split Plot Treatment, B=Blocks.

Source	df	SS
Blocks	b-1	$gt \sum_j (\bar{y}_j - \bar{\bar{y}})^2$
Whole Plot Trt	t-1	$rg \sum_i (\bar{y}_i - \bar{\bar{y}})^2$
B × WP	(b-1)(t-1)	$g \sum_{ij} (\bar{y}_{ij} - \bar{y}_i - \bar{y}_j + \bar{\bar{y}})^2$
(Whole Plot Error)		
Split Plot Trt	g-1	$bt \sum_k (\bar{y}_k - \bar{\bar{y}})^2$
SP Trt × WP Trt	(g-1)(t-1)	$b \sum_{ik} (\bar{y}_{ij} - \bar{y}_i - \bar{y}_j + \bar{\bar{y}})^2$
B × SP Trt	(b-1)(g-1)	$t \sum_{jk} (\bar{y}_{jk} - \bar{y}_j - \bar{y}_k + \bar{\bar{y}})^2$
B × SP Trt × WP Trt	(b-1)(g-1)(t-1)	$\sum_{ijk} (y_{ijk} - \bar{y}_{ij} - \bar{y}_{ik} - \bar{y}_{jk}$ $+ \bar{y}_i + \bar{y}_j + \bar{y}_k - \bar{\bar{y}})^2$
Total	grt-1	

Table 5.6. Data Layout for split plot design with RCB on whole plot treatments. T = Whole Plot Treatment, G = Split Plot Treatment, R = Replication

Replication (Blocks)

1   · · ·   r

Source	df	SS	MS	F	p-value
Rep	1	6.961	6.961		
Trt	3	14.775	4.925	19.811	0.018
Trt × Rep	3	0.746	0.2486		
Variety	5	2.071	0.414	$\frac{.414}{0.369} = 1.122$	.451
Trt × Variety	15	1.526	0.102	$\frac{.102}{.104} = 0.977$	.518
Variety × Rep	5	1.849	0.369		
Trt × Variety × Rep	15	1.562	0.104		

The anova tells us that there are significant differences in the WP treatments, but there is nothing going on at the split plot level. If we had pooled the split plot error terms into one split plot error with 20 degrees of freedom, the split plot tests would have showed us a slight significance in varieties (Exercise 5.8)

Note that this design is most appropriate if we are more interested in getting good information on the variety differences, and less interested in the treatment differences, as there will be greater precision on the split plot varieties than the whole plot treatments. However, this is also a case of rejoice - you have no choice - since it would be very labor intensive to put the fertilizer treatments at the split plot level. (The fertilizer is usually applied by machine to larger plots. Also, if they were at the split plot level there would be concern about the treatments (spreading fertilizer) being confined to their own plots.

‖

5.3.2 Expected Squares and F-tests

Calculation of the expected mean squares should, by now, be routine. There are really no surprises in the calculations, so most of the details are left to an

Table 5.7. Expected mean squares for a split plot design with the whole plots in a RCB, with data layout as in Table 5.6.

Source	df	EMS
Blocks	$b-1$	$\sigma_\delta^2 + g\sigma_\varepsilon^2 + t\sigma_{\beta\gamma}^2 + gt\sigma_\beta^2$
Whole Plot Trt	$t-1$	$\sigma_\delta^2 + g\sigma_\varepsilon^2 + \frac{bg}{t-1}\sum_i \tau_i^2$
Blocks × WP Trts	$(b-1)(t-1)$	$\sigma_\delta^2 + g\sigma_\varepsilon^2$
Split Plot Trt	$g-1$	$\sigma_\delta^2 + t\sigma_{\beta\gamma}^2 + \frac{bt}{g-1}\sum_k \gamma_k^2$
Split Plot Trt × Whole Plot Trt	$(g-1)(t-1)$	$\sigma_\delta^2 + \frac{b}{(g-1)(t-1)}\sum_{ik}(\tau\gamma)_{ik}^2$
Blocks × SP Trt	$(b-1)(g-1)$	$\sigma_\delta^2 + t\sigma_{\beta\gamma}^2$
Blocks × SP Trt × WP Trts	$(b-1)(g-1)(t-1)$	σ_δ^2
Total	bgt-1	

exercise. We do two calculations, however, to show how the split plot RCB balances the treatments. Using model (5.14) and the definition of the sums of squares, we have

$$E(\text{SS(WP Trts)}) = bg\sum_i[\tau_i + \bar{\beta} + \bar{\varepsilon}_i + (\bar{\beta}\gamma) + \bar{\delta}_i - \bar{\beta} - \bar{\bar{\varepsilon}} - (\bar{\beta}\gamma) - \bar{\delta}]^2$$

$$= bg\sum_i[\tau_i^2 + (\bar{\varepsilon}_i - \bar{\bar{\varepsilon}})^2 + (\bar{\delta}_i - \bar{\delta})^2],$$

where we see that the terms involving β and $(\beta\gamma)$ have cancelled. Note that in Table 5.7 these variance terms do not appear above the line, except in the sum of squares for blocks. Similarly, in the sum of squares for the split plot treatment, the terms involving β also cancel (since these factors are crossed). However,

$$E(\text{SS(SP Trts)}) = bt\sum_k[\bar{\bar{\varepsilon}} + \gamma_k + (\bar{\beta}\gamma)_k + \bar{\delta}_k - \bar{\bar{\varepsilon}} - (\bar{\beta}\gamma) - \bar{\delta}]^2$$

$$= bt\sum_k[\gamma_k^2 + ((\bar{\beta}\gamma)_k - (\bar{\beta}\gamma))^2 + (\bar{\delta}_k - \bar{\delta})^2],$$

so the whole plot error disappears from the split plot treatment, but the variance due to the interaction, $\sigma_{\beta\gamma}^2$, remains.

Using the model (5.14) we obtain the expected mean squares of Table 5.7, which are quite similar to those of Table 5.3 with two notable exceptions. Here there are variances due to Blocks and the Block × SP treatment interaction, the latter appearing below the line. (Note that σ_β^2 does not appear below the line). Also recall that the whole plot error term here is, in fact, the Block × WP treatment interaction.

The presence of $\sigma_{\beta\gamma}^2$ results in two separate error terms below the line. From the expected mean squares we see that there are two distinct F-ratios:

(5.16) $\dfrac{\text{MS(Split Plot Trt)}}{\text{MS(Blocks × SP Trt)}}$ and $\dfrac{\text{MS(Split Plot Trt × Whole Plot)}}{\text{MS(Blocks × SP Trt × WP Trts)}}$, which, unfortunately, could have the effect of reducing degrees of freedom and hence power of these tests.

Note: Pooling Blocks × SP Trt with Blocks × SP Trt × WP Trts requires the assumption that $\sigma^2_{\beta\gamma} = 0$, that is, no interaction between blocks and the split plot treatment.

If the plots are randomly assigned to the split plot treatments, the assumption that $\sigma^2_{\beta\gamma} = 0$ becomes more reasonable. If this assumption cannot be made, one could adopt the strategy of first testing $H_0 : \sigma^2_{\beta\gamma} = 0$ and then pooling if this test is accepted. However, then we are in the possible situation of compounding the error terms.

The best strategy? If degrees of freedom are a concern, we can pool the error terms and see from the expected mean squares that we are being conservative. The presence of $\sigma^2_{\beta\gamma}$ will result in an error term that is a bit larger than the correct one; so if we do reject H_0 we can be somewhat comfortable with the decision. Of course, we are sacrificing a bit of power, which can result in the commission of a Type II error (not finding a true difference).

Consequences of pooling split plot errors

The EMS in Table 5.7 show us how to do the remaining tests. The null hypothesis of no whole plot treatment effect is

$$H_0 : \tau_i = 0 \text{ for all } i$$

and is tested by

$$F_{t-1,(b-1)(t-1)} = \frac{\text{MS(WP Trts)}}{\text{MS(Blocks} \times \text{WP Trts)}}.$$

At the split plot level, the split plot main effect $H_0 : \gamma_k = 0$, for all k, and the interaction $H_0 : (\tau\gamma)_{ik} = 0$, for all i and k, are tested, respectively, with the mean square ratios in (5.16).

Note that if we actually pool the split plot error, and recalling what we know about nested and crossed factors, we have

$$\text{SS(Blocks} \times \text{SP Trt)} + \text{SS(Blocks} \times \text{SP Trt} \times \text{WP Trts)}$$
$$= \text{SS(Blocks} \times \text{SP Trt (}in \text{ WP Trts))},$$

which is exactly the split plot error term in Table 5.3. Of course, there the split plot treatment was not crossed with replication except in the whole plots, so the error term could not be separated as above. In the RCB split plot, the split plot treatment is crossed with blocks, which leads to the separation of the error terms.

5.3.3 Estimating Contrasts

We now turn to estimating the variances of the contrasts in the RCB split plot. Most of the calculations are quite similar to those done in Section 5.2.3, so we will suppress many of the details, leaving them to Exercise 5.35. As before, we consider the four cases:

(1) Whole Plot Means: $\sum_i a_i \tau_i$, where $\sum_i a_i = 0$.
(2) Split Plot Means: $\sum_k a_k \gamma_k$, where $\sum_k a_k = 0$.
(3) Interaction Means, Same Level of Whole Plot: $\sum_k a_k (\tau\gamma)_{ik}$, where $\sum_k a_k = 0$.
(4) Interaction Means, Different Whole Plot Level: $\sum_{ik} a_{ik} (\tau\gamma)_{ik}$, where $\sum_{ik} a_{ik} = 0$.

Since the treatment design in the split plot RCB is the same as in the split plot CRD, the least squares estimates of the treatment means are exactly the same, and are given in (5.5).

Case (1): Whole Plot Comparisons

When we calculated $\mathrm{Var}(\sum_i a_i \bar{Y}_i)$ for the CRD split plot design, we noted that the whole plot means were independent. Here, the whole plot means in the same block are correlated, and if we take this into account we have

$$(5.17) \qquad \mathrm{Var}\left(\sum_i a_i \bar{Y}_i\right) = \frac{\sigma_\delta^2 + g\sigma_\varepsilon^2}{bg} \sum_i a_i^2.$$

Again, replication at the split plot level has less of an effect here, in that it only cuts down a piece of the variance. From the anova EMS we see that $E(\mathrm{MS(WP\ Error)}) = \sigma_\delta^2 + g\sigma_\varepsilon^2$, giving us an estimate of this variance.

Case (2): Split Plot Comparisons

In calculating the variance of $\sum_k a_k \bar{y}_k$, we again have to take account of the correlation, and doing so results in

$$(5.18) \qquad \mathrm{Var}\left(\sum_k a_k \bar{Y}_k\right) = \frac{\sigma_\delta^2 + t\sigma_{\beta\gamma}^2}{bt} \sum_k a_k^2,$$

where, as in the CRD case, the contrast has eliminated the whole plot error from the variance. From Table 5.7 we see that this variance can be estimated with $\mathrm{MS(Blocks \times SP\ Trts)}$ or, with the assumption that $\sigma_{\beta\gamma}^2 = 0$, we can use a pooled error term.

We now look at Cases (3) and (4), the interaction contrasts within and between levels of the whole plots. The calculations for the RCB split plot turn out to be a bit trickier than those in the CRD split plot, and we will have to work harder for variance estimates. Some of the details are left to Exercise 5.35.

Case (3): Interaction Mean Comparisons: Same Whole Plot _____

Calculation of the variance of $\sum_k a_k \bar{y}_{ik}$, within the same level of the whole plot treatment, works out similarly to the CRD split plot, where we use the fact that $\sum_k a_k = 0$ to simplify the variance formula. The end result is

$$(5.19) \qquad \operatorname{Var}\left(\sum_k a_k \bar{Y}_{ik}\right) = \frac{\sigma_\delta^2 + \sigma_{\beta\gamma}^2}{b} \sum_k a_k^2,$$

where, again, we see that the whole plot variance σ_ε^2 does not appear in the interaction contrasts *within* a whole plot., but now we do see the appearance of the variance due to the split plot treatment by block interaction.

Although the error variance calculation was rather straightforward, we are in a bit of trouble in that, from Table 5.7 there is no obvious estimate of $\sigma_\delta^2 + \sigma_{\beta\gamma}^2$. We will deal with this at the end of the section.

Case (4): Interaction Mean Comparisons: Different Whole Plot ___

Finally, we turn to Case (4) and calculate the variance of a contrast between interaction means at different levels of a whole plot. Although the calculations are straightforward (if arduous), we again run into trouble in the estimation of the variance. The variance of a general interaction contrast is given by

$$\operatorname{Var}\left(\sum_{ik} a_{ik} \bar{Y}_{ik}\right) = \frac{\sigma_\delta^2}{b} \sum_{ik} a_{ik}^2 + \frac{\sigma_{\beta\gamma}^2}{b} \sum_k \left(\sum_i a_{ik}\right)^2 + \frac{\sigma_\varepsilon^2}{b} \sum_i \left(\sum_k a_{ik}\right)^2.$$
(5.20)

This is a rather nasty expression, and again we cannot directly estimate the variances using the mean squares in Table 5.7. Note that if we could assume $\sum_i a_{ik} = \sum_k a_{ik} = 0$, then we would have no problem in estimating the variance, but these terms will be zero only for certain contrasts (a product construction will do it).

Estimating the Contrast Variances _____

We first make a few observations about the contrast variances that we have just calculated Note the implications of these contrast variances:

(i) In (5.20), the multipliers of $\sigma_{\beta\gamma}^2$ and σ_ε^2 are zero if $\sum_i a_{ik} = 0$ and $\sum_k a_{ik} = 0$, respectively. That is, if the a_{ik} define a contrast in the whole plot means, then $\sigma_{\beta\gamma}^2$ disappears, while if they define a contrast in the split plot means, then σ_ε^2 disappears. We are then left with a variance that only depends on σ_δ^2, which can be estimated with MS(Blocks × SP Trt × WP Trts).

(ii) If the a_{ik} come from a product construction, that is, start with a_i and a_k satisfying $\sum_i a_i = 0$ and $\sum_k a_k = 0$, and the interaction contrast is $\sum_{ik} a_i a_k \bar{Y}_{ik}$, then the variance only depends on σ_δ^2

(iii) In (5.20), if there is only one value of i, then the variance reduces to that of Case (3), as it must.

(iv) Under the assumption that $\sigma^2_{\beta\gamma} = 0$, the expressions for the contrast variances in Cases (3) and (4) here become identical to the CRD case.

Finally, how do we proceed with estimation? Of course, the optimal route would be to estimate the variance components in (5.19) and (5.20) and then obtain (or approximate) the distribution. However, we are going to take a simpler route and proceed in a manner similar to other texts.

That is, we make the unjustified assumption that $\sigma^2_{\beta\gamma} = 0$, which does two things:

> Watch out for this assumption

(1) The variance estimation is now identical to that in the CRD split plot.

(2) From Table 5.7, we can pool the two error terms to get one term with $t(b-1)(g-1)$ degrees of freedom that estimates σ^2_δ.

The effect of this strategy may ultimately be conservative, in that the $\sigma^2_{\beta\gamma}$ appears in the expected mean squares and, if not zero, would result in an inflated variance. With the assumption that $\sigma^2_{\beta\gamma} = 0$, the variance (5.20) can be rewritten as (see (5.13) and Exercise 5.33)

$$(5.21) \quad \mathrm{Var}\left(\sum_{ik} a_{ik}\bar{Y}_{ik}\right) = \frac{\sigma^2_\delta}{b}\sum_{ik}(a_{ik} - \bar{a}_i)^2 + \frac{\sigma^2_\delta + g\sigma^2_\varepsilon}{bg}\sum_i\left(\sum_k a_{ik}\right)^2.$$

which is a better form for use with Satterthwaite's approximation.

We continue with Example 5.4 to illustrate the contrast inferences.

Example 5.5. VARIETY SPLIT PLOT CONCLUDED   The anova for the data of Example 5.3 (dataset `VarietySP`), identifying the factorial structure of the treatments and pooling the split plot error, is

Source	df	SS	MS	F	p-value
Rep	1	6.961	6.961		
Trt	3	14.775	4.925	19.811	0.018
P	1	12.140	12.140	48.836	0.006
K	1	1.261	1.261	5.073	0.110
P × K	1	1.374	1.374	5.526	0.100
Trt × Rep	3	0.746	0.249		
Variety	5	2.071	0.414	2.429	0.071
Trt × Variety	15	1.526	0.102	0.597	0.845
P × Variety	5	0.346	0.069	0.406	0.839
K × Variety	5	0.888	0.178	1.042	0.421
P × K × Variety	5	0.291	0.058	0.342	0.882
Split Plot Error	20	3.411	0.170		

Using the pooled split plot error gave a more significant test on Variety than using separate errors, but other than that, the anova here is quite similar to the previous one. The interaction means are

	pk	pK	Pk	PK
A	3.56	3.57	4.54	5.17
G	3.78	3.33	4.34	4.64
K	3.59	3.78	4.42	5.00
N	4.20	4.38	4.55	5.47
O	3.96	4.05	4.30	5.64
R	3.73	3.62	4.68	4.87

and two interactions of interest examine the effect of Potassium (K) on the A and G varieties:

Main Effect of K vs. A and G				Interaction of K vs. A and G			
	pk	pK	Pk	PK		pk	pK	Pk	PK

	pk	pK	Pk	PK
A	1	−1	1	−1
G	1	−1	1	−1

	pk	pK	Pk	PK
A	1	−1	1	−1
G	−1	1	−1	1

The first contrast, on the main effect, put us in Case (4), while the second contrast, dealing with the interaction, is Case (3). See Exercise 5.8   ‖

5.4 Estimating Effects

We digress a bit here and examine the difference between estimating contrasts of means (which we have been doing) and *effects*. That is, from (5.5) we see that the interaction means are $\bar{y}_{ik}$ but the interaction effects (sometimes called *least squares means*) are $(\hat{\tau\gamma})_{ik} = \bar{y}_{ik} - \bar{y}_i - \bar{y}_k + \bar{\bar{y}}$, and lead to different contrast estimates.

First, note that for estimating μ, τ_i, or γ_k, contrasts in cell means or effects are exactly the same (Exercise 5.11), but the interaction contrasts can differ. That is

$$\text{Cell mean contrast} : \sum_{ik} a_{ik} \bar{y}_{ik},$$

$$\text{Effect contrast} : \sum_{ik} a_{ik} (\hat{\tau\gamma})_{ik},$$

where $\bar{y}_{ik}$ estimates $\mathrm{E}\bar{Y}_{ik} = \mu + \tau_i + \gamma_j + (\tau\gamma)_{ik}$, the cell mean, and $(\hat{\tau\gamma})_{ik}$ estimates $\mathrm{E}(\hat{\tau\gamma})_{ik} = (\tau\gamma)_{ik}$, the cell effect. It is common to only give variances for contrasts in the cell means, but we will see that there can be an advantage to using the effect contrasts. (Note that if the a_{ik} satisfy $\sum_i a_{ik} = \sum_k a_{ik} = 0$, as would happen if they came from a product construction, then the two contrasts are identical.)

The choice of whether to estimate the cell mean or the cell effect is one to be left to the experimenter. What we are concerned with here is the resulting inference; in particular, how do the variances differ between mean estimates and effect estimates.

Note: We find that if we estimate the effects rather than the means we can typically do this with greater precision.

We look at the case where the whole plot treatments are in a CRD which, as we will see, also covers the RCB case (Exercise 5.36).

From (5.1), the variance of an interaction contrast using the least squares estimates (5.5) is

$$\text{Var}\left(\sum_{ik} a_{ik}(\hat{\tau\gamma})_{ik}\right) = \text{Var}\left(\sum_{ik} a_{ik}(\bar{y}_{ik} - \bar{y}_i - \bar{y}_k + \bar{\bar{y}})\right)$$

(5.22)
$$= \text{Var}\left(\sum_{ik} a_{ik}(\bar{y}_{ik} - \bar{y}_i - \bar{y}_k)\right) \qquad [\text{since } \sum_{ik} a_{ik} = 0]$$

$$= \text{Var}\left(\sum_{ik} a_{ik}(\bar{\varepsilon}_i + \bar{\delta}_{ik} - \bar{\varepsilon}_i - \bar{\delta}_i - \bar{\bar{\varepsilon}} - \bar{\delta}_k)\right) \qquad [\text{from (5.1)}].$$

Notice that the terms with $\bar{\varepsilon}_i$ cancel, and the term $\bar{\bar{\varepsilon}}$ vanishes because $\sum_{ik} a_{ik} = 0$. Thus, there is *no contribution* from the whole plot error in this contrast variance. Compare this to (5.12), where the whole plot error contribution only vanishes if $\bar{a}_i = 0$. We do not need this condition for the whole plot error to disappear from the variance (5.22), and it is in this sense that we can say that the interaction effect is always estimated with greater precision than the cell means.

Continuing, we have

$$\text{Var}\left(\sum_{ik} a_{ik}(\hat{\tau\gamma})_{ik}\right) = \text{Var}\left(\sum_{ik} a_{ik}(\bar{\delta}_{ik} - \bar{\delta}_i - \bar{\delta}_k)\right)$$

(5.23)
$$= \text{Var}\left(\sum_i \left[\sum_k a_{ik}\bar{\delta}_{ik} - g\bar{a}_i\bar{\delta}_i\right] - t\sum_k \bar{a}_k\bar{\delta}_k\right).$$

We now expand the variance and deal with each piece separately. The calculations are a bit painful and are summarized in Exercise 5.36. The result is

$$\text{Var}\left(\sum_{ik} a_{ik}(\hat{\tau\gamma})_{ik}\right) = \frac{\sigma_\delta^2}{r}\left[\sum_{ik} a_{ik}^2 - \sum_i \sum_k (\bar{a}_i + \bar{a}_k)^2\right]$$

(5.24)
$$= \frac{\sigma_\delta^2}{r}\sum_{ik} [a_{ik} - (\bar{a}_i + \bar{a}_k)]^2.$$

Example 5.6. DIET SPLIT PLOT CONCLUDED   The first interaction contrast looked at in Example 5.2 was

	AM	PM	
1	1	-1	0
2	0	0	0
3	0	0	0
4	0	0	0
	$\frac{1}{4}$	$-\frac{1}{4}$	

where the margins give the values of $\bar{a}_i$ and $\bar{a}_k$. For this contrast we obtain the value $\sum_{ik}[a_{ik} - (\bar{a}_i + \bar{a}_k)]^2 = 3/2$, leading to an effect variance of $(16.17/3)(3/2)$, smaller than $(16.17)(2/3)$, the variance of the contrast in cell means.   ‖

We have done all of our calculations for the case of having the whole plots in a CRD. However, in one of the more fortuitous occurrences in statistical design, if the whole plots are in an RCB the calculations are exactly the same. To see this, from (5.14), the treatment design is a twoway crossed design (with the fixed effects τ_i, γ_k and $(\tau\gamma)_{ik}$, and everything else in error). Thus, the least squares estimate of $(\hat{\tau\gamma})_{ik}$ is $\bar{y}_{ik} - \bar{y}_i - \bar{y}_k + \bar{\bar{y}}$. If $\sum_{ik} a_{ik} = 0$, it can be shown (Exercise 5.36) that, under model (5.14), $\text{Var}(\sum_{ik} a_{ik}(\hat{\tau\gamma})_{ik})$ is given by (5.24).

> **Note:** One important difference about estimating interaction effect contrasts rather than mean contrasts is that we do not run into the difficulty in estimating the variance that occurred in Sections 5.2.3 and 5.3.3.

That is, since the variance is only a function of the parameter σ_δ^2, we always have a mean square to estimate it and do not have to resort to using Satterthwaite's approximation.

5.5 Splitting Twice

In a split plot design one treatment (the split plot treatment) is randomized in the levels of another treatment (the whole plot treatment). There is, of course, no reason to stop at just one of these – we can randomize another treatment to the levels of the split plot treatment, creating a *split split plot* design. Some things to note:

(1) The split in the experiment should be dictated by the physical constraints and perhaps the desire for greater accuracy in the measurement of a particular treatment.

(2) We will be a little less formal here (and in much of what follows). Although we will indicate tests with EMS and give many contrast variances, we will not formally verify Cochran's Theorem.

(3) Of course, we could split more than twice and, for example, have a split split split plot design. But we will stop at two splits; the pattern should be clear.

(4) We will look at two cases, one where the whole plots are in a CRD, and another where they are in an RCB. There are surprises in both cases.

CRD on the Whole Plots

Example 5.7. OZONE SPLIT SPLIT PLOT   An experiment is conducted to test the effect of ozone gas on plants. The researcher assigned two environmental chambers to each of four levels of ozone (a total of eight chambers). Six varieties of plants were placed in each chamber. At the end of the experiment the plants were removed and analyzed. Data were recorded for two locations on each plant - one at the root (R) and one at the top (T). A schematic of the data looks like

Here the whole plot treatments are the levels of ozone, which are in a CRD with two observations per treatment (the two chambers). Variety is the split plot treatment, as it is randomized in the levels of ozone. Finally, location is the *split split* plot treatment, and here we are physically splitting the unit – taking two observations from each plant. The anova for this experiment is

Source	df
Ozone	3
Whole Plot Error (Chambers in Ozone)	4
Variety	5
V × O	15
Split Plot Error (V × C in O)	20
Location	1
L × V	5
L × O	3
L × V × O	15
Split Split Plot Error (L × C in V × O)	24
Total	95

This is a split split plot design with the whole plots in a oneway CRD. Note the following:

(1) The three treatments (L, V, O) are crossed.
(2) There is a nesting of the random factor (Chambers) within the whole plot treatment.
(3) The first two error terms (WP and SP) are the same as in Section 5.2. The split split plot error, formed from the residuals, reflect the interaction of Location with the random factor Chambers, but this interaction is nested in V × O.

‖

A model for the split split plot experiment with the whole plots in a CRD is

(5.25)
$$Y_{ijk\ell} = \mu + \tau_i + \varepsilon_{ij} + \gamma_k + (\tau\gamma)_{ik} + \delta_{ijk}$$
$$+ \psi_\ell + (\tau\psi)_{i\ell} + (\gamma\psi)_{k\ell} + (\tau\gamma\psi)_{ik\ell} + \omega_{ijk\ell},$$

where $i = 1, \ldots, t$, $j = 1, \ldots, r$, $k = 1, \ldots, g$, $\ell = 1, \ldots, s$ and

$$Y_{ijk} = \text{response,}$$
$$\mu = \text{overall mean effect,}$$
$$\tau_i = \text{whole plot treatment,}$$
$$\varepsilon_{ij} = \text{whole plot error, } \varepsilon_{ij} \overset{\text{iid}}{\sim} N(0, \sigma_v^2 e),$$
$$\gamma_k = \text{split plot treatments,}$$
$$(\tau\gamma)_{ik} = \text{interaction (at split plot level),}$$
$$\delta_{ijk} = \text{split plot error, } \overset{\text{iid}}{\sim} N(0, \sigma_\delta^2),$$
$$\text{independent of } \varepsilon_{ij},$$
$$\psi_\ell = \text{split split plot treatment,}$$
$$(\tau\psi)_{i\ell}, (\gamma\psi)_{k\ell}, (\tau\gamma\psi)_{ik\ell} = \text{interactions (at split split plot level),}$$

Table 5.8. Anova for split plot design with RCB on whole plot treatments. T = Whole Plot Treatment, G = Split Plot Treatment, R = Replication.

Source	df	SS
Whole Plot (WP) Trt	t-1	$srg \sum_i (\bar{y}_i - \bar{\bar{y}})^2$
WP Error	t(r-1)	$sg \sum_{ij} (\bar{y}_{ij} - \bar{y}_i)^2$
(Replication in WP)		
Split Plot (SP) Trt	g-1	$srt \sum_k (\bar{y}_k - \bar{\bar{y}})^2$
SP Trt × WP Trt	(g-1)(t-1)	$sr \sum_{ik} (\bar{y}_{ik} - \bar{y}_i - \bar{y}_k + \bar{\bar{y}})^2$
SP Error	t(g-1)(r-1)	$s \sum_i \left[\sum_{jk} (y_{ijk} - \bar{y}_{ij} - \bar{y}_{ik} + \bar{y}_i)^2 \right]$
SP Trt × Reps in WP		
Split Split Plot (SSP) Trt	s-1	$rgt \sum_\ell (\bar{y}_\ell - \bar{\bar{y}})^2$
SSP Trt × WP Trt	(s-1)(t-1)	$rg \sum_{i\ell} (\bar{y}_{i\ell} - \bar{y}_i - \bar{y}_\ell + \bar{\bar{y}})^2$
SSP Trt × SP Trt	(s-1)(g-1)	$rt \sum_{k\ell} (\bar{y}_{k\ell} - \bar{y}_k - \bar{y}_\ell + \bar{\bar{y}})^2$
SSP Trt × SP Trt × WP Trt	(s-1)(g-1)(t-1)	$r \sum_{ik\ell} (\bar{y}_{ik\ell} - \bar{y}_{ik} - \bar{y}_{i\ell} - \bar{y}_{kj\ell} + \bar{y}_i + \bar{y}_k + \bar{y}_\ell - \bar{\bar{y}})^2$
SSP Error	(s-1)(r-1)tg	$\sum_{ik} \left[\sum_{j\ell} (y_{ijk\ell} - \bar{y}_{ijk} - \bar{y}_{ik\ell} + \bar{y}_{ik})^2 \right]$
(SSP Trt × Reps) in (SP Trt × WP Trt)		
Total	sgrt-1	

$$\omega_{ijk\ell} = \text{split split plot error,} \overset{iid}{\sim} N(0, \sigma_\omega^2),$$
$$\text{independent of } \varepsilon_{ij} \text{ and } \delta_{ijk}.$$

Here, for identifiability considerations we invoke the parameter constraints

$$\bar{\tau} = \bar{\gamma} = (\bar{\tau\gamma})_i = (\bar{\tau\gamma})_k = 0,$$
$$\bar{\psi} = (\bar{\tau\psi})_i = (\bar{\tau\psi})_\ell = (\bar{\gamma\psi})_k = (\bar{\gamma\psi})_\ell = 0,$$
$$(\bar{\tau\gamma\psi})_{ik} = (\bar{\tau\gamma\psi})_{i\ell} = (\bar{\tau\gamma\psi})_{k\ell} = 0,$$

where we recall again that this is merely a renaming of the effects and does not signify any change in the real parameter space.

The full anova for the split split plot model, identifying all terms, is given in Table 5.8. Note that the split split plot error is a pooling of the split split plot × rep interactions – this is not a fourway interaction. Compare this to the split plot error, which represents a pooling of the split plot × rep interaction over the whole plots.

The whole plot error comes from the replication of the whole plot treatments, just as in any CRD, and the split plot errors come from the respective interactions – reminiscent of an RCB – but with an added nesting structure. If either $k = 1$ or $s = 1$, the design collapses back to an ordinary split plot design.

Table 5.9. Expected mean squares for a split split plot design with the whole plots in a CRD, with model (5.25).

Source	df	EMS
Whole Plot Trt	$t-1$	$\sigma_\omega^2 + s\sigma_\delta^2 + sg\sigma_\varepsilon^2 + \frac{srg}{t-1}\sum_i \tau_i^2$
Replication (in WP)	$t(r-1)$	$\sigma_\omega^2 + s\sigma_\delta^2 + sg\sigma_\varepsilon^2$
Split Plot Trt	$g-1$	$\sigma_\omega^2 + s\sigma_\delta^2 + \frac{srt}{g-1}\sum_k \gamma_k^2$
SP Trt × WP Trt	$(g-1)(t-1)$	$\sigma_\omega^2 + s\sigma_\delta^2 + \frac{sr}{(g-1)(t-1)}\sum_{ik}(\tau\gamma)_{ik}^2$
SP Trt × Reps (in WP)	$t(g-1)(r-1)$	$\sigma_\omega^2 + s\sigma_\delta^2$
Split Split Plot Trt	$s-1$	$\sigma_\omega^2 + \frac{rgt}{s-1}\sum_k \psi_\ell^2$
SSP Trt × WP Trt	$(s-1)(t-1)$	$\sigma_\omega^2 + \frac{rg}{(s-1)(t-1)}\sum_{i\ell}(\tau\psi)_{i\ell}^2$
SSP Trt × SP Trt	$(s-1)(g-1)$	$\sigma_\omega^2 + \frac{rt}{(s-1)(g-1)}\sum_{k\ell}(\gamma\psi)_{k\ell}^2$
SSP Trt × SP Trt × WP Trt	$(s-1)(g-1)(t-1)$	$\sigma_\omega^2 + \frac{r}{(s-1)(g-1)(t-1)}\sum_{ik\ell}(\tau\gamma\psi)_{ik\ell}^2$
(SSP Trt × Reps) in (SP Trt × WP Trt)	$tg(s-1)(r-1)$	σ_ω^2
Total	$sgrt-1$	

The expected mean squares for the split split plot experiment mimic those of the split plot, and are given in Table 5.9. We suppress details of the calculations, as they are quite similar to (but a bit more tedious than) previous EMS calculations.

Example 5.8. OZONE SPLIT SPLIT PLOT CONTINUED   The anova for the Ozone chamber experiment is

Source	df	SS	MS	F	p-value
Ozone	3	0.254	0.085	17.465	0.009
Whole Plot Error	4	0.019	0.005		
Variety	5	0.153	0.031	1.045	0.419
Ozone × Variety	15	0.169	0.011	0.384	0.968
Split Plot Error	20	0.586	0.029		
Location	1	0.026	0.026	0.645	0.429
Ozone × Location	3	0.185	0.061	1.532	0.232
Variety × Location	5	0.281	0.056	1.393	0.262
Ozone × Variety × Location	15	0.693	0.046	1.145	0.373
Split Split Plot Error	24	0.967	0.040		

In this "classical" split split plot analysis, at each level of the design we use the respective error term as the denominator in the test. From Table 5.9 we see that these are the appropriate tests, with no further assumptions needed on the effects.   ‖

Finally, we look at contrasts in this split split plot design. Suppressing the details, which are similar to those in Section 5.2.3, we have:

(1) Interaction mean SSP × WP:

$$\text{Var}\left(\sum_{i\ell} a_{i\ell}\bar{Y}_{i\ell}\right) = \frac{\sigma_\omega^2}{rg}\sum_{i\ell} a_{i\ell}^2 + \frac{\sigma_\delta^2 + g\sigma_\varepsilon^2}{rg}\sum_i\left(\sum_\ell a_{i\ell}\right)^2,$$

where we note that $\sum_\ell a_{i\ell} = 0$ if we have a contrast within the level of the whole plots, and, of course, this happens if we are comparing within one level of the WP treatment. This variance can be estimated without the need for the Satterthwaite approximation.

(2) Interaction mean: SSP × SP:

$$\text{Var}\left(\sum_{k\ell} a_{k\ell}\bar{Y}_{k\ell}\right) = \frac{\sigma_\omega^2}{rt}\sum_{k\ell} a_{k\ell}^2 + \frac{\sigma_\delta^2}{rt}\sum_k\left(\sum_\ell a_{k\ell}\right)^2,$$

where the second term is zero if we are balanced in the SP treatment or are in one level. The estimation of this variance may require the Satterthwaite approximation.

(3) SSP means:

$$\text{Var}\left(\sum_\ell a_\ell\bar{Y}_\ell\right) = \frac{\sigma_\omega^2}{rtg}\sum_\ell a_\ell^2 + \frac{\sigma_\delta^2 + g\sigma_\varepsilon^2}{rtg}\left(\sum_\ell a_\ell\right)^2,$$

where the last term is always zero for a contrast. This variance can be estimated without the need for the Satterthwaite approximation.

RCB on the Whole Plots

We next consider the case of having the whole plot treatments in an RCB. As happened in Section 5.3, the error structure becomes a bit more complicated.

Example 5.9. RCB SPLIT SPLIT PLOT   Three different washing solutions are being compared to study their effectiveness in retarding bacteria growth in 5-gallon milk containers. The analysis is done in three laboratories, which act as blocks, and the bacteria in each container is measured on four different days using two different assays. This is a split split plot design, but here the whole plot treatments (solutions) are in an RCB. This brings on added complications, as we saw in Section 5.3, because now the error terms are not as straightforward.

A schematic of the design is

	Solution		
	1	2	3
	Assay	Assay	Assay
	1 2	1 2	1 2
Lab 1	1 x x	1 x x	1 x x
	Day 2 x x	Day 2 x x	Day 2 x x
	3 x x	3 x x	3 x x
	4 x x	4 x x	4 x x

⋮   ⋮

	Solution		
	1	2	3
	Assay	Assay	Assay
	1 2	1 2	1 2
Lab 3	1 x x	1 x x	1 x x
	Day 2 x x	Day 2 x x	Day 2 x x
	3 x x	3 x x	3 x x
	4 x x	4 x x	4 x x

Note that, in contrast to having the whole plots in a CRD, here all of the treatments are crossed, and the chambers (the whole plot units) are nested within the whole plot treatment. An anova (source and df) for this experiment is

Source	df
Lab (Blocks)	2
Solution	2
Whole Plot Error (Solution × Lab)	4
Day	3
D × S	6
D × Lab	6
D × S × Lab	12
Assay	1
A × S	2
A × D	3
A × S × D	6
A × Lab	2
A × S × Lab	4
A × D × Lab	6
A × S × D × Lab	12
Total	71

where we see that, through the split plot level, this looks like the designs of Section 5.3. At the split split level, the error terms are even more abundant, and, as might be suspected, the formal test of any effect is with its interaction with Lab (blocks). ‖

A model for the split split plot experiment with the whole plots in an RCB is

$$
\begin{aligned}
Y_{ijk\ell} = \ & \mu + \tau_i + \beta_j + \varepsilon_{ij} + \gamma_k + (\tau\gamma)_{ik} + (\beta\gamma)_{jk} + \delta_{ijk} \\
& + \psi_\ell + (\tau\psi)_{i\ell} + (\gamma\psi)_{k\ell} + (\tau\gamma\psi)_{ik\ell} \\
& + (\beta\psi)_{j\ell} + (\beta\tau\psi)_{ij\ell} + (\beta\gamma\psi)_{jk\ell} + \omega_{ijk\ell},
\end{aligned}
$$
(5.26)

where $i = 1,\ldots,t$, $j = 1,\ldots,r$, $k = 1,\ldots,g$, $\ell = 1,\ldots,s$ and

$$
\begin{aligned}
Y_{ijk} \ = \ & \text{response,} \\
\mu \ = \ & \text{overall mean effect,} \\
\tau_i \ = \ & \text{whole plot treatment,} \\
\beta_j \ = \ & \text{blocks, } \beta_j \overset{\text{iid}}{\sim} N(0,\sigma_\beta^2), \\
\varepsilon_{ij} \ = \ & \text{whole plot error, } \varepsilon_{ij} \overset{\text{iid}}{\sim} N(0,\sigma_\varepsilon^2), \\
\gamma_k \ = \ & \text{split plot treatments,} \\
(\tau\gamma)_{ik} \ = \ & \text{fixed interaction (at split plot level),} \\
(\beta\gamma)_{jk} \ = \ & \text{random interaction (at split plot level),} \\
(\beta\gamma)_{jk} \overset{\text{iid}}{\sim} \ & N(0,\sigma_{\beta\gamma}^2), \\
\delta_{ijk} \ = \ & \text{split plot error, } \overset{\text{iid}}{\sim} N(0,\sigma_\delta^2), \\
\psi_\ell \ = \ & \text{split split plot treatment,} \\
(\tau\psi)_{i\ell}, (\gamma\psi)_{k\ell}, (\tau\gamma\psi)_{ik\ell} \ = \ & \text{fixed interactions (at split split plot level),} \\
(\beta\psi)_{ij\ell}, (\beta\tau\psi)_{ij\ell}, (\beta\gamma\psi)_{jk\ell} \ = \ & \text{random interactions (at split split plot level),} \\
(\beta\psi)_{ij\ell} \overset{\text{iid}}{\sim} N(0,\sigma_{\beta\psi}^2), \quad (\beta\tau\psi)_{ij\ell} & \overset{\text{iid}}{\sim} N(0,\sigma_{\beta\tau\psi}^2), \quad (\beta\gamma\psi)_{jk\ell} \overset{\text{iid}}{\sim} N(0,\sigma_{\beta\gamma\psi}^2), \\
\omega_{ijk\ell} \ = \ & \text{split split plot error, } \overset{\text{iid}}{\sim} N(0,\sigma_\omega^2),
\end{aligned}
$$

where we assume independence between all errors. For identifiability considerations we invoke the parameter constraints

$$
\bar{\tau} = \bar{\gamma} = (\bar{\tau\gamma})_i = (\bar{\tau\gamma})_k = 0,
$$
$$
\bar{\psi} = (\bar{\tau\psi})_i = (\bar{\tau\psi})_\ell = (\bar{\gamma\psi})_k = (\bar{\gamma\psi})_\ell = 0,
$$
$$
(\bar{\tau\gamma\psi})_{ik} = (\bar{\tau\gamma\psi})_{i\ell} = (\bar{\tau\gamma\psi})_{k\ell} = 0.
$$

The first thing that strikes us is the large number of error terms, which we need to wind our way through. First note that the three terms that are designated as error (ε, δ, and ω) are themselves interaction terms. For example, ε_{ij} is the WP $\times$ Block interaction, and could well have been designated $(\tau\beta)_{ij}$.

Table 5.10. Expected mean squares for a split split plot design with the whole plots in a RCB, with model (5.26)

Source	df	EMS
Blocks (B)	$b-1$	$\sigma_\omega^2 + s\sigma_\delta^2 + sg\sigma_\varepsilon^2 + st\sigma_{\beta\gamma}^2 + gt\sigma_\beta^2$
WP Trt (T)	$t-1$	$\sigma_\omega^2 + s\sigma_\delta^2 + sg\sigma_\varepsilon^2 + \frac{sbg}{t-1}\sum_i \tau_i^2$
B × T	$(b-1)(t-1)$	$\sigma_\omega^2 + s\sigma_\delta^2 + sg\sigma_\varepsilon^2$
SP Trt (G)	$g-1$	$\sigma_\omega^2 + s\sigma_\delta^2 + st\sigma_{\beta\gamma}^2 + \frac{sbt}{g-1}\sum_k \gamma_k^2$
G × T	$(g-1)(t-1)$	$\sigma_\omega^2 + s\sigma_\delta^2 + \frac{sb}{(g-1)(t-1)}\sum_{ik}(\tau\gamma)_{ik}^2$
G × B	$(b-1)(g-1)$	$\sigma_\omega^2 + s\sigma_\delta^2 + st\sigma_{\beta\gamma}^2$
G × T × B	$(b-1)(g-1)(t-1)$	$\sigma_\omega^2 + s\sigma_\delta^2$
SSP Trt (S)	$s-1$	$\sigma_\omega^2 + t\sigma_{\beta\gamma\psi}^2 + g\sigma_{\beta\tau\psi}^2$ $+tg\sigma_{\beta\psi}^2 + \frac{rtg}{s-1}\sum_\ell \psi_\ell^2$
S × T	$(s-1)(t-1)$	$\sigma_\omega^2 + g\sigma_{\beta\tau\psi}^2 + \frac{rg}{(s-1)(t-1)}\sum_{i\ell}(\tau\psi)_{i\ell}^2$
S × G	$(s-1)(g-1)$	$\sigma_\omega^2 + t\sigma_{\beta\gamma\psi}^2 + \frac{rt}{(s-1)(g-1)}\sum_{k\ell}(\gamma\psi)_{k\ell}^2$
S × G × T	$(s-1)(t-1)(g-1)$	$\sigma_\omega^2 + \frac{r}{(s-1)(g-1)(t-1)}\sum_{ik\ell}(\tau\gamma\psi)_{ik\ell}^2$
S × B	$(s-1)(b-1)$	$\sigma_\omega^2 + t\sigma_{\beta\gamma\psi}^2 + g\sigma_{\beta\tau\psi}^2 + tg\sigma_{\beta\psi}^2$
S × T × B	$(s-1)(t-1)(b-1)$	$\sigma_\omega^2 + g\sigma_{\beta\tau\psi}^2$
S × G × B	$(s-1)(g-1)(b-1)$	$\sigma_\omega^2 + t\sigma_{\beta\gamma\psi}^2$
S × G× T × B	$(s-1)(t-1)(g-1)(b-1)$	σ_ω^2
Total	bgts-1	

The other random terms are all interactions with blocks, and it is important to keep track of them so that we can deduce the proper error terms for estimation and testing.

Typical analyses of these designs, and default computer analyses, will present one error at each level. However, to do this requires some assumptions, which are evident from calculating the expected mean squares (Table 5.10), but are also evident if we recall the EMS of Section 5.3. There, we could pool the errors at the split plot level only if there was no interaction between the split plot treatment and the blocks. From Table 5.10, we see that not only do we require no interaction at the split plot level ($\sigma_{\beta\gamma}^2 = 0$), in order to have one split split plot error there must be no interaction involving the split split plot treatment and the blocks, that is, we must assume that $\sigma_{\beta\psi}^2 = \sigma_{\beta\tau\psi}^2 = \sigma_{\beta\gamma\psi}^2 = 0$. Note, however, that the individual tests resulting from Table 5.10 are valid without any assumptions about the interaction terms.

Example 5.10. SOLUTION DATA SPLIT SPLIT   The standard anova for the solutions data looks like

Source	df	SS	MS	F	p-value
Lab	2	81.492	40.746		
Solution	2	402.82	201.41	88.426	0.0004
Whole Plot Error	4	9.11	2.28		
Day	3	2023.75	674.58	42.210	< .0001
Solution × Day	6	118.23	19.71	1.233	0.3356
Split Plot Error	18	287.67	15.98		
Assay	1	2720.64	2720.64	137.890	< .0001
Solution × Assay	2	186.96	93.48	4.738	0.0184
Day × Assay	3	1904.52	634.84	32.176	< .0001
Solution × Day × Assay	6	71.95	11.99	0.608	0.7210
Split Split Plot Error	24	473.53	19.73		

where there is only one error at each level. We know that both the Split Plot Error and Split Split Plot Error represent a pooled error term. At the split level we must assume that there is no Day × Lab interaction, and at the split split level we must assume that all interactions involving Assay and Lab are zero (see Exercise 5.15).   ‖

The variances of treatment contrasts in the RCB split split plot are also a bit more complicated than their CRD counterparts. We have:

(1) Interaction mean SSP × WP:

$$\text{Var}\left(\sum_{i\ell} a_{i\ell}\bar{Y}_{i\ell}\right) = V_1 \sum_{i\ell} a_{i\ell}^2 + V_2 \sum_i \left(\sum_\ell a_{i\ell}\right)^2,$$

where

$$V_1 = \frac{\sigma_\omega^2 + g\sigma_{\beta\tau\psi}^2}{rg} \text{ and } V_2 = \frac{\sigma_\delta^2 + \sigma_{\beta\gamma\psi}^2 + g\sigma_{\beta\psi}^2 + g\sigma_\varepsilon^2}{rg}$$

and we note that $\sum_\ell a_{i\ell} = 0$ if we have a contrast within the level of the whole plots. Of course, this happens if we are comparing within one level of the WP treatment. If the interaction variance components are assumed to be zero, this variance can be estimated without the need for the Satterthwaite approximation.

(2) Interaction mean: SSP × SP:

$$\text{Var}\left(\sum_{k\ell} a_{k\ell}\bar{Y}_{k\ell}\right) = V_1 \sum_{k\ell} a_{k\ell}^2 + V_2 \sum_k \left(\sum_\ell a_{k\ell}\right)^2,$$

where

$$V_1 = \frac{\sigma_\omega^2 + t\sigma_{\beta\gamma\psi}^2}{rt} \text{ and } V_2 = \frac{\sigma_\delta^2 + \sigma_{\beta\tau\psi}^2 + t\sigma_{\beta\psi}^2 + t\sigma_{\beta\gamma}^2}{rg}$$

where the second term is zero if we are balanced in the SP treatment or are in one level. The estimation of this variance may require the Satterthwaite approximation.

(3) SSP means:

$$\text{Var}\left(\sum_\ell a_\ell \bar{Y}_\ell\right) = V_1 \sum_\ell a_\ell^2,$$

where

$$V_1 = \frac{\sigma_\omega^2 + g\sigma_{\beta\tau\psi}^2 + tg\sigma_{\beta\gamma\psi}^2 + tg\sigma_{\beta\psi}^2}{rtg},$$

where V_1 can be estimated using the S × B interaction. If we assume that the interaction variance components in V_1 are all zero, then we can estimate V_1 using the Split Split Plot Error.

The split split plot design also implies a restriction on the randomization in the experiment. To better understand this we look at the following example.

Example 5.11. RANDOMIZATION PATTERNS

Consider an experiment with three crossed factors, A, B, and C, each at three levels. Keeping the treatment design as a crossed experiment, we can have different experiment designs. In particular, we look at a CRD, a split plot, and a split split plot. The following picture illustrates possible randomization of the first nine observations.

The complete randomization of the CRD allows the observations to be taken in any order. In the split designs, once the whole plot is chosen, the randomization proceeds in there. If $B \times C$ is the split plot treatment then randomization is unrestricted within the whole plot. In the split split plot design the split split treatment is randomized within the levels of the split plot treatment. ‖

5.6 Variations on a Theme

Here we briefly look at three variations of the split plot design. The *strip plot design* reflects a specific type of randomization, one which originated in agricultural situations, while the *crossover design* is a useful variation of the split plot that is more common in experiments on human subjects. The third variation, the *repeated measures design* brings in a new error structure and, in this sense, should be considered a different design.

5.6.1 Strip Plots

The strip plot design, which also has other names, is an experimental design with two crossed treatments. (It is different from the crossed blocks design, discussed in Section 3.6.2.) Other names for this design are *split block*, which may still be the most common but is somewhat of a misnomer, or *crisscross*, which is descriptively the most accurate, but perhaps too whimsical for statisticians.

Example 5.12. STRIP PLOT (SPLIT BLOCK) EXPERIMENT   A sugar planter wanted to determine the effect of potassium and phosphorus fertilizers on the yield of sugarcane. He established an experiment with the following factors:

Factor A = Potassium (K) Levels = 0, 25 kg/ha, 50 kg/ha,
Factor B = Phosphorus (P) Levels = 25 kg/ha, 50 kg/ha.

He decided to use a design with three blocks. Because he wanted to use farm-scale equipment to apply the chemicals, he assigned the potassium rates to 3 strips within blocks. He then assigned phosphorus to strips of 2 plots at right angles to the potassium strips within the blocks. The field plan and yields (kg/plot) are

Block

	I			II			III		
	K3	K1	K2	K1	K3	K2	K2	K1	K3
P1	56	32	49	38	62	50	63	54	68
P2	67	54	58	52	72	64	54	44	51

In the strip plot design, within each block the potassium (K) is randomized on the vertical strips, and the phosphorus (P) is randomized on horizontal strips. The anova for the sugarcane experiment is

Source	df
Blocks	2
K	2
K x B	4
P	1
P x B	2
K x P	2
K x P x B	4
Total	17

Note the difference between this design and a split plot design. If, for example, we ran this as a split plot with K as the whole plot treatment and P as the split plot treatment, then we would have applied one level of K to the entire plot, and then randomized P in the same way. ‖

The treatment design of the strip plot is a twoway crossed design, with model

$$(5.27) \quad Y_{ijk} = \mu + \tau_i + \beta_j + (\beta\tau)_{ij} + \gamma_k + (\beta\gamma)_{jk} + (\tau\gamma)_{ik} + (\beta\tau\gamma)_{ijk} + \varepsilon_{ijk},$$

where $i = 1, \ldots, t$, $j = 1, \ldots, r$, $k = 1, \ldots, g$, τ and γ are the treatments, and

$$\beta_j \sim N(0, \sigma_\beta^2), \quad \varepsilon_{ij} \sim N(0, \sigma_\varepsilon^2),$$

all independent. The Treatment × Block interactions will be the respective error terms. The anova and EMS for this experiment is given in Table 5.11.

where we see that the F-tests for the treatments are against the respective interactions.

The strip plot design actually has three experimental units, which are illustrated in Figure 5.1. Each of the treatments is applied to distinct experimental units, and the interaction treatment has the intersection as its experimental

Table 5.11. Expected mean squares for a strip plot design.

Source	df	EMS
Blocks	$b-1$	$\sigma_\varepsilon^2 + \sigma_{\beta\tau\gamma}^2 + t\sigma_{\beta\gamma}^2 + g\sigma_{\tau\beta}^2 + tg\sigma_\beta^2$
T	$t-1$	$\sigma_\varepsilon^2 + \sigma_{\beta\tau\gamma}^2 + g\sigma_{\tau\beta}^2 + \frac{rg}{t-1}\sum_i \tau_i^2$
Blocks × T	$(b-1)(t-1)$	$\sigma_\varepsilon^2 + \sigma_{\beta\tau\gamma}^2 + g\sigma_{\tau\beta}^2$
G	$g-1$	$\sigma_\varepsilon^2 + \sigma_{\beta\tau\gamma}^2 + t\sigma_{\beta\gamma}^2 + \frac{rt}{g-1}\sum_k \gamma_k^2$
Blocks × G	$(b-1)(g-1)$	$\sigma_\varepsilon^2 + \sigma_{\beta\tau\gamma}^2 + t\sigma_{\beta\gamma}^2$
T × G	$(g-1)(t-1)$	$\sigma_\varepsilon^2 + \sigma_{\beta\tau\gamma}^2 + \frac{r}{(t-1)(g-1)}\sum_{ik} \tau\gamma_{ik}^2$
Blocks × T × G	$(b-1)(t-1)(g-1)$	$\sigma_\varepsilon^2 + \sigma_{\beta\tau\gamma}^2$
Total	bgt-1	

Fig. 5.1. The three experimental units of a strip plot design. The treatments are applied to rows and columns, and the interaction experimental unit is the intersection.

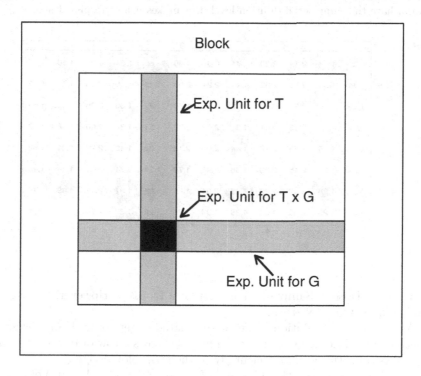

unit. It is also the case that the correlation structure is different for the treatments and interaction, with higher correlation in the interaction (since those observations are in the same row and column (see Exercise 5.41).

Although strip plot designs were originally developed to accommodate treatments applied with farm-scale equipment, the next example shows that these designs have a place in the 21st century.

Example 5.13. STRIP PLOT BIOASSAY   David Lansky, of Precision Bioassay, shares this experimental design. Companies use bioassays to measure the functional activity of protein products because the chemical analytic methods are not completely sufficient. Biological assays, which measure function, are typically complex, slow, costly, and yield low-precision estimates of activity. Imagine using a corn field to measure each batch of fertilizer!

Cells grown in culture are often sensitive to subtle features in the environment, and may grow better on one side of the plate than another. Thus, if a design such as a CRD is used, we must use randomization to control this variation and avoid grouped dilution or serial dilution when constructing

Fig. 5.2. One block of the strip plot bioassay experiment of Example 5.13 showing the "field" layout. The rows indicate the samples (reference, half, double, or one), and the columns indicate the serial dilution level (dose). Note that all samples in a column have the same serial dilution level; this makes it a strip plot design.

		block : C											
ref2	A	1.21	1.27	2.14	2.20	1.44	2.07	1.89	1.32	1.31	2.09	1.50	1.59
hlf2	B	1.34	1.33	2.35	2.32	1.41	2.24	2.03	1.28	1.40	2.43	1.73	1.75
one2	C	1.22	1.35	2.21	2.16	1.36	2.38	1.88	1.30	1.36	2.08	1.56	1.58
dub2	D	1.27	1.38	2.25	2.41	1.38	2.02	1.72	1.21	1.32	2.06	1.38	1.52
hlf1	E	1.37	1.45	2.36	2.26	1.44	2.06	2.18	1.30	1.42	2.26	1.81	1.62
dub1	F	1.24	1.35	2.18	2.36	1.35	1.90	1.73	1.23	1.31	2.11	1.46	1.45
ref1	G	1.27	1.38	2.42	2.22	1.43	2.04	1.85	1.34	1.37	2.11	1.49	1.48
one1	H	1.28	1.20	2.02	1.93	1.35	1.93	1.88	1.26	1.32	1.93	1.44	1.51
		1	2	3	4	5	6	7	8	9	10	11	12

the factorial Dose × Sample design used for a bioassay. However, this can be handled in a strip plot design.

As samples (or dilutions) are often treated together with multichannel pipettes, this places a solution *simultaneously* across a row or down a column. This physical setup leads us naturally to the strip plot design.

Figure 5.2 shows the "field layout" of one block (a 96-well plate) of a three block experiment from a cell culture bioassay run in a strip plot design using three blocks (96-well plates), four samples and 12 doses of each sample. The response is a measurement of optical density, the color intensity of a dye, and it is probably best to take logs before any analysis. The four samples are reference, one, half, and dub. These are all sugar, and the samples one, half, and dub are made from reference with relative concentrations of 1, 1/2, and 2. There are two rows on each assay plate randomly assigned to each dilution. One column on each plate is randomly assigned to each of the 12 serial dilutions. The serial dilutions are done using a multichannel pipette, separately on each plate, which applies the dilution to the entire column *simultaneously*.

Note that there are 12 columns corresponding to the dilutions, and 8 rows for two replications of the four treatments. The anova, as in Table 5.11, can be partitioned in different ways, and here is one option.

Source	df
Blocks	2
Dilutions (Columns)	11
D × B	22
Rows	7
Samples	3
Rows(in Samples)	4
Rows × Blocks	14
S × B	6
Rows(in Samples) × B	8
D × Rows	77
D × S	33
D × Rows(in Samples)	44
D × Rows × B	154
D × S× B	66
D × Rows(in Samples)× B	88
Total	287

Note that this partitioning of the sum of squares implicitly assumes that rows are nested in samples but crossed with blocks (see Exercise 5.18).   ‖

5.6.2 Crossover Designs

The crossover design combines a bit of everything - RCB, Strip plot, Latin square, and provides a means of getting tighter control on differences at the cost of an assumption on the order of treatments. In this section we look at two different crossover designs.

Simple Crossover Designs

The simplest case, which is the most popular implementation, is the two-period (P_1, P_2) Simple Crossover Design (SCOD) with two groups (G_1, G_2) and two treatments (T_1, T_2) which can be displayed as

$$
\begin{array}{c|c|c|}
 & G_1 & G_2 \\
\hline
P_1 & T_1 & T_2 \\
\hline
P_2 & T_2 & T_1 \\
\hline
\end{array}
\quad \text{or} \quad
\begin{array}{c|c|c|}
 & T_1 & T_2 \\
\hline
P_1 & G_1 & G_2 \\
\hline
P_2 & G_2 & G_1 \\
\hline
\end{array}
$$

From the picture the name of the design should be obvious. The groups are "crossed over" to the other treatment so each group receives both treatments, but in opposite orders. Thus, each group is its own control, allowing us to use fewer subjects and still get good precision (which makes this design quite popular in drug studies involving human subjects). So we are saving observations but retaining tight comparisons (greater precision) for the treatment differences – what did we give up?

The treatments are given in both orders to control for any order effect, with a *washout period* between treatments. During this period, it is assumed that the effect of the first treatment "washes out", and the subject starts the second period without any residual effect from the first treatment, an assumption that there is no *carryover* effect. That is, we assume that the groups start P_2 equivalent to the start of P_1. It is important to note that this is an assumption about interactions, not main effects, as we will see below.

The SCOD is, in fact, a split plot design with the subjects as the whole plots (subjects are the experimental units) and the treatment as the split plot treatment. The whole plot treatment is the order in which the treatments are given, and with s subjects in each group we have the anova

Source	df
Order	1
Subjects (in Order)	$s - 2$
Period	1
P × O (Treatments)	1
P × Subjects (in Order)	$s - 2$

Note: This is a split plot design with a CRD on the whole plots.

The treatment test is totally confounded with the Period × Order interaction (same sum of squares), so to use this design to test treatment, we need to assume that this interaction is zero.

Example 5.14. EXERCISE CROSSOVER DESIGN   The following study was done by Belko *et al.* (1984) to study the effects of aerobic exercise and weight loss on riboflavin requirements of moderately obese marginally deficient young women. The study was conducted as a two-period crossover design with 12 subjects allocated to one of two exercise sequences (NE/E or E/NE, where NE=no exercise, E=exercise in the form of 50 minutes of aerobics 5 days per week) in a CRD with 6 subjects per sequence. A two week baseline period preceded the actual crossover. All subjects received a diet with 0.8 mg riboflavin per 1000 kcal. One response variable of interest was the urinary excretion of riboflavin (UrRibo) expressed as a percentage of intake. The data look like

Order	Subject	UrRibo	Order	Subject	UrRibo
NE	1	29.5	E	7	14.0
Period 1	⋮	⋮		⋮	⋮
	6	20.4		12	15.0
E	1	31.6	NE	7	26.3
Period 2	⋮	⋮		⋮	⋮
	6	11.3		12	27.8

and are given in dataset Riboflavin.

The anova is

Source	df	SS	MS	F	p-value
Order	1	217.20	217.20	1.489	0.250
Subjects (in Order)	10	1457.95	145.79		
Period	1	3.53	3.53	0.059	0.813
Period × Order (Trt)	1	835.44	835.44	14.034	0.004
Split Plot Error	10	595.28	59.53		

where we see that the effect of order is not significant, so the effect of being first was the same for each treatment. The Period × Order interaction is quite significant, and this effect is the same as the Treatment effect. Thus, if we are willing to assume that there is no Period × Order interaction, we can conclude that there is a significant treatment effect. Of course, this is an assumption that we must make, as there is no way to test these effects separately.   ‖

Some observations about the SCOD:

(1) The test on order is testing the carryover effect. Both groups have received both treatments, so the only difference is the order in which the treatments are given. If order is significant, this means that the carryover from T_1 is different from the carryover from T_2. (Nonsignificance does not mean there is no carryover, but, rather, equal carryover.)

(2) The carryover, or order, test, is at the whole plot level, so it is the test of lesser precision in this design.

(3) The conclusions about treatments are affected by order
  (a) If order is not significant, then we can make conclusions about T_1 versus T_2.
  (b) If order is significant, then we can only make conclusions about T_1 after T_2 versus T_2 after T_1. In this case, some have recommended discarding Period 2 and doing a two-sample t-test, but this violates our rule of never discarding data.

(4) All of the tests in the SCOD anova are, in fact, t-tests, so the entire SCOD can be done this way (see Exercise 5.22).

There is a connection between crossover designs and Latin squares that is really more about balancing the order than Latin squares. Suppose that we rewrote the data layout for the experiment of Example 5.14 in the following way:

	Subjects 1 7			Subjects 2 8				Subjects 6 12	
Period 1	NE	E	Period 1	NE	E	......	Period 1	NE	E
Period 2	E	NE	Period 2	E	NE		Period 2	E	NE

Here we see that we can rearrange the data into 2×2 Latin squares. In fact, as the periods are the same, this is a Latin rectangle (see the discussion at Example 3.15). This, of course, gets us nothing as the analysis is the same as before – identifying the squares just renames some variability but the analysis is the same.

Remember, the analysis is dictated by the experiment design that was run. Just because we can write the data layout as a Latin square does not mean we should analyze it that way. In fact, unless it was run as a Latin square, such an analysis would be wrong.

What this does show, however, is that the order is balanced in that the same number of subjects get order NE-E as get order E-NE. Without this balance we could not separate the order effect from the period effect, as well as the treatment effect. Moreover, we could examine the partitioning of the sums of squares to see, in each design (crossover with or without Latin square) where the degrees of freedom go, and for future reference, what might be the better way of running the experiment (see Exercise 5.21).

Three-Period Crossover

We briefly move out of the realm of the simple crossover design and discuss three-period crossovers, where there are three treatments. As we will see, the crossover design starts to get unwieldy here, and moving beyond three treatments is probably not a good idea. The necessity of accommodating multiple washout periods and the assumption of no carryover effect become tenuous as the number of treatments increases. However, the three period crossover is a reasonable extension.

The consideration of balancing for order becomes more important when there are more than two treatments. In a crossover design with 3 treatments there are $3 \times 2 \times 1 = 6$ orders which, with six subjects, we could arrange as

		Subjects 1 2 3 4 5 6
	1	a b c c a b
Period	2	b c a b c a
	3	c a b a b c

Here we have balanced all six orders, by giving one to each subject. Note that with four treatment there are 12 orders, and with 5 treatments there are 60 orders, so things quickly get out of hand.

Example 5.15. THREE-PERIOD CROSSOVER

Six subjects have been recruited for an alcohol-drug study, in which three drugs were investigated for mitigating effects of alcohol on reaction time The three drugs (A,B,C) will be given to the subjects in three time periods, with order balanced as in the above diagram. The data are in dataset `Alcohol` and result in the following anova

Source	df	SS
Order	5	6252.4
Period	2	1053.8
Period × Order	10	13056.2
Drug	2	2276.8
Residual	8	10779.4

where we see that the effect of the treatment is now only a piece of the Period × Order interaction. Note that in this design Subjects and Order are completely confounded and the design is, in fact, an RCB and not a split plot.   ‖

Notice that the design on the first three subjects is a 3×3 Latin square, as is the design on the second three subjects. Moreover, these Latin squares are orthogonal (see Section 3.6.3). We can examine the effectiveness of using Latin squares, as we did for the simple crossover, and we leave that to Exercise 5.23.

Whole plot treatments can also be accommodated in the three-period crossover. If there are multiple subjects and the order is modeled as a whole plot treatment, we are back in the case of a split plot design. Also, there can be an actual whole plot treatment, and we again have a split plot design. See Exercise 5.24 for these variations.

Note: For the inferences to be valid, the covariance structure of the crossover designs must be the same as was assumed for the split plot.

That is, we assume that there is equicorrelation in the split plot treatment responses (see Technical Note 5.8.2).

For the two-period crossover design this is satisfied, but seems less reasonable assumption for the three-period crossover. This is because as the observations get further apart in time, we might expect the correlation to decrease. As we look at the next variation, the equicorrelation assumption seems even less reasonable.

5.6.3 Repeated Measures

In a repeated measures design, we typically take multiple measurements on a subject over time. If any treatment is applied to the subjects, they immediately become the whole plots, and the treatment "Time" is the split plot treatment.

Example 5.16. HYPERTENSION   Levey *et al.* (1995) did an experiment to compare the blood pressure responses of 22 white men with moderate hypertension to two six-week metabolic diets. The diets were low in sodium (Na), but varied in calcium (Ca) intake from food, with one diet averaging 1,400 mg/day (High Ca) and the other averaging 400 mg/day (Low Ca). A portion of these data are in dataset **Hypertension**, and look like

Treatment	Subject	Time		
		1	2	3
	1	133	141	100
HighCa	⋮	⋮	⋮	⋮
	5	171	142	128
	6	104	139	153
LowCa	⋮	⋮	⋮	⋮
	10	147	167	157

Blood pressure was measured at the beginning, midpoint, and end of the six-week study.

It is clear that this is a split plot design with Subjects as the whole plots, Diet being the whole plot treatment, and "Time" being the split plot treatment. The 3 measurements on each subject are obviously correlated, however it is difficult to assume that the correlation structure is as in Section 5.2.1, that is, that the correlation between any two observations (within a subject) are equicorrelated.   ‖

For measurements over time, it is often more reasonable is to assume that the correlation decreases over time, that is, if k indexes time, a plausible correlation model is

$$(5.28) \qquad \mathrm{Corr}(Y_{ijk}, Y_{ijk'}) = \rho^{|k-k'|} ,$$

so the correlation decreases as the observations are further apart in time. This is known as an *AR(1)* (Auto Regressive) correlation structure.

Although this is a plausible correlation structure for a time-dependent response, this covariance structure is too general to satisfy Cochran's Theorem, and thus the ratios of mean squares do not have F-distributions. There are a number of ways to approach this problem:

(1) Note that if there are only two repeated measures, then there is equicorrelation (as there is only one correlation), Cochran's theorem is satisfied,

and the usual F-tests are valid. This suggests that if there are more than two repeated measures, we can look at contrasts and obtain valid tests (see Exercise 5.27).

(2) We can use an approximate F-test. Such tests are usually conservative, but are quite simple to implement.

(3) The multivariate procedure known as *Hotelling's T^2* will provide a valid test against any covariance structure. However, there is typically a substantial loss of power when moving to multivariate tests.

(4) The repeated measures can be summarized into one measurement such as an average, or a slope of a regression line. Since there is now only one measurement per subject, the correlation problem disappears.

(5) In the design of the experiment, different subjects could be used in each time group. This eliminates the correlation problem as the observations are now independent but can lead to a large variance increase, as each subject is no longer his own control.

Huynh and Feldt (1970) have derived general conditions for valid F-ratios in repeated measures designs, and we look at those and other F approximations and tests in Miscellanea 5.9.2.

Example 5.17. HYPERTENSION CONTINUED   For the experiment of Example 5.16, the anova is

Source	df	SS	MS	F	p-value
Treatment	1	1153.2	1153.2	1.5981	0.2418
Whole Plot Error	8	5772.9	721.6		
Time	2	343.3	171.6		
Trt × Time	2	5028.2	2514.1		
Split Plot Error	16	1771.9	110.7		

The whole plot test is a valid F-test, as the design is a CRD on the whole plots. Recall that the whole plot analysis is done on the sums over the split plot treatment (see Section 5.1.2). At the split plot level, we can take one of the approaches that is outlined above.

As an illustration, suppose that the experiment is most interested in assessing the change in blood pressure over time, as a function of the treatments. To summarize this, we can fit a linear regression to each subject and use the slope coefficients as input. Note that we are not assuming that the response is linear, but only using the linear regression to give a summary of the trend.

The ten slopes are

Slopes for each subject

	HighCa					LowCa			
1	2	3	4	5	6	7	8	9	10
−16.5	−14.0	−3.5	−7.5	−21.5	24.5	25.5	18.5	22.0	5.0

and the oneway CRD anova on the slopes yields

Source	df	SS	MS	F	p-value
Treatments	1	2512.23	2512.23	41.619	0.0002
Within	8	482.90	60.36		

This is a valid anova – the subjects are independent and the normality assumption is not unreasonable. Slope estimates are averages, so we are a bit more comfortable with this assumption although, in this case, we only have three observations in each slope. The anova on the slopes is very significant, showing that the responses to the diets, when measured in blood pressure trend, are significantly different.   ‖

5.7 Exercises

Essential

5.1 The whole plot factor A is allocated in a CRD (3 observations/treatment), and the split plot factor B is in a CRD within A.

		A					
		0			1		
	Obs.	1	2	3	1	2	3
B	0	12	15	12	14	12	17
	1	11	13	10	12	11	14

(a) Analyze the effects of A and B using a standard split plot analysis.
(b) Analyze the effects of A and B using the appropriate paired and two-sample t-tests.
(c) Comment on the similarities or differences in parts (a) and (b).

5.2 Referring to Section 5.1.2:
(a) Recreate the three anova tables corresponding to the complete split plot analysis, the above the line analysis, and the below the line analysis. The data are in dataset `Diet`. In particular, explain why the sums of squares in the oneway CRD are half of what they are in the complete split plot analysis
(b) Show that the sum of squares due to Time is equal to $n\bar{d}^2$, where $\bar{d}$ is the mean of the differences and $n = 24$. Why should we use 24 when there are 12 differences?
(c) Calculate the anova table for an RCB where the subjects are blocks, and Time is the treatment (so we ignore Diets). For this design the Time × Subject interaction has 11 degrees of freedom. Show (numerically, or algebraically if you really love algebra) that

$$\text{SS(Time} \times \text{Subject)} = \text{SS(Diets} \times \text{Time)} + \text{SS(Time} \times \text{Subjects(in Diets))}.$$

5.3 Brogan and Kutner (1980) analyze data from a prospective randomized surgical trial allocating cirrhotic patients to one of two treatments: to a nonselective shunt (standard operation) or to a selective shunt (new operation). The response variable is the maximal rate of urea synthesis, where low values are associated with poor live function. The response was measured before and after surgery. The data are given in dataset **Shunt**, and look like

Subject	Treatment	Pre	Post
1	Selective	51	48
⋮	⋮	⋮	⋮
8	Selective	42	54
9	Nonselective	34	16
⋮	⋮	⋮	⋮
21	Nonselective	43	32

(a) Analyze these data in a split plot anova. Are you comfortable with the assumptions?

(b) Each of the three anova tests can be done as a t-test (Section 5.1.2). Verify the following:

    (i) The whole plot test is equal to a two-sample t-test with response equal to Pre + Post.

    (ii) The interaction test is equal to a two-sample t-test with response equal to Pre − Post.

    (iii) The split plot test can be done as a two-sample t-test with response equal to Pre − Post in one group and Post − Pre in the other group. However, this is not equal to the anova test unless the group sizes are equal. Explain the discrepancy.

5.4 The Federal Plant Soils Nutrition Laboratory conducted a study on zinc uptake in nutrient solution culture under controlled greenhouse conditions. Four covered tanks (large water baths) were maintained at constant temperature, two tanks at 10°C and two tanks at 20°C. The top of the tank, and the top of the pots, were covered so that the roots of the plants, growing in the nutrient solution, were maintained at the 10° or 20° temperature. The air temperature for all tops was maintained at a constant temperature for all tanks. Within each tank there were six pots with four barley plants in each pot. The nutrient solution in each pot was maintained at constant levels of all essential nutrients except zinc (0, 0.03, 0.33 μM $ZnSO_4$). The four barley plants in each pot were harvested after 20 days and the fraction of total zinc in the plant that had been translocated from the roots to the tops was determined.

(a) Write the model for a split plot analysis.

(b) Write the anova table (source, df and EMS), and indicate all tests.

(c) Describe a set of contrasts that address the factors under study.

5.5 An experimenter in the College of Medicine and the University of Florida was interested in comparing the performance of a new type of catheter (a triple lumin catheter) to the standard type. (A catheter is a tube that can be inserted into a body cavity duct to allow drainage or injection of fluids). The response to be measured is the pressure inside the catheter (measured in millimeters of Hg), which is measured by sensors placed at two points along the catheter,

Distal and Proximate. For each patient the measurements were to be taken at two organ ducts, one near the pancreas and one in the biliary duct of the liver.

For each patient, the treatment design is a $2 \times 2 \times 2$ factorial: Organ Ducts $\times$ Catheter Type $\times$ Sensor Location. For the 30 patients available for the study, each patient will have these 8 pressure measurements taken at each treatment combination during their surgery. There are a number of ways to carry out the randomization - here are three:

(1) Randomize throughout the $2 \times 2 \times 2$ factorial.
(2) Choose an organ duct, then randomize throughout Catheter $\times$ Sensor combinations.
(3) Choose a catheter type, randomize the organs within catheter, then randomize sensor in organ.

(a) Identify each of the three designs.
(b) For each design, write the anova table and indicate all tests.
(c) Which design would you recommend if the experimenter is equally interested in all treatment effects? Explain.
(d) Which design would you recommend if the experimenter is mainly interested in sensor effects? Explain.
(e) In fact, the experimenter did method (3) (without consulting a statistician), and was mainly interested in the effect of catheters. Was this a good choice of design? If not, suggest a better one and defend your choice.

5.6 Referring to Example 5.2, for each of the following four interaction contrasts, give estimates of variance and 90% confidence intervals. Use Satterthwaite where needed.

Different WP			Interaction			Interaction			Different WP		
	AM	PM		AM	PM		AM	PM		AM	PM
1	1	0	1	1	0	1	1	−1	1	−1	−1
2	0	−1	2	0	−1	2	0	0	2	−1	−1
3	0	0	3	−1	0	3	−1	1	3	−1	−1
4	0	0	4	0	1	4	0	0	4	3	3

5.7 In Section 5.2.1 we saw the correlation structure for the split plot design with whole plots in a CRD, and it is relatively simple. The correlation in the split plot with an RCB on the whole plots is a bit more complicated. For that design, show that:
(a) $\text{Cov}(Y_{ijk}, Y_{ij'k}) = 0$ unless $j = j'$.
(b) Verify the following table of covariances:

	Same WP	Different WP
Same SP	−	$\sigma_\beta^2 + \sigma_{\beta\gamma}^2$
Different SP	$\sigma_\beta^2 + \sigma_\varepsilon^2$	σ_β^2

Which of these covariances do you expect to be greater? Explain.

5.8 Referring to Example 5.5:
(a) Verify the anova table.
(b) Perform the F-tests at the split plot level using the pooled split plot error. Compare the results of your tests to those obtained in the example. Discuss the assumptions that you need to do this.
(c) Test the null hypotheses corresponding to the contrasts given in the example.

(d) The variety "Narragannsut" is the most recently developed. Give the contrast coefficients for comparing Narragannsut against the average of the other varieties, and give the contrast for testing whether this effect interacts with the levels of P.

(e) Perform the tests for the contrasts in (c).

5.9 A dataset presented by Hicks (1993) has to do with the quality of electrical components that are baked during manufacture. The data are

	Oven Temp ° F	
Baking	580	620
Time 5	217, 188	229, 160
(minutes) 15	175, 195	155, 161

(a) The above data layout does not tell us the design. If this experiment were done as a CRD, explain in detail how the observations would have to be taken. For example, if the first observation was at Oven Temp $= 620$ and Baking Time $= 15$, what has to be done before the next observation, at Oven Temp $= 580$ and Baking Time $= 15$ is taken. Is this a good way to run the experiment?

(b) A more informative data layout is

	580	620			580	620
Rep 1 5	217	229	Rep 2 5	188	160
15	175	155	15	195	161

which suggests that the experiment was blocked. Explain how to take the data if this is a (i) RCB or (ii) split plot. If the experiment were run as a split plot, what is the better choice for the whole plot treatment (from a practical standpoint).

(c) Write out the anova table (Source and df) for all three designs. Show the correspondence between the terms in the anovas. What design would you recommend and why?

(d) For each of the three designs (CRD, RCB, split plot) calculate the correlation between (i) two observations in the same Time, different Temp (ii) two observations in the same Temp, different Time. Does this information change your answer in (b)?

5.10 A variation on the data in Exercise 5.9 results in the following split-split plot design. Suppose that the experiment, detailed in part (b) of Exercise 5.9, was run three separate times, once in each of three labs. The labs now become the whole plot treatment, and Time and Temperature are now the split plot and split split plot treatment, respectively.

(a) Write down the anova (source and df) for the experiment done in three labs, where each lab does the experiment detailed in part (b) of Exercise 5.9. Are the whole plot treatments in a CRD or in an RCB? Indicate how you would do the F-tests.

(b) Using dataset **Oven**, analyze the data. Give a confidence interval for the Time $\times$ Temp interaction.

5.11 Referring to the discussion at the beginning of Section 5.4:

(a) Show that for estimating μ, τ_i, or γ_k, contrasts in cell means or effects are exactly the same.

(b) Referring to Example 5.2, calculate the variance of all four contrasts when using (i) cell means and (ii) effects. Are the effect variances always smaller? Explain.

(c) Referring to Example 5.5, calculate the variance of both contrasts when using (i) cell means and (ii) effects. Are the effect variances always smaller? Explain.

5.12 Referring to Examples 5.7 and 5.8:

(a) Reproduce the anova table (the data are in dataset `OzoneSSP`).

(b) Show how to estimate the variance of a Ozone $\times$ Location interaction contrast and a Variety $\times$ Location interaction contrast. Give examples of contrasts that only use the split split plot error term.

(c) The split split plot error can also be calculated as a pooled split plot error. For each level of Variety, the design is a split plot (WP = Ozone, SP = Location). For each of these six designs, calculate the split plot error, and show that pooling these errors yields the split split plot error of the original design.

5.13 Recall that in a microarray experiment, the expression levels of genes are measured through the use of "probes" or "probe sets" (Exercise 3.13). The probes are sections of the gene, and two probes within the same gene might display different expression levels. An experimenter wants to exploit this phenomenon in the following sense. If we measure the same genes on a number of different species (or subjects or varieties), and detect a Variety $\times$ Probe interaction within a gene, this may be evidence of an allele difference (for example, Aa versus AA). Thus, an experiment was designed to test this effect.

The following schematic illustrates such a design, with varieties as the whole plot treatment, genes as the split plot treatment, and probes at the split split plot level.

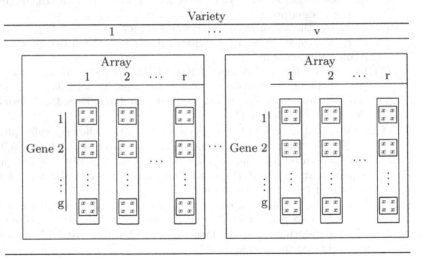

We see that Variety is the whole plot treatment, applied to the arrays, which are the whole plots. The Genes are the split plot treatment, as they are within each array. Furthermore, the Probes, where the probe set for each gene is represented by the small box with four xs, is *nested* within the genes (as each gene has its own set of probes) but crossed with the varieties.

(a) If there are v Varieties, r Arrays, g Genes, and p Probes, a model for this analysis is

$$Y_{ijk\ell} = \mu + V_i + \varepsilon_{ij} + G_k + (VG)_{ik} + \delta_{ijk} + P_{k\ell} + (VP)_{ik\ell} + \xi_{ijk\ell}$$

$$i = 1, \ldots, v, \quad j = 1, \ldots, r, \quad i = k, \ldots, n, \quad \ell = 1, \ldots, p.$$

Show that the anova for this layout is

Source	df
V	$v - 1$
Whole Plot Error	$v(r - 1)$
G	$g - 1$
G × V	$(g - 1)(v - 1)$
(G × A) in V	$(g - 1)(r - 1)v$
P in G	$(p - 1)g$
(P × V) in G	$(p - 1)(v - 1)g$
(P in G) × (A in V)	$(p - 1)(r - 1)gv$

Indicate how you would perform the tests (what are the appropriate denominators?)

(b) The data can be found in dataset **ArraySSP**. Verify the following anova table (recall that with expression data we typically take logs).

Source	df	SS	MS	F	p-value
Variety	2	201.007	100.503	607.53	0.0001
Whole Plot Error	3	0.496	0.165		
Gene	9	109.689	12.188	92.958	< .0001
Variety × Gene	18	75.953	4.220	32.184	< .0001
Split Plot Error	27	3.540	0.131		
Probe	30	9.6115	0.3204	2.2834	0.0015
Variety × Probe	60	13.5920	0.2265	1.6145	0.0195
Split Split Plot Error	90	12.6281	0.1403		

(c) The significance of the Variety × Probe interaction indicates that there may be an allele difference for some genes. Derive the error term that would be used for estimating an interaction contrast between probes in two different varieties

(d) Calculate the expected mean squares for the anova, to justify the tests. (They should be similar to Table 5.9.)

5.14 Referring to Example 5.8:

(a) Estimate the variance of a *difference* in two means (*i*) at the split split plot level within, (*ii*) at the split split plot level within the same level of the split plot treatment, (*iii*) at the split split plot level but different levels of the split plot treatment

(b) Estimate the variance of a general contrast in the split split plot means, and the interaction means (SSP × WP and SSP × SP). Give examples of SSP × SP contrasts where Satterthwaite is and is not needed.

5.15 Referring to Example 5.9:

(a) Write out a model equation for the split split plot design.

(b) Analyze these data as in Example 5.10. That is, verify the anova table (use the dataset **SolutionSSP**).

(c) Analyze these data according to the EMS of Table 5.10, That is, do not pool the error terms but rather use the individual errors. Is the analysis in part (b) justified, that is, are you willing to conclude that the appropriate interaction terms are zero?

5.16 Referring to Example 5.10:

(a) Estimate the variance of a *difference* in two means (i) at the split split plot level within, (ii) at the split split plot level within the same level of the split plot treatment, (iii) at the split split plot level but different levels of the split plot treatment.

(b) Estimate the variance of a general contrast in the split split plot means, and the interaction means (SSP × WP and SSP × SP). Give examples of SSP × SP contrasts where Satterthwaite is and is not needed.

5.17 Referring to Example 5.12:

(a) Explain why if K is ignored, this is an RCB on P, and if P is ignored, this is an RCB on K.

(b) Analyze the above data (in dataset **Sugarcane**) according to the strip plot design. What are your conclusions?

(c) The experimenter would like to know which treatment combination gives the highest yield. What is your answer? Can you justify it?

(d) If P were randomized within the levels of K this would be a split plot design. Reanalyze the data as a split plot, producing the anova table and relevant tests. Do any conclusions change?

5.18 Referring to Example 5.13 and using the data in dataset **Bioassay**:

(a) If we only identify Rows, and not Samples, this is a simple strip plot design. Verify the following anova table. The data are in dataset **Bioassay**, and logs should be take before any analysis.

Source	df	SS
Blocks	2	0.0295
Dilution (Columns)	11	12.8960
Block × Dilution	22	0.2684
Rows	7	0.2714
Block × Rows	14	0.1827
Dilution × Rows	77	0.2936
Dilution × Rows × Block	154	0.3387

Now identify the factor "Samples" and partition the sums of squares further according to the table in Example 5.13.

(b) The second anova table in part (a) actually assumes that Rows are nested in Samples, but crossed with Blocks. It is, perhaps, more realistic to assume that Rows are nested in Samples × Blocks, which would result from modeling Rows as a random factor. Under this assumption, the anova table in Example 5.13 is incorrect, and the partitioning of the sums of squares should be as follows:

Source	df
Blocks	2
Dilutions (Columns)	11
D × B	22
Samples	3
Samples × Blocks	6
Rows(in Samples × Blocks)	12
D × Samples	33
D × Samples × Blocks	66
D × Rows(in Samples × Blocks)	132
Total	287

To calculate the sums of squares for this table, verify (numerically or algebraically) the relationships

$$\text{SS}(\text{Rows(in S} \times \text{B)}) = \text{SS}(\text{Rows(in S)}) + \text{SS}(\text{Rows(in S)} \times \text{B}),$$
$$\text{SS}(\text{D} \times \text{Rows(in S} \times \text{B)}) = \text{SS}(\text{D} \times \text{Rows(in S)}) + \text{SS}(\text{D} \times \text{Rows(in S)} \times \text{B}).$$

Explain why it is reasonable to refer to the term D × Rows(in Samples × Blocks) as a within error.

(c) Unfortunately, the response in this experiment is highly nonlinear, even after taking logs, so the ordinary linear-type anova tests are not appropriate. However, we can do pairwise comparisons of the sample dilutions 1/2, 1, and 2, each against the reference. Do these tests and justify your choice of error term.

(d) One appropriate analysis is to model the dilution effect with a four parameter logistic curve (see Ratkowsky and Reedy 1986 for an interpretation of the parameters)

$$y = \beta_0 + \frac{\beta_1}{1 + \exp(\beta_2 - \beta_3 x)}.$$

The factor "Dilutions" is ordered in the amount of dilution. The dataset **Bioassay** has the correct ordering of the dilutions from 1 to 12, but realize that this will not match up with the data presented in Figure 5.2, which is a field layout with the dilutions randomized. Fit the logistic curves to each of the samples, and evaluate whether the asymptotes are the same (see Miscellanea 5.9.2, but note that this is *not* a repeated measures design).

5.19 Referring to Example 5.14:

(a) Verify the anova table. Show that the sum of squares for Period × Order is the same as the sum of squares for treatments.

(b) Is there any evidence of carryover effects?

(c) Is there any evidence of a trend due to time? Explain.

(d) Compare the mean effect of exercise versus no exercise on urinary riboflavin. Discuss the reasonableness of the assumption that this comparison is free of carryover effects?

(e) Compute the comparison of exercise versus no exercise for the first period only. Is this difference significant if we only use the error term from the first period data?

(f) Suppose 24 subjects had been available. Suggest an alternate design.

(g) Compare your alternative design to the one that was done (anova, tests). List two advantages and two disadvantages.

5.20 Referring to Exercise 5.19, there were, in addition, baseline urinary riboflavin values (UrRibo):

Subject	1	2	3	4	5	6	7	8	9	10	11	12
Baseline UrRibo	59.9	59.4	45.8	45.1	32.3	45.7	37.5	68.3	31.0	51.4	42.5	56.5

Subjects 1-6 and 7-12 correspond to Groups I and II (sequences NE/E and E/NE)

(a) Do the two groups have similar baselines? Do you think this might affect the analysis?

(b) Explain why baseline would be an appropriate covariate.

(c) Redo the analysis of Exercise 5.19 using Baseline UrRibo as a covariate. Do the baseline observations improve your test for carryover effects?

5.21 Referring to Section 5.6.2:

(a) If the experiment of Exercise 5.14 were run in Latin squares, show that the anova would be

Source	df
Squares	5
Subjects (in Squares)	6
Period	1
Period × Square	5
Period × Subject (in Squares)	6

and use dataset `Riboflavin` to fill in in the sums of squares.

(b) Show that

$$SS(Squares) = SS(Order) + SS(Squares \times Order)$$

and

$$SS(Period \times Subject (in Squares)) = SS(Period \times Order)$$
$$+SS(Period \times Squares \times Order)$$

with SS(Period × Order) = SS(Treatments). Construct an anova table with sources, df and sums of squares that contains all of the sources and shows how they are related.

(c) Based on the breakdowns of the degrees of freedom, discuss the conditions, and the variabilities of the sources, that would make it more advantageous to run the experiment in Latin squares.

5.22 Two drugs, A and B, are given to each of six subjects, with the goal of assessing the availability of the drug in the blood. The response is AUC (Area Under the Concentration curve). The data are

	Order					
	1			2		
	Subjects			Subjects		
	1	2	3	4	5	6
Period 1	A 51.9	35.1	38.6	B 50.8	41.1	39.1
Period 2	B 43.5	45.4	35.4	A 44.2	33.4	32.7

(a) Give the anova table for the this design, and do the appropriate tests. (Note that it is common to take logs of the AUC readings.)

(b) Show that the test on Order is the same as a two-sample t-test, where the response is the sum of the responses of each subject on the two treatments.

(c) Show that the test on Period is the same as a two-sample t-test, where the response is the difference of the responses of each subject on the two treatments.

(d) Show that the test on Treatments (or P × O) is the same as a two-sample t-test, where the response is the difference of the responses of each subject on the two periods.

(e) One advantage of doing the analysis as a sequence of t-tests is that we can accommodate the unequal variance situation. For each of (b)-(d) assess the assumption of equality of variances and, if you think it necessary, calculation an approximate t-statistics that does not use the equal variance assumption.

5.23 Referring to the three-period crossover design of Section 5.6.2, we noted that the orderings comprised two 3 × 3 Latin squares with data layout

		Subjects			Subjects		
		1	2	3	4	5	6
	1	a	b	c	c	a	b
Period	2	b	c	a	b	c	a
	3	c	a	b	a	b	c

Use dataset `Alcohol` to produce the anova table for this data layout and compare it to the anova in Exercise 5.15. Show how the sums of squares are related, and where the test on treatments comes from in each design. Based on the breakdowns of the degrees of freedom, discuss the conditions and the variabilities of the sources, that would make it more advantageous to run the experiment in Latin squares.

5.24 We also noted that in Example 5.15, Order and Subjects are confounded, and this can be remedied by having more than one subject in an order. Moreover, there could be a further whole plot treatment in the design.

(a) Consider adding Subjects 7-12 to the experiment under the following conditions

  (*i*) The design on Subjects 7-12 is a replication of that on Subjects 1-6, so Order is now a whole plot treatment with two subjects per order.

  (*ii*) In addition to what is done in (*i*), Subjects 1-6 receive one level of an additional whole plot treatment, while Subjects 7-12 receive a difference level of the treatment, so we have

Treatments

	1							2					
	Subjects							Subjects					
	1	2	3	4	5	6		7	8	9	10	11	12
1	a	b	c	c	a	b	1	a	b	c	c	a	b
Period 2	b	c	a	b	c	a	Period 2	b	c	a	b	c	a
3	c	a	b	a	b	c	3	c	a	b	a	b	c

For each of these designs give the anova table (source and df) and show how to test treatments. Also show how the sums of squares are connected – in particular – does the below the line analysis change?

(b) We use data from Lee *et al.* (1998) to illustrate the designs in part (a). They report a three-period crossover study on the drug Ondansetron, which is used to prevent side effects from chemotherapy and other procedures. Three variations (V_1-V_3) of the drug were used in healthy Korean male subjects randomized to received the formulations. The plasma concentrations of Ondansetron were monitored over a period of 12 hours after the administration, and the area under the curve (AUC)was reported. The researchers were particularly interested in the *bioequivalence* of the variations, that is, that there is no difference in the response to the treatments. Analyze the data, given in dataset `Ondansetron`, to see if the conclusion of bioequivalence can be supported.

5.25 Frequently, a drug is given in combination with another drug or compound. This co-administration may be done for a variety of reasons. For example, the effects of the two drugs are additive or synergistic in their pharmacologic or pharmacodynamic activity (e.g., beta blocker + diuretic for hypertension – similar efficacy with fewer side-effects); a drug plus an adjuvant to speed or improve the effects of the drug (acetaminophen/aspirin/ibuprofen + caffeine for improved analgesia); or just plain convenience as in many cold/flu products (acetaminophen + dextromethorphan + pseudoephedrine + diphenhydramine for aches, cough, congestion, and runny nose). The question is whether one compound affects the pharmacokinetics, say, of the other when co-administered versus the administration of a single product (and vice versa, possibly).

A typical design to address these questions for compounds A and B would be a 3 period crossover with the arms A, B, and A + B may or may not all be of interest. With three treatments there are $3! = 6$ possible orderings. We have 12 subjects available, and can run three orders with 4 subjects per order. For each of the following scenarios decide which orderings you would run and give the anova table (source and df) indicating the tests and the possible confounding:

(a) There is no interest in B's pharmacokinetics, only if A's are altered in B's presence.

(b) There is interest in the pharmacokinetics of both entities.

5.26 (Switchback Reversal Design) A variation on the crossover design is the switchback reversal design, where in the simplest case we have two groups of experimental units and three time periods. One group is assigned to the treatment sequence A/B/A and the other group receives the sequence B/A/B. This design is particularly effective when the response of interest has a linear trend through time. This trend would be eliminated from the within-unit variability, which would be used to assess treatment effects. This is a consequence of the fact that the treatment comparison is orthogonal to the linear component for period effects.

The following data, found in dataset `Cow`, are taken from Brandt[1] and represent the results of an experiment designed to measure the effects of two diets on milk production (in pounds) of cows. Care was taken to have the groups nearly equal (weight, milk productions, weight, etc.), and the treatments consisted of different diets. Treatment A was a diet of 1 part corn and cob meal to 1 part ground oats, while treatment B had 4 parts corn and cob meal to 1 part

---

[1] Brandt, A. E. (1938). Tests of significance in reversal or switchback trials. *Iowa. Agric. Exp. Res. Bull.* 234.

ground oats. The three treatment periods covered 105 days, with 3 periods of 35 days each. The yields of the first seven days of each period were not considered because of possible carryover effects from the previous diet.

	Group 1			Group 2		
Period	1	2	3	1	2	3
Treatment	A	B	A	B	A	B
Cow 1	433	413.7	362.9	Cow 6	671.3	610.3	596.8
⋮	⋮	⋮	⋮	⋮	⋮	⋮	⋮
5	655	616.1	494.6	10	764.4	717.6	717

The question of primary interest is whether diets A and B differ significantly with regard to milk yield, which can be answered by the contrast that assesses quadratic Period × Group interaction.

(a) Complete the following anova table (sums of squares and tests):

Source	df
Groups	1
Cows(in Groups)	8
Periods	2
Periods × Group	2
Linear Period× Group	1
Quadratic Period× Group	1
Error = Period × Cows(in Groups)	16

(b) Explain why the quadratic Period × Group interaction yields the appropriate treatment comparison.

5.27 Referring to the experiment of Example 5.16:

(a) If interest is in the level of blood pressure over the duration of the experiment, at the split plot level we could summarize the data with the averages for each subject. Do this, and compare the results to those in Example 5.17.

(b) Repeated measure split plot tests can also be done validly as sets of contrasts. Set up a reasonable set of orthogonal contrasts for the time treatment and do the analysis.

(c) It was mentioned in Section 5.6.3 that if different group of subjects were used in each time period, then the correlation problem would disappear. If this were done in the experiment in Example 5.16 the design would be a twoway CRD.

  (*i*) Explain how to run the CRD, and give the anova table (source and df).

  (*ii*) Compare the anova table in (*i*) to the one in Example 5.17. Show that the 24 df for "Within" in the CRD is partitioned into the WP error df and SP error df. Illustrate this with the numbers from the dataset **Hypertension**.

  (*iii*) Depending on the structure of the error, are there circumstances when the CRD is preferred to the repeated measures design in terms of the precision of the error estimate?

5.28 The delivery of a drug to the blood system is a function of the method of administration. A typical drug can be given orally in either tablet or liquid form, or can be administered by injection, inhalation, etc. The following data are blood concentration levels of a drug delivered by the standard control method "C", and two alternatives denoted by 1 and 2. A portion of the data appear below, with the full dataset in `Delivery`. Each row represents one subject, and the concentration level is measured over four equally spaced time periods.

	Time			
Method	1	2	3	4
C	73	69	66	61
C	64	65	59	58
⋮	⋮	⋮	⋮	⋮
1	77	77	92	83
1	79	91	96	75
⋮	⋮	⋮	⋮	⋮
2	54	65	68	68
2	70	84	79	64

(a) Explain why it is incorrect to analyze these data as a oneway CRD on delivery method.

(b) Write out the more appropriate split plot anova (source and df) indicating the tests.

(c) What covariance structure seems reasonable for these data? Does this violate the split plot assumptions?

(d) Run the anova, and partition the treatment sum of squares into orthogonal polynomial contrasts. Test the contrasts and state any conclusions.

(e) Explain why the contrast tests are valid.

(f) Use the approximations of Miscellanea 5.9.2 to attach significance levels to the usual split plot anova F-tests. How do these results agree with those in part (d)?

Accompaniment

5.29 EMS calculations:

(a) Referring to Table 5.3, fill in the details to verify the expected mean squares. (Recall Lemmas 2.16 and 3.16.)

(b) Finish the calculations of the expected mean squares started in Section 5.3.2 and verify the remaining EMS terms in Table 5.7.

5.30 Referring to the least squares criteria of (5.4):

(a) Verify that (5.30) are the least squares estimates of the effects and compare them to those in (2.8).

(b) Although we know that least squares estimates are unbiased, verify the unbiasedness of the estimates in (a) under model (5.1).

(c) Derive the least squares estimates and verify their unbiasedness using model (5.14).

5.31 For the microarray design of (5.31), a comparison of major interest is the interaction of genes and treatment. Specifically, for a given gene ℓ, we want to estimate the difference in expression levels for treatments i and i', $(GT)_{i\ell} - (GT)_{i'\ell}$.

(a) Show that the least squares estimate of $(GT)_{i\ell} - (GT)_{i'\ell}$ is

$$(\bar{y}_{i\cdot\ell} - \bar{y}_{i\cdot\cdot}) - (\bar{y}_{i'\cdot\ell} - \bar{y}_{i'\cdot\cdot}).$$

(b) Show that the variance of the estimate in part (a) is

$$\frac{g-1}{g}\left(\frac{2}{r}\right)\sigma^2,$$

where, here, σ^2 is the split plot error. (Note that the number of genes in the experiment has no effect on reducing variance; it is r, the number of true replications, that counts.)

5.32 Referring to (5.23), verify the useful expected values: (a) $E\bar{\delta}_{ik}\bar{\delta}_{k'} = \sigma_\delta^2/tr$, if $k \neq k'$, (b) $E\bar{\delta}_{ik}\bar{\delta}_k = \sigma_\delta^2/tr$, (c) $E\bar{\delta}_i\bar{\delta}_k = \sigma_\delta^2/r$.

5.33 Some fun with the Satterthwaite approximation:
(a) Show how to go from (5.12) to (5.13) using the identity

$$\sum_{i=1}^{n}(x_i - \bar{x})^2 = \sum_{i=1}^{n} x_i^2 - n(\bar{x})^2.$$

Referring to Technical Note 5.8.1:
(b) Use the fact that for a χ_ν^2 random variable $E(\chi_\nu^2) = \nu$ to verify (5.30).
(c) Show that straightforward matching of second moments yields

$$E\left(\sum_{i=1}^{k} a_i Y_i\right)^2 = E(\frac{\chi_\nu^2}{\nu})^2 = \frac{2}{\nu} + 1,$$

and hence $\hat{\nu} = 2/[(\sum_{i=1}^{k} a_i Y_i)^2 - 1]$, which could be negative.

5.34 Referring to the calculations leading to (5.8) and (5.10), here we further explore the variance and correlation of the split plot means when the whole plots are a CRD.
(a) Show that

$$\text{Corr}(\bar{Y}_k, \bar{Y}_{k'}) = \frac{\sigma_\varepsilon^2}{\sigma_\varepsilon^2 + \sigma_\delta^2} \overset{\text{def}}{=} \rho_{SP}$$

and compare this correlation to the RCB intraclass correlation ρ_B (3.10).
(b) Show that an alternate expression for the split plot contrast variance is

$$\text{Var}\left(\sum_{k} a_k \bar{Y}_k\right) = \frac{\sigma_\varepsilon^2 + \sigma_\delta^2}{tr}\left[(1 - \rho_{SP})\sum_{k} a_k^2 + \rho_{SP}\left(\sum_{k} a_k^2 + 2\sum_{k>k'} a_k a_{k'}\right)\right],$$

and using the definition of ρ_{SP} show that this expression is the same as (5.8).
(c) The variance of an interaction contrast within a whole plot is given in (5.10). Using the independence of replicates within a whole plot, show that

$$\text{Var}(\bar{Y}_{ik}) = \frac{\sigma_\varepsilon^2 + \sigma_\delta^2}{r},$$

and using (5.6) show that

$$\text{Var}\left(\sum_{k} a_k \bar{Y}_{ik}\right) = \frac{\sigma_\varepsilon^2 + \sigma_\delta^2}{r}\sum_{k} a_k^2 + 2\frac{\sigma_\varepsilon^2}{r}\sum_{k>k'} a_k a_{k'} = \frac{\sigma_\delta^2}{r}\sum_{k} a_k^2.$$

5.35 Referring to Section 5.3.3 and the calculation of contrast variances in the RCB split plot:

(a) At the whole plot level, show that

$$\text{Var}(\bar{Y}_i) = \frac{\sigma_\varepsilon^2 + \sigma_\beta^2}{b} + \frac{\sigma_\delta^2 + \sigma_{\beta\gamma}^2}{bg}, \quad \text{Cov}(\bar{Y}_i, \bar{Y}_{i'}) = \frac{\sigma_\beta^2}{b} + \frac{\sigma_{\beta\gamma}^2}{bg},$$

and hence $\text{Var}(\sum_i a_i \bar{Y}_i)$ is given by (5.17).

(b) At the split plot level, show that

$$\text{Var}(\bar{Y}_k) = \frac{\sigma_\beta^2 + \sigma_{\beta\gamma}^2}{b} + \frac{\sigma_\delta^2 + \sigma_\varepsilon^2}{bt}, \quad \text{Cov}(\bar{Y}_k, \bar{Y}_{k'}) = \frac{\sigma_\beta^2}{b} + \frac{\sigma_\varepsilon^2}{bt},$$

and hence $\text{Var}(\sum_k a_k \bar{Y}_k)$ is given by (5.18).

(c) For the interaction contrast within the same whole plot level, show that

$$\text{Var}(\bar{Y}_{ik}) = \frac{1}{b}(\sigma_\delta^2 + \sigma_{\beta\gamma}^2 + \sigma_\varepsilon^2 + \sigma_\beta^2), \quad \text{Cov}(\bar{Y}_{ik}, \bar{Y}_{ik'}) = \frac{1}{b}(\sigma_\varepsilon^2 + \sigma_\beta^2),$$

and hence $\text{Var}(\sum_k a_k \bar{Y}_{ik})$ is given by (5.19).

(d) For interaction contrasts with different whole plot levels, use the fact that

$$\text{Var}\left(\sum_{ik} a_{ik} \bar{Y}_{ik}\right) = \sum_i \text{Var}\left(\sum_k a_{ik} \bar{Y}_{ik}\right)$$

$$+ 2\sum_{i>i'} \text{Cov}\left(\sum_k a_{ik} \bar{Y}_{ik}, \sum_k a_{i'k} \bar{Y}_{i'k}\right)$$

to use part (c) to calculate the first term. Then show that

$$\text{Cov}\left(\sum_k a_{ik} \bar{Y}_{ik}, \sum_k a_{i'k} \bar{Y}_{i'k}\right) = \frac{\sigma_\beta^2}{b}\sum_{kk'} a_{ik} a_{i'k'} + \frac{\sigma_{\beta\gamma}^2}{b}\sum_k a_{ik} a_{i'k}$$

and establish (5.20).

5.36 Referring to (5.23), show that

(a)

$$\text{Var}\left(\sum_{ik} a_{ik} (\hat{\tau\gamma})_{ik}\right) = \sum_i \text{Var}\left(\sum_k a_{ik} \bar{\delta}_{ik} - g\bar{a}_i \bar{\delta}_i\right) + \text{Var}\left(t\sum_k \bar{a}_k \bar{\delta}_k\right)$$

$$- 2\text{Cov}\left(\sum_i\left[\sum_k a_{ik}\bar{\delta}_{ik} - g\bar{a}_i\bar{\delta}_i\right], t\sum_k \bar{a}_k\bar{\delta}_k\right)$$

$$= (\{1\}) + (\{2\}) - 2(\{3\})$$

and that

(b)

$$\{1\} = \frac{\sigma_\delta^2}{r}\left[\sum_{ik} a_{ik}^2 - g\sum_i \bar{a}_i^2\right], \quad \{2\} = \frac{\sigma_\delta^2}{r} t \sum_k \bar{a}_k^2,$$

and

$$\{3\} = \frac{\sigma_\delta^2}{r} \left[t \sum_k \bar{a}_k^2 - \sum_i \sum_k \bar{a}_i \bar{a}_k \right] = \frac{\sigma_\delta^2}{r} t \sum_k \bar{a}_k^2,$$

since $\sum_i \sum_k \bar{a}_i \bar{a}_k = 0$ for a contrast.

(c) Therefore, show that

$$\mathrm{Var}\left(\sum_{ik} a_{ik}(\widehat{\tau\gamma})_{ik} \right) = \frac{\sigma_\delta^2}{r} \left[\sum_{ik} a_{ik}^2 - \sum_i \sum_k (\bar{a}_i + \bar{a}_k)^2 \right]$$

$$= \frac{\sigma_\delta^2}{r} \sum_{ik} [a_{ik} - (\bar{a}_i + \bar{a}_k)]^2,$$

where, to establish the final equality, show that

$$\sum_{ik} [a_{ik} - (\bar{a}_i + \bar{a}_k)]^2 = \sum_{ik} a_{ik}^2 - \sum_k (\bar{a}_i + \bar{a}_k)^2 + 4 \sum_{ik} \bar{a}_i \bar{a}_k.$$

(d) Show that $\mathrm{Var}\left(\sum_{ik} a_{ik}(\widehat{\tau\gamma})_{ik} \right)$ is given by (5.24), where the whole plots are in an RCB.

5.37 Referring to Table 5.9, verify the expressions for the expected mean squares that are given there.

5.38 Referring to Section 5.5:
   (a) Verify the expressions for the variances of contrasts in the CRD split split plot. (The calculations are similar to those in Section 5.2.3.)
   (b) Verify the expressions for the variances of contrasts in the RCB split split plot.
   (c) (For the stouthearted) Verify the EMS in Table 5.10 for the RCB split split plot (5.26).

5.39 Referring to Technical Note 5.8.2:
   (a) Show that the covariance matrix, Σ, of the observation vector $\mathbf{Y}$ is block diagonal with tr blocks of the form

$$\sigma_\delta^2 I_g + \sigma_\varepsilon^2 J_g.$$

   (b) Establish the following properties of A_1, A_2, and C:
      (i) A_1, A_2, and C are idempotent (follows from the idempotency of $B_1 - B_3$ and $I - B_1 - B_2 + B_3$ and the facts that $B_i B_j = B_3$ and $B_4 B_4' = r I_{tg}$).
      (ii) $A_1 A_2 = A_1 C = A_2 C = 0$ and $A = A_1 + A_2 + C$ is idempotent
      (iii) $A_1 \Sigma = \sigma_\delta^2 A_1$, $A_2 \Sigma = \sigma_\delta^2 A_2$, $C\Sigma = \sigma_\delta^2 C$.
   (c) Use the results in part (b) to prove Lemma 5.18.

5.40 (a) Use Lemma 5.18 to prove Theorem 5.19.
   (b) Formulate and prove a version of Cochran's Theorem for below the line tests when the whole plots are in an RCB. You can assume that $\sigma_{\beta\gamma}^2 = 0$.

5.41 Referring to the strip plot design of Section 5.6.1, show that
   (a) $\mathrm{Cov}(Y_{ijk}, Y_{i'j'k'}) = \begin{cases} 0 & \text{unless } j = j' \\ \sigma_\beta^2 + \sigma_{\tau\beta}^2 & \text{if } j = j', i = i', k \neq k' \\ \sigma_\beta^2 + \sigma_{\beta\gamma}^2 & \text{if } j = j', i \neq i', k = k'. \end{cases}$

(b) If there are replications within the crossed blocks, so the observations are $Y_{ijk\ell}$, where the model is the same as (5.27) except now we have $\varepsilon_{ijk\ell}$, show that the covariance of observations within the same cell is

$$\mathrm{Cov}(Y_{ijk\ell}, Y_{ijk\ell'}) = \sigma_\beta^2 + \sigma_{\beta\gamma}^2 + \sigma_{\beta\tau}^2 + \sigma_{\beta\tau\gamma}^2.$$

Referring to the strip plot design of Section 5.6.1:
(a) Verify the EMS of Table 5.11.
(b) Verify the variances of the treatment contrasts

  (i) T: $\mathrm{Var}\left(\sum_i a_i \bar{Y}_i\right) = \frac{\sigma_\varepsilon^2 + \sigma_{\beta\tau\gamma}^2 + g\sigma_{\tau\beta}^2}{rg} \sum_i a_i^2$

  (ii) G: $\mathrm{Var}\left(\sum_k a_k \bar{Y}_k\right) = \frac{\sigma_\varepsilon^2 + \sigma_{\beta\tau\gamma}^2 + t\sigma_{\beta\gamma}^2}{rg} \sum_k a_k^2$

  (iii) T $\times$ G: $\mathrm{Var}\left(\sum_{ik} a_{ik} \bar{Y}_{ik}\right) = \frac{\sigma_\varepsilon^2 + \sigma_{\beta\tau\gamma}^2}{r} \sum_{ik} a_{ik}^2 + \frac{\sigma_{\beta\gamma}^2}{r} \sum_k \left(\sum_i a_{ik}\right)^2$
  $+ \frac{\sigma_{\tau\beta}^2}{r} \sum_i \left(\sum_k a_{ik}\right)^2.$

5.8 Technical Notes

5.8.1 Satterthwaite's Approximation

From Technical Note 2.8.2 we have seen that the sum of independent χ^2 random variables is again χ^2, and together with the developments in Cochran's Theorem this has lead to the F-tests and confidence intervals in the anova. However, in obtaining contrast variances in the split plot design, we find that the error variance is not a simple sum of χ^2 random variables.

Specifically, from Theorem 2.18, we know that if W_i are independent χ^2 random variables, then so is $\sum_i W_i$. Now consider Case (4) of the CRD split plot design, specifically (5.13). There we will estimate the variance with

$$\frac{\mathrm{MS(SP\ error)}}{r} \sum_{ik} (a_{ik} - \bar{a}_i)^2 + \frac{\mathrm{MS(WP\ error)}}{rg} \sum_i \left(\sum_k a_{ik}\right)^2$$

(5.29) $$= A \times \mathrm{MS(SP\ error)} + B \times \mathrm{MS(WP\ error)},$$

where

$$A = \frac{1}{r} \sum_{ik} (a_{ik} - \bar{a}_i)^2 \text{ and } B = \frac{1}{rg} \sum_i \left(\sum_k a_{ik}\right)^2$$

and $\mathrm{MS(SP\ error)}/\sigma_\delta^2$ and $\mathrm{MS(WP\ error)}/(\sigma_\delta^2 + g\sigma_\varepsilon^2)$ are χ^2 random variables divided by their degrees of freedom.

Although $\sum_i W_i$ is χ^2, this is not the case for $\sum_i a_i W_i$, where the a_is are known (nonequal) constants, and that is what we have in (5.29). This distribution is, in general, quite difficult to obtain. It does seem reasonable, however, to assume that a χ_ν^2, for some value of ν, will provide a good approximation. Furthermore, the method of moments can be very useful in obtaining approximations to such distributions of statistics.

For these situations, the approximation of Satterthwaite (1946) does a good job, and is still used today. He was interested in approximating the denominator of a t-statistic, but solved a more general case which can best be stated as follows: For given constants $a_1, \ldots, a_k$ and independent mean squares $M_1, \ldots, M_k$ with degrees of freedom $\nu_1, \ldots, \nu_k$, find a value of ν so that

$$\sum_{i=1}^{k} a_i M_i \sim \frac{\chi_\nu^2}{\nu} \qquad \text{(approximately)}.$$

A first moment match yields the following (Exercise 5.33):

(5.30) $$\mathrm{E}\left(\sum_{i=1}^{k} a_i M_i\right) = \sum_{i=1}^{k} a_i \mathrm{E}(M_i) = \sum_{i=1}^{k} a_i,$$

$$\mathrm{E}(\chi_\nu^2/\nu) = 1,$$

which tells us that we need to have $\sum_{i=1}^{k} a_i = 1$, which does not help us to estimate ν (and, as we will see, is not really a constraint).

So next we match second moments, but Satterthwaite found that he had to do this in a clever way. Straightforward matching of the second moments yields a possibly negative estimate of ν, but the following argument succeeds. Write

$$\mathrm{E}\left(\sum a_i M_i\right)^2 = \mathrm{Var}\left(\sum a_i M_i\right) + \left(\mathrm{E}\sum a_i M_i\right)^2$$

$$= \left(\mathrm{E}\sum a_i M_i\right)^2 \left[\frac{\mathrm{Var}(\sum a_i M_i)}{(\mathrm{E}\sum a_i M_i)^2} + 1\right].$$

Now use the fact that $\mathrm{E}\sum a_i M_i = 1$ on the expectation *outside* of the square brackets, but *not* on the expectation inside the square brackets. Since $\mathrm{E}(\chi_\nu^2/\nu)^2 = (2/\nu) + 1$, equating these expressions we obtain

$$\nu = \frac{2(\mathrm{E}\sum a_i M_i)^2}{\mathrm{Var}(\sum a_i M_i)}.$$

Finally, use the fact that $M_1, \ldots, M_k$ are independent random variables to write

$$\mathrm{Var}\left(\sum a_i M_i\right) = \sum a_i^2 \mathrm{Var}\, M_i = 2\sum \frac{a_i^2 (\mathrm{E} M_i)^2}{\nu_i},$$

where in the last equality we use the fact that $\mathrm{Var}\, M_i = 2(\mathrm{E} M_i)^2/\nu_i$. Substituting this expression for the variance and removing the expectations, we obtain Satterthwaite's estimator

$$\hat\nu = \frac{\left(\sum a_i M_i\right)^2}{\sum \frac{a_i^2}{\nu_i} M_i^2}.$$

Finally, note that the constraint $\sum_i a_i = 1$ really does not matter, as we can divide the numerator and denominator above by $(\sum_i a_i)^2$ without changing the value of $\hat\nu$, and obtain $\sum_i a_i = 1$. This approximation is quite good as long as all of the a_i are positive.

For more details on the Satterthwaite approximation see Casella and Berger (2001, Section 7.2).

5.8.2 Cochran's Theorem for Split Plot Designs

In this section we give the details of how Cochran's Theorem (Theorem 2.20) applies to split plot designs.

We first note two things:

(1) For a split plot with the whole plots in a CRD, above the line the tests are from the CRD. Thus, we can apply Cochran's Theorem for CRDs (Technical Note 2.8.3) to justify those tests.

(2) Similarly, if the whole plot treatments are in an RCB, we can apply Cochran's Theorem for RCBs (Technical Note 3.8.3).

Thus, we only need give the details for the below the line tests, for the split plot treatment and the split plot - whole plot treatment interaction. Here we will give the details for the case of the whole plot treatments in a CRD, leaving the RCB case to Exercise 5.40. We have model (5.1)

$$Y_{ijk} = \mu + \tau_i + \varepsilon_{ij} + \gamma_k + (\tau\gamma)_{ik} + \delta_{ijk},$$

with

$$\mathrm{Var}(Y_{ijk}) = \sigma_\delta^2 + \sigma_\varepsilon^2, \quad \mathrm{Cov}(Y_{ijk}, Y_{ijk'}) = \sigma_\varepsilon^2,$$

with all other covariances equal to zero (that is, there is only correlation between the split plot observations within the same whole plot rep). To have the covariance matrix in a nice form, we order the data vector **y** as

$$\mathbf{Y}' = \{Yijk\}'$$
$$= (Y_{111}, \cdots, Y_{11g}, \cdots, Y_{t11}, \cdots, Y_{t1g}, \cdots, Y_{1r1}, \cdots, Y_{1rg}, \cdots, Y_{tr1}, \cdots, Y_{trg}),$$

which results in a block diagonal covariance matrix (which we write as "BD(·)") with tr blocks:

$$\mathrm{Cov}(\mathbf{Y}) = \mathrm{BD}(\sigma_\delta^2 I_g + \sigma_\varepsilon^2 J_g).$$

See Exercise 5.39.

We next define four matrices (note that these are not the same as the matrices defined in Technical Note 3.8.2 because of the ordering of the **Y** vector)

$$B_1 = \frac{1}{t} \begin{pmatrix} I_g \\ \vdots \\ I_g \end{pmatrix}_{tg \times g} (I_g \ \cdots \ I_g)_{g \times tg}$$

$$B_2 = \frac{1}{g} \begin{pmatrix} \mathbf{1}_{g \times 1} & 0 & \cdots & 0 \\ 0 & \mathbf{1}_{g \times 1} & \cdots & 0 \\ \vdots & \vdots & \vdots & \vdots \\ 0 & 0 & \cdots & \mathbf{1}_{g \times 1} \end{pmatrix}_{tg \times t} \begin{pmatrix} \mathbf{1}_{1 \times g} & 0 & \cdots & 0 \\ 0 & \mathbf{1}_{1 \times g} & \cdots & 0 \\ \vdots & \vdots & \vdots & \vdots \\ 0 & 0 & \cdots & \mathbf{1}_{1 \times g} \end{pmatrix}_{t \times tg},$$

$$B_3 = \frac{1}{tg} J_{tg}, \qquad B_4 = \mathrm{BD}(I_g \cdots I_g)_{g \times gr} \quad t \text{ times},$$

and now we write the sums of squares in terms of these matrices and **Y**. Define the matrices

$$A_1 = \frac{1}{r} B_4'(B_1 - B_3)B_4, \quad A_2 = \frac{1}{r} B_4'(I - B_1 - B_2 + B_3)B_4,$$

and

$$C = \mathrm{BD}(I - C_1 - C_2 + C_3),$$

where

$$C_1 = \frac{1}{r} \begin{pmatrix} I_g \\ \vdots \\ I_g \end{pmatrix}_{rg \times g} (I_g \; \cdots \; I_g)_{g \times rg}$$

$$C_2 = \frac{1}{g} \begin{pmatrix} \mathbf{1}_{g \times 1} & 0 & \cdots & 0 \\ 0 & \mathbf{1}_{g \times 1} & \cdots & 0 \\ \vdots & \vdots & \vdots & \vdots \\ 0 & 0 & \cdots & \mathbf{1}_{g \times 1} \end{pmatrix}_{rg \times r} \begin{pmatrix} \mathbf{1}_{1 \times g} & 0 & \cdots & 0 \\ 0 & \mathbf{1}_{1 \times g} & \cdots & 0 \\ \vdots & \vdots & \vdots & \vdots \\ 0 & 0 & \cdots & \mathbf{1}_{1 \times g} \end{pmatrix}_{r \times rg},$$

$$C_3 = \frac{1}{rg} J_{rg}.$$

Then

$$\mathbf{Y}' A_1 \mathbf{Y} = \text{SS(SP Trts)}, \quad \mathbf{Y}' A_2 \mathbf{Y} = \text{SS(SP Trts} \times \text{WP Trts)},$$
$$\mathbf{Y}' C \mathbf{Y} = \text{SS(SP Error)}.$$

Here is the application of Cochran's Theorem to the split plot design. The proofs are similar to the other Cochran results, and are left to an exercise (Exercise 5.39).

Lemma 5.18. *Let* $\mathbf{Y} \sim \text{N}(0, \Sigma)$, *where* $\Sigma = \text{BD}(\sigma_\delta^2 I_g + \sigma_\varepsilon^2 J_g)$. *Then*

$$A_1^* = \frac{1}{\sigma_\delta^2} A_1, \quad A_2^* = \frac{1}{\sigma_\delta^2} A_2, \quad C^* = \frac{1}{\sigma_\delta^2} C, \text{ and } A^* = A_1^* + A_2^* + C^*$$

satisfy the assumptions of Theorem 2.20 (Cochran's Theorem). That is,
(1) $A^* \Sigma$ *is idempotent.*
(2) $A_1^* \Sigma$, $A_2^* \Sigma$, *and* $C^* \Sigma$ *are idempotent.*
(3) $A_1^* \Sigma A_2^* = 0$, $A_1^* \Sigma C^* = 0$, $A_2^* \Sigma C^* = 0$.

Using this lemma we can establish the following theorem.

Theorem 5.19 (Cochran's Theorem for Split Plots - Whole Plots in CRD). *Under the split plot anova model (5.1),* $\mathbf{Y} = \{Y_{ij}\}$ *is multivariate normal with* $\Sigma = \text{BD}(\sigma_\delta^2 I + \sigma_\varepsilon^2 J)$.
(1) Under $H_0 : \gamma_k - \bar{\gamma} = 0$, *for all* k

$$\frac{\text{MS(SP Trt)}}{\text{MS(SP Error)}} = \frac{\mathbf{Y}' A_1^* \mathbf{Y}/r_1}{\mathbf{Y}' C^* \mathbf{Y}/r_C} \sim F_{r_1, r_C}$$

where $r_1 = tr(A_1)$ *and* $r_C = tr(C)$.
(2) Under $H_0 : (\tau\gamma)_{ik} = 0$ *for all* i *and* k

$$\frac{\text{MS(SP} \times \text{WP)}}{\text{MS(SP Error)}} = \frac{\mathbf{Y}' A_2^* \mathbf{Y}/r_2}{\mathbf{Y}' C^* \mathbf{Y}/r_C} \sim F_{r_2, r_C}$$

where $r_2 = tr(A_2)$
(3) For any contrast $(a_{11}, \ldots, a_{tg})$,

$$\frac{\sum_{ik} a_{ik} \bar{Y}_{ik} - \sum_{ik} a_{ik} \tau \gamma_{ik}}{\sqrt{\frac{\text{MS(SP Error)}}{r} \sum_{ik} a_{ik}^2}} \sim t_{r_C}.$$

5.9 Miscellanea

5.9.1 Microarray Design II

Here we go a bit further into the design of oligonucleotide microarrays, the single-dye system made by `Affymetrix`. Recall that the experimental unit is the RNA (actually the subject from which it was taken), as that is where the treatment is applied. (For plants this could be growing conditions or varieties, for humans this could be disease status.) On the chip are the genes, and here we are splitting the experimental unit to get the expression level of all of the genes. Thus, the genes are a split plot treatment, and the microarray design is a simple split plot.

As in any split plot design, the whole plot treatments can have many different design – this is the design on the subjects - and can be a CRD, RCB, or something else. Assuming a oneway CRD for the whole plots, a model for the analysis is

(5.31)
$$y_{ijk} = \mu + T_i + A_{ij} + G_k + (GT)_{ik} + \varepsilon_{ijk},$$
$$i = 1, \ldots, t, \quad j = 1, \ldots, r, \quad k = 1, \ldots, g,$$

where y_{ijk} is the log expression level of gene k on subject j in treatment i, T_i is the treatment, A_{ij} are the microarrays in the treatments, and G_k are the genes. We assume that each treatment has r subjects (microarrays). The anova is

Source	df
Treatments	$t-1$
Whole Plot Error	$t(r-1)$
Genes	$g-1$
Gene × Treatment	$(t-1)(g-1)$
Split Plot Error	$t(g-1)(r-1)$

The whole plot error comes from the mean square for Arrays in Treatments, and the split plot error is Gene × Array in Treatment. Note that the microarrays must be replicated otherwise there are no tests!

Typically, the factor of most interest is the Gene × Treatment interaction, because that signifies that the genes are reacting differently to the treatments, and this could help the experimenter find genes that are connected to a disease state, or some other trait. It is fortunate that this test is at the split-plot level and is thus more precise.

5.9.2 Beyond Cochran

As mentioned in Section 5.6.3 the covariance structures of the repeated measures anova are outside the assumptions of Cochran's Theorem. As a result, at the split plot level, the ratios of mean squares do not have an F distribution under the appropriate null hypotheses. Here we look at a variety of methods to address this problem.

Huynh/Feldt Conditions

Huynh and Feldt (1970) give necessary and sufficient conditions on the error structure under which we can obtain valid F-ratios. They first give their conditions for the case of one whole plot treatment and extend it to the split plot model.

Theorem 5.20. *Huynh/Feldt Conditions*
(1) (One Whole Plot Treatment) For the model

$$Y_{ij} = \mu + \tau_i + \beta_j + \varepsilon_{ij}, \quad i = 1, \ldots, t, \quad j = 1, \ldots b,$$

where $Y_j = (Y_{ij}, Y_{2j}, \ldots, Y_{tj})$ are the observations on subject (or block) j over the treatments (time in a repeated measures design) and have a multivariate normal distribution. Let $\Sigma_{t \times t} = \{\sigma_{rs}\}$ denote the common covariance matrix of the Y_js. Under the null hypothesis H_0 : no treatment effect, the ratio of mean squares $\mathrm{MS}(\mathrm{Trt})/\mathrm{MS}(\mathrm{Trt} \times \mathrm{Block})$ has an F distribution if and only if for some constants α_r, α_s, and $\lambda > 0$

$$\sigma_{rs} = \begin{cases} \alpha_r + \alpha_s & \text{if } r \neq s \\ \alpha_r + \alpha_s + \lambda & \text{if } r = s. \end{cases}$$

(2) (Split Plot Design) In the split plot model, the F-tests at the split plot level are valid F-tests if and only if the covariance matrix condition in (1) is satisfied within each level of the whole plot treatment. The αs may differ, but there must be a common λ.

This theorem gives us a richer covariance structure than Theorem 5.19, which would require $\alpha_r = \alpha_s$ for all r and s. However, it does not get us to covariance structures such as AR(1), which remain popular for modeling trends over time.

Multivariate Analysis

The split plot model is very similar to the model used for the multivariate technique of *Profile Analysis*, the only difference being the covariance structures. In profile analysis the covariance is allowed to be general. Thus, the test statistic for profile analysis, *Hotelling's T^2*, will provide an exact test in the repeated measures case. (Indeed, Hotelling's T^2 will provide an exact test for any covariance structure, at the cost of a decrease in power.) For example, for model (5.1), but allowing the covariance structure to be totally general, we can test $H_0 : \gamma_k = 0$ for all k with the statistic

$$T^2 = rt\bar{Y}'M(M'SM)^{-1}M'\bar{Y},$$

where Y' = the vector of split plot means, S is the sample covariance matrix of the split plot means, that is, $S = \{s_{kk'}\}$, where

$$s_{kk'} = \frac{1}{t(r-1)} \sum_i \sum_{kk'} \sum_j (Y_{ijk} - \bar{Y}_{i\cdot k})(Y_{ijk'} - \bar{Y}_{i\cdot k'}),$$

and

$$M = \begin{pmatrix} 1 & 0 & \cdots & \cdots & 0 \\ -1 & 1 & \cdots & \cdots & 0 \\ 0 & -1 & 1 & \cdots & 0 \\ \vdots & \vdots & \vdots & \vdots & \vdots \\ 0 & 0 & \cdots & \cdots & -1 \end{pmatrix}_{g \times g-1}.$$

Under H_0,

$$\frac{rt - t - g + 2}{t(r-1)(g-1)} T^2 \sim F_{g-1, rt-t-g+2}.$$

Thus, we can get an exact test in the repeated measures design, but we give up a lot. Since we are estimating the entire covariance matrix, with $g(g+1)/2$ parameters, we are losing a great deal of power. In the split plot model, or in the AR(1) covariance model, we only have to estimate a two or three parameters. In general, it is somewhat of a losing proposition to estimate the entire covariance matrix; we typically want to model it with a small number of parameters, otherwise too much power is lost. However, it is nice to know that an exact test does exist.

Approximate F-tests

In the notation of Section 5.2, with similar approximations holding for the other split plot designs,

$$\frac{\text{MS(Split Plot Trt)}}{\text{MS(Split Plot Error)}} \overset{\text{approx}}{\sim} F_{1, t(r-1))}$$

and

$$\frac{\text{MS(SP Trt} \times \text{WP Trt)}}{\text{MS(Split Plot Error)}} \overset{\text{approx}}{\sim} F_{t-1, t(r-1)}.$$

Geisser and Greenhouse (1958) have shown that the approximation is conservative, which implies that the cutoff points will be greater than those of the true distribution (whatever it is) and thus rejection with the approximation will imply rejection against the true cutoff. Thus, the approximate test will maintain the nominal α -level (the Type I error rate will be $\leq \alpha$), but the approximate test will be less powerful than using the true cutoff point.

Summarizing the Repeated Measure

As was illustrated in Example 5.17, another approach to testing repeated measures is through the use of a summary statistic. This can be a most powerful approach, but relies on both the expertise of the experimenter and the existence of a summary measure that is meaningful.

Reducing the repeated measure to a summary statistic eliminates the correlation problem as there is now only one split plot measurement for each subject (whole plot). In addition, it effectively eliminates the split plot since, with one observation per subject, the design is now only on the whole plot treatments. However, if the summary measure is a good one, there will be a variance reduction and a corresponding increase in power. Also, as the summary statistics are usually some sort of average, we would expect the Central Limit Theorem to be on our side.

Typical summary statistics are:

(1) Means. Summarizing each subject with the mean over time reduces the analysis to comparing average levels over time, which might be a quantity of interest.

(2) Contrasts. We can do a full set of orthogonal contrasts or a few contrasts of interest. Since the repeated measure is usually time, polynomial contrasts are usually applicable.

(3) Coefficient estimates. For each subject we could fit a curve such as $y = \beta_0 + \beta_1 x$, or a growth curve such as the logistic curve $y = \frac{\beta_0}{1+\beta_1 \exp{-\beta_2 x}}$, and use the coefficient estimate of each subject as the dependent variable in the regression. In the case of linear regression, β_1 measures a rate of change, which is often a meaningful quantity. The logistic curve is a bit trickier, but could also be meaningful. For example, β_0 is the asymptote, which is the total growth, and β_2 is a relative growth rate.

The drawback to this approach, other than the problem of finding a summary measure, is that we may end up doing many analyses if we have a number of summary measures. Thus, in such a situation, we will have to adjust the conclusions to account for multiplicity of tests(Miscellanea 2.9.1).

For more details on Hotelling's T^2 and the F approximations, see Morrison (2005), and for the full picture on the analysis of repeated measures see Davis (2002).

Confounding in Blocks

Would you think me impertinent if I were to put your theories to a more severe test?

Dr. Watson to Sherlock Holmes
The Sign of Four

It is easy to conduct an experiment in such a way that no useful inferences can be made...

William Cochran and Gertrude Cox
Experimental Designs

6.1 Introduction

Thus far, all of the designs we have looked at have been *complete* in that every treatment has appeared in every block. This is the best situation and gives us the best information for treatment comparisons. However, there are many situations where we cannot put every treatment in every block (often due to time, money, or physical constraints of the experiment). For example, a microarray experiment using a two-dye chip is restricted to two treatments per block (microarray). In these cases the design becomes *incomplete* in that not every treatment is in every block.

If the design is incomplete, then we immediately are faced with the fact that treatment comparisons are confounded with block effects which, of course, will cause problems. There is the obvious problem that the block difference may affect treatment comparisons, and we also have the problem that block variances could creep into the variance of a treatment comparison. The point of this chapter is to see how to deal with incomplete designs so that we can mitigate these problems.

Example 6.1. DIET AND BLOOD PRESSURE   A study was to be conducted to assess the effects of diet on blood pressure in African-American males. Three factors are to be measured:

A = amount of fruits and vegetables in the diet (low/high),
B = amount of fat in the diet (low/high),
C = amount of dairy products in the diet. (low/high)

There are eight treatment combinations to be arranged in a 2^3 factorial, where each treatment combination will be administered as follows: A subject will have a baseline blood pressure reading taken, then will be fed (at a laboratory) according to one of the eight diet plans (treatments). After three weeks, another blood pressure reading will be taken.

Unfortunately, administering the diet plans is very labor-intensive, and only four treatment combinations can be run at one time. Thus, the experiment will be run in two blocks, each lasting three weeks.

Notation for describing treatment combinations	**Notation**: For factors of two levels, we denote the "high" level by the corresponding lower case letter, and the "low" level by the absence of the letter. Thus the treatment combination abc denotes the combination where each treatment is at the high level, and bc denotes the combination where A is at the low level and the other two are at the high level. The treatment combination with all factors at the low level is denoted by (1). (See also Example 2.11.)

The following design was decided upon:

Block

1			2	
a	b		(1)	ab
c	abc		ac	bc

With eight subjects per treatment combination, the anova table looks like

Source	df
Blocks	1
Trts	6
T × B	0
Within	56
Total	63

where there are only 6 degrees of freedom for treatments because of the confounding with blocks. Typically, the treatment tests are against the T × B interaction, but there are no degrees of freedom to estimate it. Thus, our only recourse is to use the within error, which brings along the assumption of no T × B interaction.

Formally, we would consider the 56 degrees of freedom to be wasted because the within variation is only useful for testing the T × B interactions, which we are hoping is not significant. However, especially with human subjects, we are often concerned with the size of the within variability, so there is some merit to having eight subjects. However, a better design, if possible, would be to use only four subjects for each treatment combination, run the experiment twice, and thus be able to estimate the T × B interaction (Exercise 6.1).

If we break down the 7 degrees of freedom for treatments we can study the confounding in blocks. We look at the treatment combinations corresponding to the component main effects and interactions, and here is where we can see the confounding. Writing out the contrasts we get

		Effect						
Block	Trt. Comb	A	B	C	AB	AC	BC	ABC
1	a	+	−	−	−	−	+	+
1	b	−	+	−	−	+	−	+
1	c	−	−	+	+	−	−	+
1	abc	+	+	+	+	+	+	+
2	(1)	−	−	−	+	+	+	−
2	ab	+	+	−	+	−	−	−
2	ac	+	−	+	−	+	−	−
2	bc	−	+	+	−	−	+	−

where we see that the ABC interaction is confounded with blocks, as Block 1 has all high levels and Block 2 has all low levels. Every other effect is balanced between the blocks in that there are two high levels and two low levels in each block, so no other effect is confounded with blocks.   ‖

So we see why the above anova table has only 6 degrees of freedom for treatments, as the block sum of squares is *exactly* the sum of squares due to the threeway interaction. By moving from an RCB to the above incomplete design, we are able to estimate all of the treatment effects except the threeway interaction. Although this is not bad, in that we only lost information on one contrast, this design is really not very good for two reasons: (*i*) We are wasting a lot of degrees of freedom by having eight subjects per cell. (*ii*) We cannot estimate the T × B interaction. The following design does better on these points.

Example 6.2. DIET AND BLOOD PRESSURE CONTINUED   The design in Example 6.1 confounded the ABC interaction with blocks, not allowing us to estimate it. One way around this problem, assuming that we are still limited to running only four treatment combinations at one time, is to replicate the experiment and confound another effect with blocks. That is, to run the following experiment

	Run 1		Run 2	
	Block		Block	
	1	2	3	4

a	b		(1)	ab		(1)	a		b	c
c	abc		ac	bc		bc	abc		ab	ac

In the second run the *BC* interaction is confounded with Blocks (Exercise 6.2), and so if both runs are done, we can do the following anova (using four subjects per treatment combination) and estimate all effects:

Source	df
Runs	1
Blocks (Within Runs)	2
Block 1 vs. Block 2 (ABC)	1
Block 3 vs. Block 4 (BC)	1
Treatments	7
A	1
B	1
C	1
AB	1
AC	1
BC	1
ABC	1
T × B (Residual)	5
Within	48
Total	63

Note that the *BC* and *ABC* effects have only partial information, as they are each estimated only with one run. The sums of squares are adjusted for blocks (they are fitted after blocks). Here we can actually use the more proper error term, the T × B interaction, to test the treatment effects. Of course, because of the small number of degrees of freedom for this term, we would be looking to pool it with the within error.

> This pooling can result in error underestimation

‖

Now for this second design we see that in each run we lost some information on one of the factors (by confounding that factor with the blocks) . In Run 1 we lost information on the threeway interaction, and in Run 2 we lost information on the *BC* interaction. We lost the information because the block contrast and the effect contrast were exactly the same, with the "+" in one block and the "−" in the other. If we think about this a little harder, we might think of doing this for each of the treatment combinations, that is, set up pairs of blocks so that each pair confounds one of the seven treatment effects.

The effects are

	Effect						
Trt. Comb	A	B	C	AB	AC	BC	ABC
(1)	−	−	−	+	+	+	−
a	+	−	−	−	−	+	+
b	−	+	−	−	+	−	+
c	−	−	+	+	−	−	+
ab	+	+	−	+	−	−	−
ac	+	−	+	−	+	−	−
bc	−	+	+	−	−	+	−
abc	+	+	+	+	+	+	+

Example 6.3. BALANCING THE WHOLE THING   For each of the above seven effects we can set up a pair of blocks that are confounded with that effect. This would result in the following design:

Block Pair	Confounded Effect	Block Pair	Confounded Effect
a ab ac abc (1) b c bc	A	a b ac bc (1) c ab abc	AB
b ab bc abc * (1) a c ac *	B	a c ab bc (1) b ac abc	AC
c ac bc abc * (1) a b ab *	C	b c ab ac * (1) a bc abc *	BC
		a b c abc (1) ab ac bc	ABC

Although we have written the blocks in pairs, to better understand the confounding pattern, the design to be run will randomize among the fourteen blocks. If we could have run a complete design, an RCB, then we could put all eight treatment combinations inside a block.

Note that if we could collapse each of the seven pairs into one block, each of those seven blocks would be a complete block, and the experiment could be run as an RCB. With 56 experimental units, one for each treatment combination, in Table 6.1 we look at the anova tables from running the experiment in 14 blocks as above, in 7 complete blocks.

Comparing the anova tables in Table 6.1 we see that seven degrees of freedom moved from the Trt × Block interaction in the RCB to blocks in the

Table 6.1. The anova tables from running the experiment in Example 6.3 in 14 incomplete blocks (left) and seven complete blocks (right).

Fourteen Incomplete Blocks		Seven Complete Blocks	
Source	df	Source	df
Blocks	13	Blocks	6
Trts	7	Trts	7
Trts × Blocks	$7 \times 5 = 35$	Trts × Blocks	$7 \times 6 = 42$
Total	55	Total	55

incomplete design. This can be explained by first looking at the six blocks that are marked with a * above. These are the only blocks that can be used to estimate the A × Block interaction. This is the same for all of the treatment effects – each can only be estimated from six blocks. Thus, the interaction degrees of freedom are 7×5, since each effect only is estimated from 6 blocks yielding 5 degrees of freedom for blocks to estimate interactions (see Exercise 6.3).

‖

The design in Example 6.3 is, in fact, a balanced incomplete block design, the workhorse of incomplete designs and the subject of the next section.

6.2 Balanced Incomplete Block Designs

In the previous section we saw that if all treatments cannot appear in every block, then confounding results in a loss of information. A *Balanced Incomplete Block Design (BIBD)* addresses this problem by balancing the confounding so that

Advantages of the BIBD

‖ (1) Every treatment is estimated with the same variance
‖ (2) Every *contrast* of treatments is estimated with the same variance.
‖ (3) The contrast variance is free of the block variance component.

It is (2) and (3) that are most important, as the BIBD balances the confounding so that we still can get good estimates of contrasts. This is accomplished by balancing pairs of treatments in the incomplete blocks

Definition 6.4. A *balanced incomplete block design (BIBD)* with t treatments and b blocks satisfies:

(1) Each block has k treatments ($k < t$).

(2) Each treatment appears in r blocks ($r < b$).

(3) Every pair of treatments appears together λ times.

Note that every pair of treatments appears together the same number of times in a BIBD. This requirement can be relaxed, but then all contrasts will not have the same variance.

Example 6.5. VERY SIMPLE BIBD   If there are $t = 4$ treatments and we only can have $k = 3$ treatments per block, we need $\binom{4}{3} = 4$ blocks for a BIBD. If we denote the treatments A, B, C, D, the BIBD is

Block

1	2	3	4
A	A	A	B
B	B	C	C
C	D	D	D

where every treatment is in $r = 3$ blocks and every pair of treatments appears together $\lambda = 2$ times.   ‖

The BIBD is characterized by the five numbers (t, k, b, r, λ), where

$$t > k, \quad \text{otherwise we could do an RCB,}$$

$$\lambda < r < b, \quad \text{with equality if we have an RCB.}$$

Note also that the requirements of the BIBD result in the following two equations, that every BIBD must satisfy:

$$rt = bk,$$
$$\lambda(t - 1) = r(k - 1).$$

> BIBD
> defining
> equations

There is nothing mysterious about these equations, they merely reflect counting the same thing in two different ways. That is,

$$rt = \# \text{ blocks/trt} \times \# \text{ trts} = \# \text{ Experimental Units,}$$
$$bk = \# \text{ blocks} \times \# \text{ trts/block} = \# \text{ Experimental Units,}$$

so the fact that $rt = bk$ is just a result of counting the number of experimental units two different ways. Similarly, but a bit more complicated, consider any one treatment, say treatment A. Then, since A occurs in r blocks, each time with $k - 1$ other treatments,

$$\# \text{ of Exp. Units in blocks containing } A = r(k - 1).$$

On the other hand, since A occurs λ times with each of $t - 1$ other treatments,

$$\# \text{ of Exp. Units in blocks containing } A = \lambda(t - 1),$$

and so $r(k - 1) = \lambda(t - 1)$.

Example 6.6. VERY SIMPLE BIBD CONTINUED   If $t = 4$ and $k = 3$, then $4r = 3b$. If we start looking for integer solutions of this equation, we see that $r = 1$ or $r = 2$ does not work, but $r = 3$ does, with $b = 4$. We then have $\lambda \times 3 = 3 \times 2$, so $\lambda = 2$, which is the design in Example 6.5.   ‖

Note that, in an RCB, we have $b = r$ and $t = k$, so all of these restrictions are immediately satisfied (see Exercise 6.4).

6.2.1 Model and Distribution Assumptions

The model that we use for the BIBD is essentially equivalent to the RCB model (3.5), that is

$$(6.1) \qquad Y_{ij} = \mu + \tau_i + \beta_j + \varepsilon_{ij},$$

where

(1) the random variables $\varepsilon_{ij} \sim$ iid $N(0, \sigma_\varepsilon^2)$ (normal errors with equal variances),
(2) the random variables $\beta_1, \ldots, \beta_b$, are iid $N(0, \sigma_\beta^2)$ and are independent of ε_{ij} for all i, j.

The difference between the BIBD model and the RCB model lies in the index set. For the RCB model we had $i = 1, \ldots, t$ and $j = 1, \ldots, b$. However, in the BIBD we do not have every treatment in every block, so we have to be more careful about how we define the index set (as we were for Latin square designs in Section 3.6.3). Here, the index set for the BIBD can be defined in two ways:

$$\{i, j\} : i = 1, \ldots, t, j \in J_i = \text{ the blocks } j \text{ that contain the treatment } i$$
$$\text{– there are } r \text{ such indices}$$
$$\{i, j\} : j = 1, \ldots, b, i \in I_j = \text{ the treatments } i \text{ that are in block } j$$
$$\text{– there are } k \text{ such indices.}$$

So, for example,

$$\bar{y}_i = \frac{1}{r} \sum_{j \in J_i} y_{ij}, \quad \bar{y}_j = \frac{1}{k} \sum_{i \in I_j} y_{ij}, \text{ and } \bar{\bar{y}} = \frac{1}{rt} \sum_{i=1}^{t} \sum_{j \in J_i} y_{ij}$$

are, respectively, the mean of treatment i, the mean of block j, and the grand mean.

As the BIBD is not a balanced design, we do not expect the least squares estimates to be the cell means. However, they are close to what we might expect, given the unbalance of the design. The least squares estimates of τ_i are (Exercise 6.29)

$$(6.2) \qquad \hat{\tau}_i = \frac{k}{\lambda t} \left(r \bar{y}_i - \sum_{j \in J_i} \bar{y}_j \right)$$

and, by virtue of being least squares, are unbiased estimators of τ_i.

This estimate is somewhat intuitive, as the first term in the parentheses is the total response from treatment i, and what is subtracted is the total of the blocks containing treatment i, which seems to be the appropriate centering term. However, if we operated strictly on intuition, we might come up with the estimate

$$\hat{\tau}_i^I = \bar{y}_i - \frac{1}{r}\sum_{j \in J_i} \bar{y}_j = \frac{1}{r}\left(r\bar{y}_i - \sum_{j \in J_i} \bar{y}_j\right),$$

where we use "I" for "intuition". Although there is nothing terribly wrong with this estimate, it is biased (but has a smaller variance than the least squares estimate $\hat{\tau}_i$).

Example 6.7. SIMPLE ESTIMATES Hicks (1993) gave the following data from a BIBD used to measure current flow through four different television tubes (A, B, C, D), where the blocks are different days.

	Treatment			
	A	B	C	D
Block 1	2	-	20	7
Block 2	-	32	14	3
Block 3	4	13	31	-
Block 4	0	23	-	11

Using these data, our estimates of the effect of treatment B are

$$\hat{\tau}_B = \frac{3}{2 \times 4}\left(68 - \frac{131}{3}\right) = 9.13,$$

$$\hat{\tau}_B^I = \frac{1}{3}\left(68 - \frac{131}{3}\right) = 8.11.$$

$\parallel$

To maintain unbiasedness, the least squares estimate uses the factor $k/\lambda t$ rather than $1/r$, which of course seems more natural since treatment i appears r times. From the fact that $\lambda(t-1) = r(k-1)$, we can write

(6.3) $$\frac{\lambda t}{k} = r\left[\frac{k-1}{k}\frac{t}{t-1}\right],$$

and it can be shown that, since $k < t$ in a BIBD, the factor in square brackets is always less than 1, yielding the larger magnitude of the least square estimates (Exercise 6.30). Note that in an RCB $k = t$ so there is equality in (6.3).

The anova table for a BIBD has the form

Source	df	SS	MS	F
Blocks	$b-1$	SS(Blocks)	MS(Blocks)	
Treatments	$t-1$	SS(Trts)	MS(Trts)	$\dfrac{\text{MS(Trts)}}{\text{MS(T} \times \text{B)}}$
T $\times$ B	$bk-b-t+1$	SS(T $\times$ B)	MS(T $\times$ B)	
Total	$bk-1$	SS(Total)		

where we see that the test on treatments is the same as in the RCB. We also note that under model (6.1), MS(T $\times$ B) is an unbiased estimator of σ_ε^2.

6.2.2 Estimating Contrasts

Although the BIBD is an unbalanced design, it is sufficiently balanced so that the variance of a treatment contrast is free from block effects. This is an enormous advantage, and allows the BIBD to achieve good precision in contrast estimation.

The variance of $\hat{\tau}_i$ is

$$\text{Var}(\hat{\tau}_i) = \left(\frac{k}{\lambda t}\right)^2 \text{Var}\left(r\bar{Y}_i - \sum_{j \in J_i} \bar{Y}_j\right)$$

$$= \left(\frac{k}{\lambda t}\right)^2 \text{Var}\left(\sum_{j \in J_i}\left[Y_{ij} - \frac{1}{k}\sum_{i' \in I_j} Y_{i'j}\right]\right)$$

$$= \left(\frac{k}{\lambda t}\right)^2 \sum_{j \in J_i} \text{Var}\left(Y_{ij} - \frac{1}{k}\sum_{i' \in I_j} Y_{i'j}\right),$$

where we use the fact that observations in different blocks are independent. It is then straightforward to show (Exercise 6.31) that

(6.4)
$$\text{Var}\left(Y_{ij} - \frac{1}{k}\sum_{i' \in I_j} Y_{i'j}\right) = r\left(1 - \frac{1}{k}\right)\sigma_\varepsilon^2$$

and thus

(6.5)
$$\text{Var}(\hat{\tau}_i) = \left(\frac{k}{\lambda t}\right)^2 r\left(1 - \frac{1}{k}\right)\sigma_\varepsilon^2 = \frac{k}{\lambda t}\left(\frac{t-1}{t}\right)\sigma_\varepsilon^2,$$

where we used the fact that $r(k-1) = \lambda(t-1)$.

The parameter λ is very important Note the important role played by λ, the number of times that the pairs appear together. It is in the denominator of the variance estimate, so increasing λ leads to more precise variance estimates. If this were an RCB, then $k/\lambda t = 1/b$, which would give the best precision.

Note: As a consequence of the BIBD, the variance of $\hat{\tau}_i$ is free of σ_β^2, the block variance. This is the real advantage of this design.

Finally, using (6.5), we can derive the variance of a contrast. We start with

$$\text{Var}\left(\sum_{i=1}^{t} a_i\hat{\tau}_i\right) = \sum_{i=1}^{t} a_i^2 \text{Var}(\hat{\tau}_i) + 2\sum_{i>i'} \text{Cov}(\hat{\tau}_i, \hat{\tau}_{i'}),$$

and using the fact that (Exercise 6.32)

(6.6) $$\text{Cov}(\hat{\tau}_i, \hat{\tau}_{i'}) = -\frac{k}{\lambda t^2}\sigma_\varepsilon^2,$$

we have

(6.7) $$\text{Var}\left(\sum_{i=1}^{t} a_i\hat{\tau}_i\right) = \frac{k}{\lambda t}\sigma_\varepsilon^2 \sum_{i=1}^{t} a_i^2.$$

As we have done in previous chapters, we can get the distribution of the contrast estimate based on model (6.1), but here, due to the algebraic overload, we skip the details. The contrast $\sum_{i=1}^{t} a_i\hat{\tau}_i$ satisfies

(6.8) $$\frac{\sum_{i=1}^{t} a_i\hat{\tau}_i - \sum_{i=1}^{t} a_i\tau_i}{\sqrt{\frac{k}{\lambda t}\text{MS}(\text{T} \times \text{B})}} \sim t_{bk-b-t+1},$$

from which we can get both tests of hypotheses and confidence intervals.

Example 6.8. ORTHOPEDIC BIBD   Berkowitz *et al.* (2005) report the results of a study done to assess the strength of self-tapping screws used in surgical procedures on bone. In particular, there is interest in whether the depth of insertion of the screw has a relationship to strength. Five treatments (1-5) were considered, each corresponding to a relative depth of placement of the screw (-1mm, 0mm, 1mm, 2mm, 3mm). The screws were inserted into blocks of synthetic bone, called *bone coupon* and a machine known as an *Instron* was used to measure the force (in Newtons) needed to extract the screw. Because of the size of the bone coupon and the disruptive nature of the extraction, only three screws could be inserted in a particular bone coupon.

The experiment was run as a BIBD with $t = 5$ and $k = 3$, with anova

Source	df	SS	MS	F	p-value
Block	9	1,097,669	12,1963	13.892	
Trt	4	2,033,580	508,395	57.906	< .0001
Residuals	16	140,474	8,780		

The anova shows that the treatments (depth of screws) are significantly different. Berkowitz *et al.* (2005) report the cell means, and then use a multiple comparison procedure to find the best treatment.

Treatment Mean Pullout Strength (N)

1	2	3	4	5
1316.94	1503.11	1941.83	1928.41	2045.55

From the development in Section 6.2.2 it should be clear that the variance of the cell means depends on σ^2_β, the block variance, and comparisons should be based on the least squares estimates, where contrast variances would only depend on σ^2_ε, estimated by the MS(Residual). This is pursued in Exercise 6.11.   ‖

Generalizations

As a generalization of a BIBD, there is the PBIBD, the *Partially* Balanced Incomplete Block Design. Here we will just give a very brief introduction; more detail can be found in Dean and Voss (1999, Chapter 11), who discuss PBIBDs and the more general group divisible designs.

In the BIBD, the parameter λ describes how many times a particular treatment is paired with all others. As we have seen, this is a very important parameter of the design, as it plays a key role in the variance of the treatment effect estimate; see (6.7). In a PBIBD, there are two values of λ defining two *associate classes* of treatments. If $\lambda_1 > \lambda_2$, then there is better information on the treatments in the λ_1 class.

This design is useful when a BIBD is too costly, or if some treatments are more important than others. The analysis is a bit more complicated than the BIBD.

6.3 Fractional Factorial Designs

At the beginning of this chapter we saw, in Example 6.3, how a BIBD cycles through a design and confounds each effect with blocks, so in the end we can recover information about each effect. If we run only a piece of the design, however, there will be a loss of information in that some effects will be confounded, and will not be estimable. This is the idea behind *Fractional Factorial Designs*.

Introduction

In a fractional factorial design we purposely confound some effects in order to run a smaller design. The key is to understand the confounding, so that the important information is not lost.

Example 6.9. A SIMPLE FRACTIONAL FACTORIAL   For example, suppose that we have the design of Example 6.3 and only run the first block

$$\boxed{\text{a ab ac abc}}$$

Formally, this is a $1/2$ *replication of a* 2^3 *factorial.* That is, the original treatment design is a 2^3 factorial (three treatments each at 2 levels in a crossed design), and the full factorial would require $2^3 = 8$ observations. However, we are only taking $2^3/2 = 4$ observations in the $1/2$ rep.

If we actually run this design, there is a massive amount of confounding that we can summarize as follows:

	Effect						
Trt. Comb	A	B	C	AB	AC	BC	ABC
a	+	−	−	−	−	+	+
ab	+	+	−	+	−	−	−
ac	+	−	+	−	+	−	−
abc	+	+	+	+	+	+	+

First note that the A effect is confounded with blocks, and cannot be estimated. (This should be no surprise, as only the high level of A is in this block.) However, note that there is confounding of pairs of effects, as they have the same contrast. Thus, B and AB are confounded, C and AC are confounded, and BC and ABC are confounded. This means that we can use the contrast to estimate one of the confounded effects only if we assume that the other effect is zero. ‖

In this section we will only deal with factorial designs in which each treatment has two levels. More general cases, such as 3^p designs, where each factor has 3 levels, or 2^p3^r, where factors can have either two or three level, are considerably more complex, and, in practical terms, somewhat less useful. See Cox and Reid (2000, Section 6.3) for some of the theory and Dean and Voss (1999, Chapter 14) for applications and examples.

For a 2^p factorial, where p is any integer, we can take a fractional replication of size $1/2^q$, for $q < p$. This would result in 2^{p-q} observations in one block. Moreover, there are many ways to choose the composition of the blocks, depending on which effects are to be confounded.

Example 6.10. A SIMPLE FRACTIONAL FACTORIAL CONTINUED From Example 6.9 it should be clear that it is very important to understand the confounding in a fractional replicate. For example, if there is interest in B, the above design is only useful if we assume that the AB interaction is zero. However, consider the fraction

$$\boxed{\text{(1) b ac abc}}$$

Here, the B effect is confounded with the ABC interaction, which we might be more comfortable in assuming is equal to zero (Exercise 6.14). ‖

Some observations about fractional factorials:

(1) It is important to understand the subject matter so that the effect of confounding structure can be minimized.
(2) The statistical analysis must be carefully planned, so valid inferences can be made in the face of the confounding.
(3) Often there is no within error estimate, and tests are done against higher-order interactions. As we have seen (Section 2.5) this will lead to conservative tests and confidence intervals.

Alias Sets and Modular Arithmetic

The confounding structure of a fractional factorial can be understood using the idea of *alias sets*, which are generated through a use of modular arithmetic.

Definition 6.11. Treatment combinations that are confounded with each other are called *aliases*. An *alias set* consists of all treatment combinations that are estimated by the same contrast. The alias set containing the overall mean is called the *defining contrast*.

Alias sets and blocks are easily found using modular arithmetic, in particular arithmetic *mod 2*, in which 0 and 1 are the only values we use, and we invoke the modular identity $1 + 1 = 0$.[1]

Now, with each effect we associate a variable x_i, $i = 1, \ldots, p$ that can only take on the values 0 (for the low level) and 1 (for the high level). So, for example, if A is at the high level we would have $x_1 = 1$, and if C is at the low level we would have $x_3 = 0$.

Example 6.12. A SIMPLE FRACTIONAL FACTORIAL CONCLUDED   To confound B with ABC, we simply write

$$x_2 = x_1 + x_2 + x_3, \text{ or } x_1 + x_3 = 0,$$

The solutions to $x_1 + x_3 = 0$ generate the treatment combinations in one block, and the solutions to $x_1 + x_3 = 1$ generate the treatment combinations in the other block. Thus the blocks are

$x_1 + x_3 = 0$		$x_1 + x_3 = 1$	
000	(1)	100	a
010	b	001	c
101	ac	110	ab
111	abc	011	bc

---

[1] Modular arithmetic essentially replaces the sum with the remainder of the sum when divided by the modular base. Thus, for mod 2 arithmetic, $1 + 1 = 2$, but 2 has remainder 0 when divided by 2, so $1 + 1 = 0$. This is an oversimplification, but it is all that we need to deal with 2^p designs. See Cox and Reid (2000, Appendix B) for an introduction to modular arithmetic for fractional factorials.

Note that the equation $x_1 + x_3 = 0$ is the defining contrast, as it is satisfied by the overall mean, with low levels for all factors. The block containing this defining contrast is sometimes known as the *intrablock subgroup*.

To find the alias sets, we can do as in Example 6.9 and write out the contrasts. Although this is preferred, it could get unwieldy for large factorials. Alternatively, we can cycle through the defining equation. To do this, start with the defining contrast $x_1 + x_3 = 0$ and cycle through by adding x_is to each side. Thus,

$$x_1 + x_3 = 0, \quad \text{so } \{AC, I\} \text{ is an alias set,}$$
$$x_1 + x_1 + x_3 = x_1 + 0 \Rightarrow x_3 = x_1, \quad \text{so } \{A, C\} \text{ is an alias set,}$$
$$x_2 + x_3 = x_2 + x_1, \quad \text{so } \{BC, AB\} \text{ is an alias set,}$$
$$x_3 + x_2 + x_3 = x_3 + x_2 + x_1 \Rightarrow x_2 = x_3 + x_2 + x_1,$$
$$\text{so } \{B, ABC\} \text{ is an alias set.}$$

$$\|$$

Note a few things about these calculations:

(1) We have repeated used the fact that $2x_i = 0$.
(2) We could use *either* block in running the 1/2 rep and we would get the same information.
(3) There are, in fact, seven distinct pairs of blocks that comprise seven distinct 1/2 replicates, each with different alias sets (see Exercise 6.15).

Of course, we are not limited to running just one block. In fact, with smaller factorials it is often the case that all of the blocks are run, so we do the complete factorial in blocks.

Fig. 6.1. The full 2^3 factorial takes the observations at each corner of the cube (a). Two possible fractions are shown in (b) and (c), corresponding to the blocks in Example 6.13. Design (c) is also the fraction used in Example 6.14.

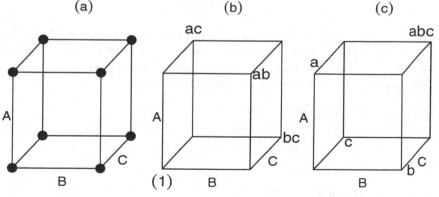

Example 6.13. COMPLETE FACTORIAL IN BLOCKS   The design in Example 6.12 is really not very good, as it confounds main effects. A better design is obtained from the defining contrast $x_1 + x_2 + x_3 = 0$ or 1 which gives the two blocks

(1)	a
ab	b
ac	c
bc	abc

The threeway interaction is confounded with blocks, and if we assume that the twoway interactions are zero, we can get a residual term to estimate the error. This results in the anova table

Source	df
Blocks	1
A	1
B	1
C	1
Residual	3

Figure 6.1 shows the relationship between the full factorial and the two blocks.
‖

There are other options in running a fractional factorial, which we look at in the following example.

Example 6.14. A CLASSIC FRACTION   In a classic agricultural experiment, a 2^3 factorial was planned where the three treatments were high and low levels of fertilizer. The factors are CaO: Lime, P: Phosphorus, and K: Potassium. A 1/2 rep was run with the following data (tons/acre):

Trt. Comb.	Yield
c	27.4
p	17.0
k	22.0
cpk	24.3

Here the defining contrast is $x_1 + x_2 + x_3 = 0$ and the alias sets are $\{I, ABC\}$, $\{A, BC\}$, $\{C, AB\}$, and $\{B, AC\}$. Thus, we can make inferences about the main effects if we assume that the twoway interactions are zero.

We can estimate the effects with contrasts as follows:

$$\text{CaO} : c + cpk - p - k = 12.7,$$
$$\text{P} : p + cpk - c - k = -8.1,$$
$$\text{K} : k + cpk - c - p = 1.9.$$

The anova for these data is

Source	df	SS
CaO	1	40.32
P	1	16.40
K	1	.90

so, unless we do something clever, we have no error estimate and hence no tests.   ‖

We have a number of choices in obtaining an error estimate

(1) We can run a second block with the same treatments, and estimate the error from the residual. This would be a good option if there is no other effect that we would like to estimate, and are satisfied with our assumptions about which effects can be assumed to be zero.
(2) We can run a second block with different treatments (using another defining equation) and estimate the error with the residual. This would be a good option if there is another effect that we would like to estimate.
(3) We can replicate the experimental unit and have a within error. This would maximize our error degrees of freedom.

Example 6.15. A CLASSIC FRACTION CONTINUED   Here we look at the three options described above for obtaining an error estimate. Consider the three experiments:

	Experiment	
1	2	3

c 27.4	c 24.0	c 27.4	(1) 25.2	c 27.4, 26.6
p 17.0	p 27.3	p 17.0	c 26.4	p 17.0, 26.8
k 22.0	k 27.5	k 22.0	p 23.9	k 22.0, 24.5
cpk 24.3	cpk 26.8	cpk 24.3	cp 26.7	cpk 24.3, 23.4

Experiment 1 allows us to estimate the main effects, and have 3 degrees of freedom for error. Experiment 2 uses a different block with defining contrast $x_3 = 0$, which confounds the CP interaction with CPK, and thus allows for the estimation of the CP interaction under the assumption that the threeway interaction is zero. That is, we can estimate the CP effect by

$$(cp - p) - (p - (1)),$$

which estimates the CP interaction if we assume that the CPK interaction is 0, as both estimates have the same contrast. We pay for this additional effect estimate by losing one degree of freedom in the residual, leaving us with two degrees of freedom for error. Finally, Experiment 3 allows the estimation of the main effects with 4 degrees of freedom for within error.

The anova for Experiment 1 is

Source	df	SS	MS	F	p-value
Blocks	1	27.75			
C	1	9.461	9.461	0.576	0.503
P	1	3.781	3.781	0.230	0.664
K	1	3.001	3.001	0.183	0.699
Residual	3	49.324	16.441		

where we see that none of the main effects are significant. For the other two anovas see Exercise 6.16   ‖

Model and Distribution Assumptions

The fractional factorial design will typically be run as a CRD (if there is only one block) or an RCB (if there is more than one block). Therefore, the model and distribution assumptions are taken from those designs; see Section 2.2 for the CRD and Section 3.2 for the RCB.

However, we will typically add assumptions about interactions, in particular that certain interactions are zero so that we can estimate the effects of interest. The form of these assumptions will be dictated by the subject matter and the design.

Typically, we will assume that most or all of the interactions are zero; this will be necessary in order to get estimates of the main effects. For example, suppose that we have the twoway model

$$(6.9) \qquad y_{ijk} = \mu + \tau_i + \beta_j + \gamma_k + \varepsilon_{ijk},$$

where τ_i and γ_k are the treatments and β_j are the blocks, and we run the 2^2 factorial in blocks of size 2:

$(1) = y_{111}$
$ab = y_{212}$

$a = y_{221}$
$b = y_{122}$

Our estimate of the A effect (τ in the model) is

$$\text{high } A - \text{low } A = \frac{1}{2}[(y_{212} + y_{221}) - (y_{111} + y_{122})],$$

and, under model (6.9), we find that (Exercise 6.33)

$$(6.10) \qquad \mathrm{E}\left(\frac{1}{2}[(Y_{212} + Y_{221}) - (Y_{111} + Y_{122})]\right) = \tau_2 - \tau_1,$$

where the β_j actually cancel out in the contrast (not just in expectation). If we were to only estimate the high level of A using $(y_{212} + y_{221})/2$, this estimate would also be unbiased, and the block effect would be zero in expectation. In the contrast the effect actually cancels, making for a more precise comparison.

We attach standard errors to these contrasts in the usual way, taking our estimate of variance from the MS(Residual) in the anova. Thus, for example,

$$\text{Var}\left(\frac{1}{2}[(Y_{212} + Y_{221}) - (Y_{111} + Y_{122})]\right) = \left(\frac{1}{4} + \frac{1}{4} + \frac{1}{4} + \frac{1}{4}\right)\sigma_\varepsilon^2.$$

Example 6.16. A CLASSIC FRACTION CONCLUDED   Finishing Example 6.15, in Experiment 1 we can estimate the P effect by

$$\frac{1}{4}[(17 + 24.3 + 27.3 + 26.8) - (27.4 + 22 + 24 + 27.5)] = -1.375,$$

with variance estimated by

$$8 \times \left(\frac{1}{4}\right)^2 16.441 = 8.2205.$$

Thus, to test H_0: no P effect, we compare $-1.375/\sqrt{8.2205} = -0.479$ to a t distribution with 3 degrees of freedom. Of course, we accept H_0. (Compare the t-statistics to the F-statistics in the anova.)   ‖

Larger Factorials, Smaller Fractions _____

For smaller fractions, which are more likely the case if the factorial is larger, we will need more than one defining equation.

Example 6.17. FRACTIONING A 2^5 FACTORIAL   For a $1/2^2 = 1/4$ replication of a 2^5 factorial, we would take $2^5/2^2 = 8$ observations. Since we are only doing $1/4$ of the design, the alias sets will contain four treatment combinations. We construct these sets with two defining contrasts

For example, suppose that we confound the mean with the fiveway interaction and a fourway interaction, so we have

(6.11) $0 = x_1 + x_2 + x_3 + x_4 + x_5$ and $0 = x_1 + x_2 + x_3 + x_4.$

Each of the contrasts can be either 0 or 1, and we now can construct the four blocks by equating the contrasts to the four combinations $(0,0)$, $(0,1)$, $(1,0)$, and $(1,1)$. For example, the intrablock subgroup satisfies (6.11) and is given by

$$(1), bc, bd, cd, abe, ace, ade, abcde.$$

Note that if each term in (6.11) is 0, the sum of the two terms will also be zero, so the sum is immediately in the alias set. Since the sum is x_5, to construct the alias sets we then cycle through the equation

$$0 = x_1 + x_2 + x_3 + x_4 + x_5 = x_1 + x_2 + x_3 + x_4 = x_5$$

by successively adding terms x_i to each piece. For example, the first alias set is $\{(1), ABCDE, ABCD, E\}$, and if we add x_1 to each piece we get the second alias set $\{A, BCDE, BCD, AE\}$. Continuing, we get the complete set of aliases

$$\{(1), ABCDE, ABCD, E\} \quad \{A, BCDE, BCD, AE\}$$
$$\{AB, CDE, CD, ABE\} \quad \{ABC, DE, D, ABCE\}$$
$$\{B, ACDE, ACD, BE\} \quad \{BC, ADE, AD, BCE\}$$
$$\{C, ABDE, ABD, CE\} \quad \{CD, ABE, AB, CDE\}.$$

Note that (6.11) is not a particularly good defining contrast, as we have confounded a main effect with the mean. For an alternative, see Exercise 6.17

‖

The pattern for even larger factorials, or smaller fractions, should now be clear. If we were to do a $1/2^3 = 1/8$ replication, we would start with three terms, and the blocks would be formed by equating these terms to either 0 or 1 in the eight combinations $(0, 0, 0)$, $(0, 0, 1)$, $(0, 1, 0)$, ..., $(1, 1, 1)$.

There is an alternate way of calculating the alias sets, based on a slight variation of the modular arithmetic explained here. We have chosen not to use that approach, as it is somewhat artificial and has the disadvantage of not being able to give the composition of the blocks. See Exercise 6.18 for more details.

6.4 Variations on a Theme

In this section we will look at a number of examples that go a little beyond the designs that we have been discussing. Although many of the examples here are in the context of microarray analysis, this merely reflects the fact that we are dealing with designs that have blocks of size two. We start with a specialized design that has some interesting properties.

Balanced Lattice Designs

As mentioned in Section 3.6.3, the existence of orthogonal Latin squares results in a specialized design which, if it can be applied, can be quite useful in practice. We start with the following definition (see, for example, Kuehl 1994).

Definition 6.18. An incomplete block design with each treatment appearing r times is *resolvable* if the blocks can be divided into r groups with each group having a complete replication of the treatments.

Table 6.2. A possible field layout for a balanced lattice square with $t = 9$ treatments.

	Rep 1				Rep 2				Rep 3				Rep 4		
	Row				Row				Row				Row		
	1	2	3		1	2	3		1	2	3		1	2	3
1	6	4	5	1	4	2	9	1	8	4	3	1	7	1	4
Col. 2	3	1	2	Col. 2	3	7	5	Col. 2	6	2	7	Col. 2	9	3	6
3	9	7	8	3	8	6	1	3	1	9	5	3	8	2	5

Complete designs, such as the RCB, are resolvable, while a BIBD is typically not resolvable. The advantage of a resolvable design is that the treatments are balanced across the blocks and it is then possible to estimate treatment contrasts free of block effects. Lattice designs are constructed to be resolvable.

We will look at two types of lattice designs. In this section we will see the properties of the more restrictive *balanced lattice square*, and we look at the less restrictive *balanced lattice* in Exercise 6.23. The major difference in these designs is that the lattice square controls for both row and column effects, while the lattice controls only one blocking factor.

Suppose that the number of treatments, t, is a square. If a set of $\sqrt{t} + 1$ orthogonal Latin squares of side t exist, then we can construct a *balanced lattice square design* for t treatments. Such designs exist for t that are powers of a prime number so, for example, we have these designs for $t = 9, 16, 25, \ldots$.

The balanced lattice square design has the property that *each pair of treatments appears once in each row and once in each column.*

Example 6.19. A BALANCED LATTICE SQUARE WITH $t = 9$ For $t = 9$ a balanced lattice square can be constructed with four replications of the 9 treatments. Once such design is given in Table 6.2. Notice that each replication contains all 9 treatments (resolvable). If we consider the rows as blocks, there are 12 blocks forming a BIBD. However, for this design there is more. If we consider the columns as blocks, there are 12 blocks forming a BIBD. As these BIBDs come from an orthogonal Latin square construction, this is how we obtain the property that each pair of treatments appears exactly once in each row and once in each column.   ‖

This highly structured design is attractive to experimenters (especially in the field) because it allows the comparison of treatments in small, manageable blocks. Moreover, if there is some incident that results in loss of data (the tractor overturns on the plants), only that small block need be redone in order to recover the entire design. There is also a variance advantage, which we will soon see.

A model for the analysis with t treatments is

$$(6.12) \qquad Y_{ij\ell m} = \mu + \tau_i + \beta_{jm} + \gamma_{\ell m} + R_m + \varepsilon_{j\ell m}, \qquad \sum_i \tau_i = 0,$$

where we have to be careful about indices. Defining $k = \sqrt{t}$ and $r = \sqrt{t} + 1$, we have

$$j = 1, \ldots k, \quad \ell = 1, \ldots k, \quad m = 1, \ldots, r.$$

The index i is redundant, but we need to set it up to keep track of the treatments. For each $i = 1, \ldots t$, we define the index set

$$i_{j\ell m} \in I_i = \{(j\ell m) : \text{the response at } (j\ell m) \text{ is from treatment } i\}.$$

To complete the model specification, we assume that rows, columns, and replications are random effects with

$$\beta_{jm} \sim \mathrm{N}(0, \sigma_\beta^2), \quad \gamma_{\ell m} \sim \mathrm{N}(0, \sigma_\gamma^2), \quad R_m \sim \mathrm{N}(0, \sigma_R^2), \quad \varepsilon_{j\ell m} \sim \mathrm{N}(0, \sigma_\varepsilon^2).$$

The anova table is

Source	df
Reps	$r - 1$
Columns	$r(k - 1)$
Rows	$r(k - 1)$
Treatments	$t - 1$
Residual	$r(k - 1)(k - 1) - (t - 1)$
Total	$rt - 1 = rk^2 - 1$

where we see that the degrees of freedom for treatments come out of the "residual" from the rows and columns. In fact,

$$\mathrm{SS}(\text{Residual}) = \mathrm{SS}(\text{Rows} \times \text{Columns in Reps}) - \mathrm{SS}(\text{Treatments}),$$

and treatments are crossed with reps but are not crossed with rows or columns.

As usual, we estimate the treatment effect τ_i with least squares. Although the derivation is straightforward, it does take effort to keep the indices straight. Moreover, as we saw with the BIBD, it is important to do the least squares derivation in order to get the correct divisor. The result is

$$(6.13) \qquad \hat{\tau}_i = \frac{t - 1}{(c - 1)^2}(\bar{y}_i - \bar{y}_{iR} - \bar{y}_{iC} + \bar{\bar{y}}),$$

where

$\bar{y}_i = $ mean of observations getting treatment i,

$\bar{y}_{iR} = $ mean of observations in rows containing treatment i,

$\bar{y}_{iC} = $ mean of observations in columns containing treatment i,

$\bar{\bar{y}} = $ mean of all observations (grand mean),

which, through virtue of being least squares, is an unbiased estimator of τ_i. The least squares estimator looks somewhat like a residual, and that is what it essentially is. Without the treatments, the Rows, Columns, and Replications will use up all of the degrees of freedom, so (as the anova table shows) the treatment effect comes out of the residual.

When evaluating a complex design such as this one, we should always ask "What do I get for all of this effort?" with the expected answer, we hope, having something to do with a variance reduction. Recall that the effort in the BIBD resulted in least squares estimates that were free of block variances, which we typically expect to be bigger than σ_ε^2. Here, we are blocking in both rows and columns, but because we have BIBDs in both directions, this results in the variance of $\hat{\tau}_i$ being free of all variance components except σ_ε^2 (Exercise 6.35). Also, the variance of any contrast $\sum_i a_i \hat{\tau}_i$ only depends on σ_ε^2. So there is a good return for our efforts.

Lattice designs were first introduced by Yates (1936, 1940). Cochran and Cox (1957) have a detailed discussion and list a number of experimental plans, although the most detailed discussion is in Federer (1955). There are more general forms of this design, for example a rectangular lattice, which is related to a PBIBD. Other developments in lattice designs can be found in Speed *et al.* (1985) and Federer (1998).

Latin Squares and Fractional Factorials

In some situations, a Latin square (Section 3.6.3) turns out to be a fractional factorial. When this is recognized it is then easy to understand the confounding structure.

Example 6.20. A MICROARRAY LATIN SQUARE

> Kerr *et al.* (2000) describe the following experiment, a microarray experiment done as a Latin square. In the experiment mRNA samples were obtained from human liver tissue, and were compared to muscle tissue. The design used two arrays in a two-dye system, and the colors were swapped according to the following diagram:

This is also a SCOD!

	Array 1	2
Dye		
Red	Liver	Muscle
Green	Muscle	Liver

with model

$$(6.14) \quad \log Y_{ijkg} = \mu + A_i + D_j + T_k + G_g + (AG)_{ig} + (TG)_{kg} + \varepsilon_{ijkg},$$

where Y_{ijkg} is the log expression level, μ is the overall mean, A_i is the array effect, D_j is the dye effect, T_k is the treatment effect and ε_{ijkg} is the error.

With 4 treatments there are 16 total effects in the model: the mean, 4 main effects, $\binom{4}{2} = 6$ twoway interactions, $\binom{4}{3} = 4$ threeway interactions, and 1 fourway interaction. However, we cannot estimate these main effects because all of the treatment combinations are not observed. Note that the index set does not run through all values because the design is not complete. We have $g = 1, \ldots, n$, but

$$(i, j, k) \in \{(1,1,1), (1,2,2), (2,1,2), (2,2,1)\}.$$

This leads to the following confounding structure:

mean ~ ADT	G ~ ADTG
A ~ DT	AG ~ DTG
D ~ AT	DG ~ ATG
T ~ AD	TG ~ ADG

Note that G is crossed with all treatments, so the right table is just a rewrite of the left table with the G effect included. Thus, the presence of G does not impact the confounding. In each table, the effects in the left column are the ones that appear in the model and are not confounded with each other. All of the main effects are confounded with twoway or fourway interactions, while the twoway interactions in the model are confounded with threeway interactions.

To have valid inferences from this design we must assume that the effects on the right side are all zero, because the sums of squares for each pair of effects are the same. Also, note that the DG effect, the Dye × Gene interaction, is not in model (6.14). This term becomes the residual error, under the assumption that there is no Dye × Gene interaction.

The three treatments are each at two levels, and to better understand the confounding we can write out the complete table of contrasts

	Effect			
Trt. Comb	Array	Dye	Tissue	D × T
1 R L*	−	−	−	+
1 R M	−	−	+	−
1 G L	−	+	−	−
1 G M*	−	+	+	+
2 R L	+	−	−	+
2 R M*	+	−	+	−
2 G L*	+	+	−	−
2 G M	+	+	+	+

where the treatment combinations marked with * are the ones that were run. These combinations are, in fact, a 1/2 rep of the 2^3 where the threeway interaction is confounded with blocks, and the confounding structure reflects the alias sets. So this Latin square is, in fact, a fractional factorial.

Continuing with the analysis, in this microarray design the main interest was in determining which genes are differentially expressed, that is, which genes react differently to the different treatments. This effect is the Treatment × Gene interaction, and for gene g it is estimated by

$$(\hat{TG})_{1g} - (\hat{TG})_{2g} =$$

$$(6.15) \qquad \frac{1}{2} \log\left(\frac{y_{111g} y_{221g}}{y_{122g} y_{212g}}\right) - \frac{1}{2n} \log\left(\prod_{g'=1}^{n} \frac{y_{111g'} y_{221g'}}{y_{122g'} y_{212g'}}\right)$$

This design allows us to estimate this effect and its error, and to test significance (see Exercise 6.24).   ‖

Reference and Loop Designs

We have already seen that a microarray experiment can be viewed as a split plot design. Now we look a bit deeper at designs for the two-dye systems, which add another layer of complexity to the design at the whole plot level. The fact that there are two dyes – measuring RNA from two experimental units on one chip – automatically implies that the experiment is an incomplete block design (unless there are only two treatments, in which case it is an RCB). And it is a very simple incomplete block in that we are forced to have only two treatments per block. Thus, if there are t treatments, to balance the design we would need $\binom{t}{2}$ microarrays.

As we know, to compare two varieties of plant, we want to grow them as close to one another as possible so that we control for all other factors in making the comparison. This is the basic principle of blocking. For the microarray, by construction, we have this principle in force, allowing us to have very good designs.

In the early days of microarray design there was much concern about *dye bias*, that is, the red and green dyes might fluoresce at different rates. To control for this, the experiment would include a dye swap, for example if treatments A and B are on one microarray with dyes red/green respectively, a second A/B microarray would be run with dyes green/red, that is, with dyes swapped. This would be done for all treatment pairs. Happily, the technology has progressed and dye bias is not the concern that it once was. However, if the experimenter can afford, it would not hurt to control for it.

Unfortunately, instead of the BIBD, another two-dye design emerged, called the *reference design*. It was much simpler that the BIBD, so it gained popularity, even though it is quite wasteful of resources. Instead of comparing two treatments on the same microarray, the reference design compares each treatment to a "reference" pool of DNA. A reference design is compared to a BIBD in Table 6.3. Reference designs probably became popular because the

Table 6.3. Experimental designs to study three varieties of interest in three blocks (microarrays) of size two. Varieties A, B, and C are of interest, while R represents a fourth variety (not of interest) to serve as a reference. Kerr and Churchill (2001a)

Reference Design		
Block 1	Block 2	Block 3
A	B	C
R	R	R

BIBD		
Block 1	Block 2	Block 3
A	B	C
B	C	A

reference can be ignored and we can treat each microarray as an observation on that treatment. But, as Kerr and Churchill (2001a) note, "Introducing a reference as an intermediate step is unnecessary and generally inefficient because it means that fully half of the data are dedicated to an extraneous sample". Also note that the BIBD gives us two observations on the varieties of interest for the same effort, and also lowers the treatment variance (see Exercise 6.34).

Example 6.21. MICROARRAY BIBD   As an example, consider this experiment on the effect of Aluminum on Zebrafish. Three treatments are to be compared: Control, $AlCl_3$(aluminum chloride), and Nano (aluminum nano particles). The treatments are applied to tanks holding the Zebrafish, RNA is then extracted from their gill tissue and subject to microarray analysis using a two-dye system.

Notice that the experimental unit is the tank, as in Exercise 1.3. The RNA from all the fish in one tank will be combined (pooled) into one sample. This, of course, does not lose degrees of freedom as the individual fish are subsamples.

With these three treatments, two designs are considered. The first is the reference design[2] and the second is a "loop" design (Kerr and Churchill 2001b; see also Simon *et al.* 2003), which allows the balancing of the dye effects. It is called a loop design because it cycles through all of the treatments, with each treatment being applied twice, once with the green dye (Cy3) and once with the red dye (Cy5). Note that in the reference design we do not have to worry about dye bias, as the bias will be constant throughout the experiment

---

[2] One typical choice for the reference RNA is a pool of all the RNA in the experiment.

Fig. 6.2. Loop designs (*a*) using three blocks, balancing dyes, and (*b*) balancing dyes with six blocks.

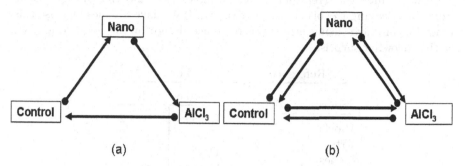

(a)   (b)

(as long as we label the treatments with the same color) and thus comparisons will be free of the dye effect.

Dye	Reference			Loop		
Green	Ref.	Ref.	Ref.	Cont.	Nano	AlCl$_3$
Red	Cont.	AlCl$_3$	Nano	Nano	AlCl$_3$	Cont.

To see why they are called loop designs, look at Figure 6.2 which shows the design as a loop. The point of the arrow represents Cy5 (red dye) and the base of the arrow represents Cy3 (green dye). Figure 6.2(*a*) is equivalent to the loop design above, and Figure 6.2(*b*) is a double loop, where all of the dyes are reversed. This latter design is looked at further in Exercise 6.26

To analyze the designs, we can first look at a one-gene model

$$(6.16) \qquad y_{ijk} = \mu + \tau_i + \beta_j + D_k + \varepsilon_{ijk},$$

where τ_i is the treatment effect, β_j is the block effect, and D_k is the dye effect. Note that for the reference design, we would not include the dye effect in the model. If we were to run these designs taking six observations and calculate the variance of a treatment difference, we would find

$$(6.17) \qquad \text{Reference Design} : \text{Var}(\hat{\tau}_i - \hat{\tau}_{i'}) = 2\sigma_\varepsilon^2 + 2\sigma_\beta^2,$$

$$(6.18) \qquad \text{Loop Design} : \text{Var}(\hat{\tau}_i - \hat{\tau}_{i'}) = \sigma_\varepsilon^2 + \frac{1}{2}\sigma_\beta^2 \, nonumber$$

which clearly shows the wastefulness of the reference design.

If we add the gene effect to the model, we have

$$(6.19) \qquad y_{ijkg} = \mu + \tau_i + \beta_j + D_k + G_g + (\tau G)_{ig} + (\beta G)_{jg} + \varepsilon_{ijkg},$$

where we include the Treatment × Gene interaction and the Block × Gene interaction, leaving the others to the residual. If we have n genes and use this model for an anova using either the reference design or the loop design, we get the following surprise

Reference			Loop	
Source	df		Source	df
Blocks	2		Blocks	2
Trts	3		Trts	2
Genes	$n-1$		Dye	1
T × G	$3(n-1)$		Genes	$n-1$
B × G	$2(n-1)$		T × G	$2(n-1)$
Residual	0		B × G	$2(n-1)$
Total	$6n-1$		Residual	$n-1$
			Total	$6n-1$

So in the reference design all of the residual degrees of freedom are "wasted" in the T × G interaction, which has a good bit of it devoted to measuring the effect of the unimportant reference treatment. Recall how the tests would be run in this anova. The gene effect would be tested with the B × G interaction, so that test is OK. However, the tests on treatment, including main effects and the T × G interaction, would be tested against the interactions of treatments and blocks. This is what would make up the residual term, which would then be used for those tests. However, we have no degrees of freedom left to estimate the residual because of the extra degrees of freedom used by the T × G interaction.

In contrast, the loop design has one fewer treatment and, consequently, there are degrees of freedom left over to test treatments and their interactions, using the residual error term. See Exercise 6.26

We also note that there is some concern that pooling the error term over all genes may not be optimal, as the genes may not all have the same variance. Procedures for dealing with this are being developed. See, for example, Cui et al. (2005).   ‖

Beyond Loops to BIBDs

Loop designs have the advantage of (i) balancing dyes and (ii) providing good comparisons between treatments that are adjacent. If there is interest in getting good comparisons between treatments that are not adjacent in the original loop, the design can be augmented. That is, blocks can be added to obtain good comparisons between additional treatments, while maintaining balance with respect to the dyes.

Example 6.22. AVOCADO GENE EXPRESSION   The experiment first described in Exercise 3.14 has a second part, which we now describe. For each avocado plant, RNA was obtained from eight different tissues:

Number	Name		Number	Name
1	medium bud		5	petal
2	small bud		6	stamen
3	leaf		7	carpel
4	sepal		8	fruit

where the numbers refer to Figure 6.3, which displays the types of designs that were considered. The goal of the experiment is to measure differential gene expression in the tissues.

Fig. 6.3. Graphical display of the designs considered in Example 6.22. The outer black arrows are a loop design, the outer black arrows plus the inner black arrows are the design actually performed, and all arrows plus the gray dotted lines are a BIBD. The point of the arrow represents Cy5 (red dye) and the base of the arrow represents Cy3 (green dye).

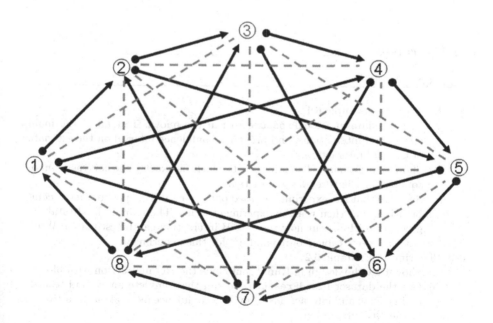

The outer black arrows are a loop design for this experiment. For treatment comparisons in the loop design, using the one-gene model (6.16), the variance of a difference is

$$(6.20) \qquad \text{Adjacent} : \mathrm{Var}(\hat{\tau}_i - \hat{\tau}_{i'}) = \sigma_\varepsilon^2 + \frac{1}{2}\sigma_\beta^2,$$

$$\text{Non-Adjacent} : \mathrm{Var}(\hat{\tau}_i - \hat{\tau}_{i'}) = \sigma_\varepsilon^2 + \sigma_\beta^2.$$

The actual design that was done was the augmented loop, where the orange and blue arrows were added to the original loop design. Adding these eight blocks allows for better comparisons, and gives us three classes of comparisons:

(i) adjacent treatments,
(ii) non-adjacent treatments with a common block, such as $\tau_2 - \tau_5$,
(iii) non-adjacent treatments with no common block, such as $\tau_2 - \tau_4$.

As might be expected as we go from $(i) \to (iii)$ the variance of a difference increases, so the augmented blocks should be chosen to decrease the variance on the more important comparisons.

We could further augment this design with the gray dashed lines in Figure 6.3, adding 12 more blocks. This new design is, in fact, a BIBD, and will provide even better (and equal) variances on all treatment contrasts. However, the disadvantage is that the BIBD is not balanced for the dye effect, so if this is a concern the BIBD may not be preferred. See Exercise 6.28 for more details.
‖

6.5 Exercises

Essential

6.1 Referring to Example 6.1
(a) Use the dataset **BloodPressure** to run the anova. Test all of the main effects and interactions, and provide a confidence interval on the estimate of the BC interaction.
(b) Show that the sum of squares due to blocks is exactly the contrast sum of squares for the threeway interaction.
(c) Suppose that the experimenter used only four subjects per treatment combination, and then ran the experiment twice. There are still 63 total degrees of freedom, but now the T × B interaction can be estimated. Write out the anova table (source and df) for this experiment.

6.2 Referring to Example 6.2:
(a) Show that the design in Run 2 confounds the BC interaction with blocks.
(b) Use the dataset **BloodPressure2** to run the complete anova and test all main effects and interactions. Provide a confidence interval on the estimate of the BC interaction.
(c) Compare you confidence interval in part (c) to that of Exercise 6.1. Explain any differences.

6.3 Referring to Example 6.3:

(a) The six blocks that could be used to estimate the A × Block interaction were marked with an asterisk (*). For each of the other six treatment × block interactions, identify the six blocks that could be used to estimate it.

(b) This design is a BIBD with $b = 14$ and $t = 8$. What are the other parameters?

(c) With eight treatments, what other BIBDs can be run? Are any others to be preferred?

6.4 Suppose that we have $t = 5$ treatments and we can only have blocks of size three.

(a) Find the smallest BIBD (fewest number of blocks). Give the values of r and λ

(b) If we could have blocks of size four, can we run a BIBD? Is there an advantage over the design in part (a)?

(c) Write out the layout of the design from part (b) showing how to arrange the treatments in blocks.

6.5 In general it can be somewhat tricky to find the layout of a BIBD. The following strategy can sometimes be used.

(a) Specific case.

   (*i*) Write out a treatment design for a 7 × 7 Latin square.

   (*ii*) Write rows 1, 2 and 4 of the 7 × 7 Latin square as its own treatment design. Is this a BIBD? If so, what are the parameters of the design?

   (*iii*) Write out the remaining rows of the 7 × 7 Latin square (3,5,6,7). If this is a BIBD what are the parameters?

(b) General case.

   (*i*) Write out the layout of a $p \times p$ Latin square (Section 3.6.3).

   (*ii*) Drop any column.

   (*iii*) Ignore rows, take the columns as blocks. Show that this design is a BIBD with $t = b = p - 1$, $r = k = p - 1$, and $\lambda = p - 2$.

6.6 A simpler version of the design in Example 6.3 occurs if we only have two treatments A and B, each with two levels in a factorial arrangement.

(a) Find the parameters of a BIBD with blocks of size 2 ($k = 2$). Explain why we must have $b = \binom{4}{2} = 6$.

(b) The six blocks are

$$
\begin{array}{cccccc}
1 & 2 & 3 & 4 & 5 & 6 \\
\boxed{\begin{array}{c}(1)\\ab\end{array}} & \boxed{\begin{array}{c}a\\b\end{array}} & \boxed{\begin{array}{c}a\\ab\end{array}} & \boxed{\begin{array}{c}(1)\\b\end{array}} & \boxed{\begin{array}{c}b\\ab\end{array}} & \boxed{\begin{array}{c}(1)\\a\end{array}}
\end{array}
$$

Each treatment combination, A, B, and AB, is confounded with a pair of blocks. Identify which blocks are confounded with each treatment combination.

(c) The anova table is

Source	df
Blocks	5
Trts	3
Trts × Blocks	3

which tells us that interactions A × Blocks, etc., are only estimated from one pair of blocks. Identify these pairs. In particular, explain why the A ×

Block interaction can only be estimated from the pair of blocks that are confounded with B, and not from the other pairs. Can you use this information to make a more general statement about estimating interactions?

6.7 An alternative to the design in Exercise 6.6 is to run the BIBD in blocks of size three.

(a) Find the parameters of the design. In particular, show that we need only four blocks.

(b) The four blocks are

1	2	3	4
(1)	(1)	(1)	a
a	a	b	b
b	ab	ab	ab

Write out the contrasts for A, B, and AB, and show that none of these effects are confounded with blocks. Write out the anova table and identify where the confounding is.

6.8 An alternative to the designs of Examples 6.1 and 6.2 is to use a split plot approach, with one of the treatments moved up to the whole plot level. (This would be considered if one factor is expected to have a big effect on the response or if there is one of lesser importance.) Suppose that this is the case for treatment C, and the experimenter is willing to trade off lesser precision on C for greater precision on A and B. The following design could be run:

Run 1				Run 2			
C				C			
Low		High		Low		High	
(1)	a	(1)	a	(1)	a	(1)	a
b	ab	b	ab	b	ab	b	ab

Note that this is a complete design – there is no confounding with blocks.

(a) Write out the anova table (source and df) and indicate how the tests will be done. Also comment on whether you would consider pooling error terms and how you would do it.

(b) Analyze the data in **BloodPressure2** according to this design. Test all main effects and interactions, and provide a confidence interval on the estimate of the BC interaction.

(c) Compare you confidence interval in part (b) to those in Exercises 6.1 and 6.2. Which one "wins"?.

6.9 Three factors, A, B, and C were each run at two levels in a BIBD. The data are

Rep 1		Rep 2		Rep 3		Rep 4	
B1	B2	B3	B4	B5	B6	B7	B8
(1) 10	a 17	(1) 11	a 8	(1) 6	b 9	a 17	(1) 9
ab 17	b 12	b 9	ab 9	a 15	ab 14	b 13	ab 15
c 9	ac 19	ac 16	c 6	bc 8	c 7	c 9	ac 17
abc 10	bc 11	abc 16	bc 2	abc 1	ac 14	abc 16	bc 14

In Rep 1 the AB interaction is confounded with blocks, which can be seen from the following contrast table:

	A B C AB		A B C AB
(1)	− − − +	a	+ − − −
ab	+ + − +	b	− + − −
c	− − + +	ac	+ − + −
abc	+ + + +	bc	− + + −

(a) For each of the other reps, find out which effect is confounded with blocks.

(b) Calculate the complete anova table and test the significance of the treatments.

(c) Estimate the main effects and give 95% confidence intervals.

6.10 Refer to the data of Example 6.7:

(a) What are the parameters of this BIBD?

(b) Show that, for this design, the least squares estimate of τ_1 (6.2) is unbiased.

(c) Verify the anova table

Source	df	SS	F	p-value
Blocks	3	6.17		
Treatments	3	880.83	4.042	.083
Residuals	5	363.17		

Note that these are the partial sums of squares, which can be produced in R by using the **drop1** command; see Exercise 2.31.

(d) Calculate the least squares estimates of τ_i and test the hypothesis H_0 : $\tau_1 = \tau_2$.

(e) Calculate the interblock estimates (see Miscellanea 6.6.1) of τ_i and compare them to the least squares estimates. Is there agreement?

6.11 Referring to Example 6.8:

(a) Give the parameters of the BIBD and show the arrangement of treatments in blocks. (Try to do this without looking at the data.)

(b) Verify the anova table. (The data are in dataset **Bone**, which is a representation of the original data.)

(c) Obtain the least squares estimates of the treatment effects and their variance.

(d) By examining pairwise difference in the least squares estimates, which treatment is the strongest? It also seems reasonable to test for a trend in the treatments, using polynomial contrasts. Do this and report your conclusion.

(e) Berkowitz *et al.* (2005) had 50 screws available and only 30 are needed for one complete run of the BIBD. They allocated the extra screws to the groups according to the following description:

A noticeable block-to-block variance was identified in the middle of the study; therefore, the same number of trials was not run for each group. It was necessary to run more trials for the groups with similar data, whereas fewer trials were run for groups that were clearly statistically different from all other groups.

Was this the best allocation of the extra observations? If so, explain why. If not, suggest a better plan.

6.12 Blood pressure is typically measured with a sphygmomanometer (the inflatable cuff put on the upper arm), but there are other means for doing this. When new blood pressure monitors are developed, their accuracy is tested against standard methods. A recent development is the easy to use *Dinamap automated oscillometric device* (see, for example, Chang *et al.* 2003 and the references there). To assess this new method, we could compare it to the standard sphygmomanometer and to a method known as intra-arterial recording, which is accurate but expensive. It is also known that blood pressure measurements could vary from the right arm to left arm, so a good design would take this into account. Therefore, we consider an experiment where the three types of measurement (S, D, I) are crossed with arms (R, L), giving six treatment combinations. Since we would consider each person a block, we necessarily are restricted to two treatments per block.

Try to do parts (a) and (b) before looking at the dataset.

(a) The design we would like to use is a BIBD. We have $t = 6$ and $k = 2$. Give the rest of the parameters for the smallest BIBD. List the treatment combinations that appear in each block.

(b) This design is, in fact, resolvable, in that we can divide it into *five* reps, where each rep contains all six treatment combinations. Do this.

(c) Data, systolic blood pressures of healthy subjects, are contained in dataset `BPMonitor` using the following code:

	Treatment					
	1	2	3	4	5	6
Monitor	S	I	D	S	I	D
ARM	R	L	R	L	R	L

The experimenters are particularly interested in whether the Dinamap gives readings equivalent[3] to the others and whether there is an interaction between monitors and arms. What can you tell them about this?

(d) Since a person has two arms, we are limited to $k = 2$ in any design. If 12 more subjects became available, how would you use them? Justify your choice.

Note: Cochran and Cox (1957) contain catalogs of design plans, including the ones asked for in this exercise.

6.13 As part of a project to understand the relationship of different genes to substantiality of crops – in particular, potatoes – a researcher investigated the effects of two factors, Photoperiod (P) and bioactive Tuber Inducing Factor solution (TIF), on gene expression levels. Each factor is at two levels (2 = high and 1 = low), and are crossed in a 2 × 2 factorial. Three separate experiments are to be run; in each genes from different tissues (leaf, root, stolen) will be used. The experiment uses an `Agilent` microarray chip, which is a two-dye system. Thus, two treatments can be applied to each array. (One treatment is labeled with red dye and the other with green dye.) With this chip there is less concern about dye bias, so the experimenter chose not to swap dyes, as in Examples 6.21 and 6.20.

For the following questions, just concentrate on one tissue, say leaf, as the same experiment design will be applied to the other tissues.

(a) The four treatment combinations are $(1), p, t, pt$. Four blocks (Arrays) were run as indicated. Here is a table of the effects, similar to that in Section 6.1. Finish the table.

---
[3] Notice that we are actually looking to accept the null hypothesis here. This puts us in the realm of *bioequivalence testing*, which can lead to a different type of testing procedure. A nice introduction to bioequivalence is given by Berger and Hsu (1996).

	Effect		
Array	Trt. Comb	P T PT	Confounded
1	(1)	− − +	T
	p	+ − −	
2	p		
	pt		
3	(1)		
	t		
4	(1)		
	pt		

(b) A BIBD with two treatment combinations per block would need six arrays. Give the parameters of the design and show how to estimate treatment differences. Can you add two arrays to the design in part (a) to have a BIBD?

(c) Sadly, we are stuck with the data collected according to the design in part (a). How can we do damage control? For example, suppose that we assume the model is of the form

$$Y_{ijk} = \mu + A_i + P_j + T_k + \varepsilon_{ijk},$$

where Y_{ijk} is the log of the expression level of the gene, A_i is the array (block) effect, and we assume that there is no interaction in the model. We can estimate treatment effects as follows: Since there is no interaction in the model, both $pt - p$ and $t - (1)$ estimate the effect of T, and these treatment combinations appear in the same array. Show that the differences $Y_{222} - Y_{221}$ and $Y_{312} - Y_{311}$ are both estimates of $T_2 - T_1$ that are free of array effects. Can you find other effect estimates that are free of block effects?

(d) The dataset PotatoLeaf1Gene has log expression level data for one gene, taken according to the design in part (a), and replicated three times. Verify that the anova is given by

Source	df	SS	MS
Rep	2	0.8647	0.4323
Treatments	3	2.0152	0.6717
Arrays	9	5.2162	0.5796
Rep × Trt	6	1.0071	0.1678
Residual	3	1.0877	0.3626

Use this information to get estimates and standard errors for the effects in part (c).

(e) The experimenter actually ran 12 arrays (three replications of a four-array experiment), and ended up with many confounded effects. If you had those twelve arrays before the data were taken, with the restriction that there can only be two treatment combinations per array, what design would you recommend? Explain why your design is better. (The three replications are for the three tissue types: leaf, root, stolen.) Can we run BIBDs?

(f) Each array actually had $11,412$ genes, and the design can be analyzed in a manner similar to that described in Miscellanea 5.9.1. Use the dataset PotatoLeaf to carry out such an analysis. (This dataset only has 150 of the 11,412 genes.) In particular, can you locate genes that have a large P or T effect?

6.14 : Referring to Example 6.10,

   (a) Show that B is confounded with ABC. List the contrast coefficients and describe all of the confounding.

   (b) Give a 1/2 replication of a 2^3 in which A is confounded with ABC, and one in which C is confounded with ABC.

   (c) Perhaps the best 1/2 replication of a 2^3 is the one given by $x_1 + x_2 + x_3 = 0$ or 1, which confounds the mean with ABC. (Example 6.13 gives the blocks.) Find the alias sets, and show that all main effects are confounded with twoway interactions, and hence are estimable if we assume that these interactions are zero.

   (d) As an alternative design, we could do a 1/4 replicate of the 2^3. One particular design would be given by the defining contrasts

$$x_1 + x_2 + x_3 = 0 \text{ or } 1, \quad x_2 + x_3 = 0 \text{ or } 1.$$

   Construct the four blocks and give the alias sets. Even if we run all four blocks, is there any reason to ever do this design?

6.15 Referring to Example 6.12:

   (a) Write out the defining equation, and find the blocks and alias sets for designs that confound (i) A with ABC and (ii) C with ABC.

   (b) Explain why there are seven distinct designs, each with its own unique alias sets. (Recall Example 6.3.)

   (c) For a 1/2 replicate of a 2^p factorial, how many distinct fractions are there?

   (d) For a $1/2^q$ replicate of a 2^p factorial, how many distinct fractions are there?

6.16 Referring to Example 6.14:

   (a) Verify that the defining contrast is $x_1 + x_2 + x_3 = 0$, give the composition of both blocks, and verify the alias sets.

   (b) For Experiment 2, verify that the defining contrast for the second block is $x_3 = 0$, and that CP is confounded with the CPK interaction. Give the composition of the two blocks and the alias sets.

   (c) Give the anova for Experiment 2, which should have sums of squares for the three main effects and the CP interaction, and 2 degrees of freedom for the residual. Are the conclusions the same as for Experiment 1?

   (d) Give the anova for Experiment 3. Are the conclusions the same as for Experiments 1 and 2?

6.17 Referring to Example 6.17:

   (a) Verify the intrablock subgroup and give the composition of the other three blocks.

   (b) Verify the alias sets.

   (c) As mentioned in the example, (6.11) is not a particularly good defining contrast as it confounds a main effect with the mean. Consider instead confounding the mean with fourway interactions using

$$0 = x_1 + x_2 + x_3 + x_4 = x_2 + x_3 + x_4 + x_5.$$

   For this defining contrast give the intrablock subgroup and the alias sets. Is this design preferable to the one given by (6.11)?

6.18 In this exercise we look at an alternative to modular arithmetic, which is fairly similar but based on multiplication rather than addition. In Example 6.10 we confounded B with ABC. We write this

$$B = ABC,$$

showing that the B effect is confounded with ABC. To determine the other alias sets, we multiply both sides of the above equation by the other effects, and use the restriction that any squared effect is equal to I, the identity. That is, $A^2 = B^2 = C^2 = I$. Thus, the first alias set is $\{B, ABC\}$ and the next is $\{AB, BC\}$.

(a) Verify that the other two alias sets are $\{I, AC\}$ and $\{A, C\}$.

(b) Show that the defining contrast (6.11) is equivalent to

$$I = ABCDE = ABCD,$$

and use this method to obtain the alias sets for this design.

6.19 A researcher conducted a 1/2 replicate of a 2^5 factorial using $x_1 + x_2 + x_3 + x_4 + x_5 = 0$ as the defining contrast. The data were

c	26	b	25	abcde 36	bde 46
cde 29	acd 35	abe	23	a	32
abd 40	bcd 28	d	35	abc 27
ace 22	e	21	bce	37	ade 39

(a) Specify the alias sets.

(b) Estimate the effects and interactions under the assumption of no interactions involving three or more factors.

(c) Give the anova table under the assumptions in (b).

(d) What treatment differences can be tested?

6.20 An animal nutrition experiment is to be performed, where the effects of four dietary factors on the growth of baby lambs are to be examined. Each factor is at two levels, high and low. They are A: vitamin A; B: protein; C: carbohydrate; and D: fat. The experimenter is restricted to only treating four lambs at a given time, so the complete factorial (2^4) is to be run in the following four blocks:

Blocks

1	2	3	4
(1)	c		
cd	d		
ab	abc		
abcd	abd		

(a) The blocks are generated by the defining contrasts $x_1 + x_2 + x_3 + x_4$ and $x_1 + x_2$. Fill in blocks 3 and 4 to complete the factorial.

(b) Find the alias sets.

(c) What model will you use for the analysis? Based on that model run the anova on the following data (also in dataset BabyLamb). Use the block structure from part (a).

Weight Gain (lbs.)

(1) 17.23	a 15.69	b 17.56	c 22.99
d 31.76	ab 21.24	ac 17.2	ad 20.92
bc 8.3	bd 16.87	cd 3.7	abc 26.61
abd 35.19	acd 6.19	bcd 5.69	abcd 6.86

(d) Write out the 1 df contrast that gives the A effect. Is it unbiased under your model? Give an estimate of the standard error of the contrast.

6.21 The *Youden square* is a misnamed design, mainly because it is not a square. This is a design with row blocks and column blocks, however, the column blocks cannot accommodate all treatments. Formally, a Youden Square is a design with

t treatment levels, c columns, r rows, and tr experimental units. Each treatment occurs once in every row, and at most once in every column.

Typical data from a Youden square experiment, with treatments A, B, C, D are

Column

	1	2	3	4
1	13(A)	10(B)	11(C)	20(D)
Row 2	16(B)	12(A)	18(D)	15(C)
3	11(C)	17(D)	11(A)	14(B)

Note that every treatment is in every row, so treatment differences are free of row effects. However, they are not free of column effects.

(a) Explain why, if we ignore columns, we have an RCB.

(b) Explain why, if we ignore rows, we have a BIBD. Identify the BIBD parameters.

(c) Write a model for the analysis identifying all terms.

(d) Which treatment contrasts are free of row effects? Which are free of column effects. Find the variances of these contrasts.

(e) Show that if we append a fourth row with treatments D C B A then we would have a BIBD. Give the parameters of the BIBD.

For more details on Youden squares see Dean and Voss (1999) or Cox and Reid (2000).

6.22 Referring to Example 6.19

(a) Verify that, in Table 6.2, each pair of treatments occurs exactly once in a row and once in a column.

(b) Verify that, if t is a square, then a BIBD exists with $k = \sqrt{t}$, $r = \sqrt{t} + 1$, and $b = \sqrt{t}(\sqrt{t} + 1)$. Relate this to the design in Table 6.2.

(c) For the anova table corresponding to the model (6.12), show that the residual degrees of freedom can be obtained by subtracting the Rep, Row, Column and Treatment df from the Total df.

6.23 A balanced lattice design is a special type of BIBD, controlling for one blocking factor. It has the BIBD properties of increased precision due to smaller blocks, and equal information on all treatment comparisons. A balanced lattice design with k treatments per block satisfies

(i) The number of treatments is k^2.

(ii)) The number of replications is $k + 1$.

(iii) Each treatment appears only once with other treatments in a sub-block.

A field layout for a balanced lattice design with 25 treatments is the following:

Arrangement of 25 Alfalfa Varieties
in Six Replications, Each with Five Blocks

Block	Rep 1	Block	Rep 2	Block	Rep 3
1	1 2 3 4 5	1	1 6 11 16 21	1	1 7 13 19 25
2	6 7 8 9 10	2	2 7 12 17 22	2	6 12 18 24 5
3	11 12 13 14 15	3	3 8 13 18 23	3	11 17 23 4 10
4	16 17 18 19 20	4	4 9 14 19 24	4	16 22 3 9 15
5	21 22 23 24 25	5	5 10 15 20 25	5	21 2 8 14 20

Block	Rep 4	Block	Rep 5	Block	Rep 6
1	1 12 23 9 20	1	1 17 8 24 15	1	1 22 18 14 10
2	6 17 3 14 25	2	6 22 13 4 20	2	6 2 23 19 15
3	11 22 8 19 5	3	11 2 18 9 25	3	11 7 3 24 20
4	16 2 13 24 10	4	16 7 23 14 5	4	16 12 8 4 25
5	21 7 18 4 15	5	21 12 3 19 10	5	21 17 13 9 5

(a) Verify that the above design is a BIBD with $t = 25$ and $k = 5$. Show that, in general, if we specify k treatments per block, then there is a BIBD (the balanced lattice) with $t = k^2$, $r = k+1$, $b = k(k+1)$, and $\lambda = 1$ (so we do not need orthogonal Latin squares here, as we did for the balanced lattice square).

(b) Write out a model and the anova table (source and df) and indicate how to test treatments. (Note that blocks are nested within reps, and treatments are crossed with reps but not with blocks.)

(c) Show that treatment contrasts are not confounded with blocks.

(d) The data for this experiment are in dataset **Lattice**. Analyze the data and estimate the variance of a treatment contrast.

(e) Use a multiple comparison procedure to find the variety with the greatest yield. State your conclusion at $\alpha = .05$.

6.24 For the Latin square microarray experiment of Example 6.20:

(a) Show that the Latin square is a 1/2 rep corresponding to the confounding equation $x_1 + x_2 + x_3 = 0$ or 1, and show that the alias sets give the confounding of the example.

(b) Verify the following anova table using the dataset **SynteniLS100**. (This is a portion of the actual data, using only 100 genes. The full dataset with 1286 genes is available at
http://www.jax.org/staff/churchill/labsite/datasets/index.html.)

Source	df	SS
Array	1	13.675
Dye	1	0.127
Treatment	1	5.577
Gene	99	87.908
A × G	99	21.550
T × G	99	46.873
Residual (D × G)	99	3.471

(c) Calculate the EMS for the anova in part (a) and indicate the tests.

(d) Run the tests and give conclusions[4]

---

[4] Kerr *et al.* (2000) were suspicious of the normality assumption and obtained their significance levels through *bootstrapping*. See Efron and Tibshirani (1993) for an introduction to bootstrapping.

(e) Verify that (6.15) is the least squares estimate of the Treatment × Gene interaction for gene g.

(f) Estimate the variance of the estimate in (c) and construct a 95% confidence interval.

6.25 Kerr *et al.* (2000) also analyzed a reference design in which mRNA samples were obtained from human liver tissue and were compared to muscle tissue. The design used two arrays in a two-dye system, but each treatment was compared to a reference tissue "Placenta". Liver and Muscle were always assigned the green dye and Placenta was assigned the red dye. A schematic of the treatment design is

	Array	
Dye	1	2
Red	Placenta	Placenta
Green	Liver	Muscle

(a) Based on the model

$$\log Y_{ijkg} = \mu + A_i + T_k + G_g + (TG)_{kg} + \varepsilon_{ijkg},$$

where Y_{ijkg} is the log expression level, μ is the overall mean, A_i is the array effect, T_k is the treatment effect, and ε_{ijkg} is the error, calculate an anova and perform the relevant tests.

Use the dataset `SynteniRef100`, a portion of the actual data, with 100 genes. The full dataset with 1286 genes is available at http://www.jax.org/staff/churchill/labsite/datasets/index.html.

(b) Is it possible, with this design, to separate the treatment effect from the dye effect? Explain. Calculate the sums of squares due to the dye effect. What other name can this sum of squares have?

(c) Show that the Treatment × Gene interaction for gene g is estimated by

$$(\hat{TG})_{1g} - (\hat{TG})_{2g} = \log\left(\frac{y_{121g}}{y_{222g}}\right) - \frac{1}{n}\log\left(\prod_{g'=1}^{n}\frac{y_{121g'}}{y_{222g'}}\right),$$

(compare to (6.15)), and find the variance of the estimate.

6.26 Referring to Example 6.21:

(a) Verify the variances in (6.17). In the loop design with three treatments, all of the treatments are adjacent, so that i and i' will appear together in the same block. In bigger loops, like the one described in Example 6.22, the treatments may not be adjacent. For this case in the loop design, show that $\text{Var}(\hat{\tau}_i - \hat{\tau}_{i'}) = \sigma_\varepsilon^2 + \sigma_\beta^2$, which is still an improvement over the reference design.

We have available six microarrays, so we can run six blocks. This will allow us to run the loop design with a dye swap, as indicated in Figure 6.2. The data for 15 genes can be found in dataset `FishGill`, where the response is log expression level.

(b) Compute the anova table for these data and perform the F-tests. All treatment comparisons are of interest. Set up confidence intervals for the contrasts Control vs. the average of AlCl3 and Nano, and AlCl3 vs. Nano. Pay attention to calculation of the variances.

(c) Write out the anova (just source and df) for a reference design using six microarrays. Is there any reason to prefer this design.

(d) With three treatments and blocks of size two, we can run a BIBD in three blocks. Give the parameters of the design. Since we have six blocks we can run two BIBDs with a dye swap. Describe this design and give the anova table assuming there are 15 genes. Is this design preferred to the one in part (b)? Explain.

6.27 Referring to Exercise 3.12, there was yet another surprise from the experimenter. When first questioned, he said that they would be using `Affymetrix` microarrays, but at the end of the consulting session he said that maybe they would use `Agilent` arrays, which are two-dye systems. This now suggests that the experiment should be run as an incomplete block design, with each microarray being a block of size two.

(a) Since there are six treatment pairs, if a subject's stem cells are divided into six parts, each treatment pair could be run on a microarray (assume one mouse per subject). Write out the anova table for this experiment.

(b) An alternative design might be a *loop* design. Show how this can be done. Can this design be balanced for dye swap?

(c) Calculate the variance of a treatment mean for the designs in (a) and (b). Which design is preferred?

6.28 Referring to Example 6.22:

(a) Verify the variances in (6.20), for the original eight-block loop design. Also calculate these variances assuming model (6.19) with n genes.

(b) Show that, for model (6.16), if we do the augmented experiment, we have

$$\text{Adjacent} : \text{Var}(\hat{\tau}_i - \hat{\tau}_{i'}) = \frac{1}{2}\sigma_\varepsilon^2 + \frac{3}{8}\sigma_\beta^2,$$

$$\text{Non-adjacent, common block} : \text{Var}(\hat{\tau}_i - \hat{\tau}_{i'}) = \frac{1}{2}\sigma_\varepsilon^2 + \frac{3}{8}\sigma_\beta^2,$$

$$\text{Non-adjacent, no common block} : \text{Var}(\hat{\tau}_i - \hat{\tau}_{i'}) = \frac{1}{2}\sigma_\varepsilon^2 + \frac{1}{2}\sigma_\beta^2.$$

(c) Show that if we add the 12 gray lines in Figure 6.3 we have a BIBD, and give the parameters of the design. Explain why the BIBD cannot balance the dye effects. How could you construct a BIBD, with a minimal number of blocks, that would balance the dye effects?

(d) Assuming no dye effect, for the BIBD using 28 blocks give the variance of a treatment contrast under model (6.16).

Accompaniment

6.29 The least squares estimate of the τ_i in (6.1) minimizes

$$\sum_i \sum_{j \in J_i} (y_{ij} - \mu - \tau_i - \beta_j)^2.$$

(a) Show that the least squares estimate of τ_i is given by (6.2). (The calculation is a bit easier if the sum is written as above for the τ_i differentiation and as $\sum_j \sum_{i \in I_j}$ for the β_j differentiation. Solve for the β_j in terms of τ_i and then solve for τ_i. Use the fact that $\sum_{j \in J_i} \sum_{i' \in I_j} \tau_{i'} = (r - \lambda)\tau_i$, which follows since treatment i is in all r blocks of J_i, treatment i' is in λ blocks, and $\sum_i \tau_i = 0$.)

(b) Show that (6.2) is unbiased for τ_i under model (6.1). In particular, show that

$$E\left(r\bar{Y}_i - \sum_{j \in J_i} \bar{Y}_j\right) = r(\mu + \tau_i) - \frac{1}{k}\sum_{j \in J_i}\sum_{i' \in I_j}(\mu + \tau_{i'}),$$

and again use the identity from part (a) about the double sum of τ_i. Then combine all the terms and use the fact that $r(k-1) = \lambda(t-1)$ to show that $E(\hat{\tau}_i) = \tau_i$.

6.30 Referring to (6.3):

(a) Show that the equation is valid using the BIBD equations.

(b) Show that the factor in square brackets is less than 1 as long as $k < t$. (Establish and use the fact that the function $x/(x-1)$ is decreasing in x if $x > 1$.) Hence, show that it is always the case in a BIBD that $\lambda t/k < r$.

(c) Explain why the naive estimator has smaller variance than the least squares estimator.

6.31 Referring to Section 6.2.2, here we will fill in the details of the variance of $\hat{\tau}_i$.

(a) Using (6.1), show that

$$\sum_{j \in J_i} \text{Var}\left(Y_{ij} - \frac{1}{k}\sum_{i' \in I_j} Y_{i'j}\right) = \sum_{j \in J_i} \text{Var}\left(\beta_j + \varepsilon_{ij} - \frac{1}{k}\sum_{i' \in I_j}(\beta_j + \varepsilon_{i'j})\right)$$

$$= \sum_{j \in J_i} \text{Var}\left(\beta_j + \varepsilon_{ij} - \beta_j - \bar{\varepsilon}_{.j}\right),$$

where we see that the β_j cancels! Also note that the range of the dot $(\cdot)$ is different for each $\bar{\varepsilon}_{.j}$, but this causes no problem.

(b) Using the fact that $\text{Var}(\varepsilon_{ij}) = \sigma_\varepsilon^2$, show that

$$\sum_{j \in J_i} \text{Var}\left(\varepsilon_{ij} - \bar{\varepsilon}_{.j}\right) = \sum_{j \in J_i}\left(1 - \frac{1}{k}\right)\sigma_\varepsilon^2 = r\left(1 - \frac{1}{k}\right)\sigma_\varepsilon^2$$

(c) Use the fact that $\lambda(t-1) = r(k-1)$ to establish (6.5).

6.32 Here we will establish the variance of a contrast in a BIBD (6.7).

(a) Show that

$$\text{Cov}(\hat{\tau}_i, \hat{\tau}_{i'})$$

$$= \left(\frac{k}{\lambda t}\right)^2 \text{Cov}\left(\sum_{j \in J_i}\left[Y_{ij} - \frac{1}{k}\sum_{\ell \in I_j}Y_{\ell j}\right], \sum_{j \in J_{i'}}\left[Y_{i'j} - \frac{1}{k}\sum_{\ell \in I_j}Y_{\ell j}\right]\right).$$

(b) In (a) show that (i) if the js do not match the covariance is zero, (ii) the js match exactly λ times, and (iii) i and i' are together λ times, which allows us to write for a particular i,

$$\text{Cov}(\hat{\tau}_i, \hat{\tau}_{i'}) = \left(\frac{k}{\lambda t}\right)^2 \lambda \text{Cov}\left(Y_{ij} - \frac{1}{k}\sum_{\ell \in I_j}Y_{\ell j}, Y_{i'j} - \frac{1}{k}\sum_{\ell \in I_j}Y_{\ell j}\right).$$

(c) Next, show that

$$Y_{ij} - \frac{1}{k}\sum_{\ell \in I_j} Y_{\ell j} = \tau_i - \bar{\tau}_{\cdot j} + \varepsilon_{ij} - \bar{\varepsilon}_{\cdot j}$$

and hence

$$\text{Cov}(\hat{\tau}_i, \hat{\tau}_{i'}) = \left(\frac{k}{\lambda t}\right)^2 \lambda \text{Cov}(\varepsilon_{ij} - \bar{\varepsilon}_{\cdot j}, \varepsilon_{i'j} - \bar{\varepsilon}_{\cdot j}).$$

(d) Show that $\text{Cov}(\varepsilon_{ij} - \bar{\varepsilon}_{\cdot j}, \varepsilon_{i'j} - \bar{\varepsilon}_{\cdot j}) = -\sigma_\varepsilon^2/k$ and hence establish (6.6) and (6.7).

6.33 Referring to the discussion at (6.10):

(a) Show that $E[(Y_{212} + Y_{221})/2] = \tau_2$, with the block effect being zero in expectation.

(b) Verify the expectation (6.10), showing that the block effect cancels. Also give the estimate of the γ effect and show that it is also unbiased.

(c) For the threeway design in Example 6.13, show that, under a no interaction model, the A effect can be unbiasedly estimated with

$$\frac{1}{4}[(a + ab + ac + abc) - ((1) + b + c + bc)]$$

and that the block effect cancels out.

(d) Suppose that in Example 6.13 we do not assume that the AC interaction is zero. Show that an unbiased estimate of this effect is

$$\frac{1}{4}\left([(abc - bc) - (ab - b)] + [(ac - c) - (a - (1))]\right).$$

(If we change the "+" to "−" we would have an unbiased estimate of the ABC effect. However, in this design ABC is confounded with blocks.)

6.34 Referring to Table 6.3 and the surrounding discussion, show that the least squares estimate of the yield difference for two varieties of interest will have variance one-third as large for the BIBD compared to that of the reference design.

6.35 Here we look into the properties of the least squares estimator (6.13) in the context of Example 6.19. Using model (6.12) applied to Example 6.19:

(a) Show that in $\hat{\tau}_i$ the parameters R_k will always cancel, so we need not be concerned with them.

(b) Using the design in Example 6.19, ignoring $\varepsilon_{j\ell m}$ (for now), and recalling that $\sum_i \tau_i = 0$, show that

$$\hat{\tau}_1 = \frac{1}{4}[4\tau_1 + \beta_{21} + \beta_{32} + \beta_{33} + \beta_{14} + \gamma_{21} + \gamma_{32} + \gamma_{13} + \gamma_{24}]$$

$$- \frac{1}{12}\left[3\tau_1 + 3(\beta_{21} + \beta_{32} + \beta_{33} + \beta_{14}) + \sum_{\ell m}\gamma_{\ell m}\right]$$

$$- \frac{1}{12}\left[3\tau_1 + \sum_{jm}\beta_{jm} + 3(\gamma_{21} + \gamma_{32} + \gamma_{13} + \gamma_{24})\right]$$

$$+ \frac{1}{36}\left[4\sum_i \tau_i + 3\sum_{jm}\beta_{jm} + 3\sum_{\ell m}\gamma_{\ell m})\right],$$

and hence show that (i) $\hat{\tau}_i$ is unbiased for τ_i and (ii) the variance of $\hat{\tau}_i$ is free of σ_β^2, σ_γ^2, and σ_R^2.

(c) We know that, in general, $\hat{\tau}_i$ is unbiased for τ_i, but we cannot yet make a general statement about the variance only depending on σ_ε^2. The proof is a nightmare of keeping track of indices, but here is an argument that will convince us that this is the case. From part (b):

   (i) Argue that $\text{Var}(\hat{\tau}_i)$ will only be a function of the $\varepsilon_{j\ell m}$.

   (ii) From (i), we can ignore the index i in the variance calculation, and substitute the variance of a row-column residual for the variance of $\hat{\tau}_i$, that is, for any i, j and ℓ,

$$\text{Var}(\hat{\tau}_i) = \frac{1}{r^2}\text{Var}\left(\sum_{m=1}^{r}(\varepsilon_{j\ell m} - \bar{\varepsilon}_{\cdot\ell m} - \bar{\varepsilon}_{j\cdot m} + \bar{\varepsilon}_{\cdot\cdot m})\right),$$

which can only depend on σ_ε^2.

6.36 Referring to Miscellanea 6.6.1, here we will do a small example to illustrate the unbiasedness of the interblock estimate $\tilde{\tau}_i$. Suppose that the following data are collected according to model (6.1):

Treatments

		1	2	3
	1	y_{11}	y_{21}	–
Block	2	y_{12}	–	y_{32}
	3	–	y_{23}	y_{33}

(a) What are the parameters of this design?

(b) Using model (6.1), ignoring the ε_{ij}, show that

$$\tilde{\tau}_1 = \frac{1}{3}(y_{11} + y_{21} + y_{12} + y_{32}) - \frac{2}{3}(y_{23} + y_{33}) = \tau_1 + \frac{2}{3}(\beta_1 + \beta_2) - \frac{4}{3}\beta_3,$$

and, thus, $\tilde{\tau}_1$ is an unbiased estimator of τ_1

(c) Similarly, show that for the intrablock estimate

$$\hat{\tau}_1 = \frac{2}{3}(y_{11} + y_{12}) - \frac{1}{3}(y_{11} + y_{21} + y_{12} + y_{32}) = \tau_1,$$

since all of the β_js cancel out. Why is this preferable to (b)?

6.6 Miscellanea

6.6.1 Interblock Information

In a BIBD we saw that the least squares estimate of τ_i is given by (6.2). There is another estimate of τ_i, based on block means only, called the *interblock* estimator. (To be precise, we should then refer to (6.2) as the *intrablock* estimate.) We will see that the interblock estimate is not as desirable as the intrablock estimate, but it can be used to improve the intrablock estimate. The interblock estimate is based on the idea that we can estimate τ_i with

$$\tilde{\tau}_i = \left(\begin{array}{c}\text{effect in blocks}\\\text{with treatment } i\end{array}\right) - \left(\begin{array}{c}\text{average}\\\text{block effect}\end{array}\right),$$

leading to the estimate

$$\tilde{\tau}_i = \frac{1}{r-\lambda}(B_i - \bar{B}),$$

with

$$B_i = \sum_{j\in J_i}\sum_{i'\in I_j} y_{i'j}, \bar{B} = \frac{1}{t}\sum_{i=1}^{t} B_i,$$

so B_i is the sum of the observations in the blocks that contain treatment i. The remarkable thing is that $\tilde{\tau}_i$ is another unbiased estimator of $\hat{\tau}_i$ (see Exercise 6.36). However, the drawback is that

$$\mathrm{Var}(\tilde{\tau}_i) = \frac{k(t-1)}{t(r-\lambda)}(\sigma_\varepsilon^2 + k\sigma_\beta^2).$$

Compare this to (6.5), where we were elated that the variance of $\hat{\tau}_i$ did not contain the block variance. Here it does, so we expect the variance of $\tilde{\tau}_i$ to be greater (possibly much greater) than that of $\hat{\tau}_i$.

It is not immediately clear how to use $\tilde{\tau}_i$. It seems reasonable to try to use a linear combination of $\tilde{\tau}_i$ and $\hat{\tau}_i$ to estimate τ_i, but we do not know what to use as weights. The classically optimal weights are inversely proportional to the variances, but the variances are unknown. Using estimated variances, it may be the case that such a linear combination will have a higher variance than just $\hat{\tau}_i$ alone. However, Brown and Cohen (1974), using techniques based on the pioneering work of Stein (1956) (who also addressed the problem of recovery of interblock information in Stein 1966), showed how to construct a combined estimator that will always dominate $\hat{\tau}_i$ as long as $b \geq 4$. Other work on combining these estimates has been done by Graybill and Weeks (1959), Shah (1970), and Bhattacharya (1980).

A

Designs Illustrated

In this appendix we give a small catalog of designs, to be used as a quick reference. We consider the case of two treatments A and B, each at two levels, with 16 observations available. We use the notation of Chapter 6, and denote the four treatment combinations by (1), a, b, and ab. The layouts indicate a possible randomization of the treatments – notice the restrictions as we move between designs. Note also that as we move between designs the error structure, and hence the degrees of freedom for the tests, changes. But in the end each design has 15 total degrees of freedom.

This appendix is based on the notes of Prof. Carl Lowe of Cornell University.

A.1 Completely Randomized Design (CRD)

Layout

a	(1)	b	a
b	a	ab	b
a	ab	(1)	ab
(1)	b	ab	(1)

Anova

Source	df
Treatments	3
A	1
B	1
A × B	1
Within Error	12
Total	15

Notes: The within error is a "pure error" in the sense that it represents the variation in responses subjected to identical experimental conditions and gives an error estimate that is not dependent on the model. This is a difficult design to run as each treatment combination has to be an independent replication, so the experimental conditions must be reconstructed each time. A disadvantage is that there is only one "block", that is, the entire experiment is run under one condition (plot of land, time, etc.).

A.2 Randomized Complete Block Design (RCB) - No Subsampling———

Layout

Block 2

| 1 | a | (1) | b | ab |

| 2 | b | a | ab | (1) |

| 3 | a | ab | (1) | b |

| 4 | (1) | b | ab | a |

Anova

Source	df
Blocks	3
Treatments	3
T × B	9
Total	15

Notes: This design is typically easier to run than a CRD. Here we pooled all of the treatment-block interactions into one term with 9 df. There is no test on the interaction in this model.

A.3 Randomized Complete Block Design (RCB) - Subsampling———

Layout

Block 1 | a | (1) | b | ab | a | ab | (1) | b |

2 | b | a | ab | (1) | (1) | b | ab | a |

Anova

Source	df
Blocks	1
Treatments	3
T × B	3
Within Error	8
Total	15

Notes: The test on treatments is not as good as Design A.2. The within error can be used to teat the significance of T × B if the observations within a block are true replications and not technical replications (see Section 3.5).

A.4 Latin Square———————————————————

Layout
Columns

Rows	1	2	3	4
1	(1)	b	a	ab
2	ab	a	b	(1)
3	b	(1)	ab	a
4	a	ab	(1)	b

Anova

Source	df
Rows	3
Columns	3
Treatments	3
Residual	6
Total	15

Notes: The design controls two gradients, but assumptions of no interactions are needed for a good test on treatments (it can be conservative – see Section 3.6.3).

A.5 Split Plot - CRD on Whole Plots

<u>Layout</u>

A
Low | High

| (1) b | | a ab |

| b (1) | | a ab |

| b (1) | | ab a |

| (1) b | | a ab |

<u>Anova</u>

Source	df
A	1
Reps (in A)	6
B	1
A × B	1
(B × Reps) in A	6
Total	15

Notes: The CRD is on the whole plot treatment A, with the split plot treatment B randomized on the whole plots. The whole plot error, Reps (in A), tests A while the split plot error, (B × Reps) in A, tests everything below the line.

A.6 Split Plot - RCB on Whole Plots

<u>Layout</u>
A

Low   High

1 | (1) b | a ab |

Block 2 | b (1) | a ab |

3 | (1) b | ab a |

4 | (1) b | ab a |

<u>Anova</u>

Source	df
Blocks	3
A	1
A × Blocks	3
B	1
A × B	1
Split Plot Error	6
Total	15

Notes: This is an RCB on the whole plots (A), with B randomized within the levels of A. Note that all factors are crossed, in contrast to Design A.5. Also recall that to have one split plot error requires assumptions about the Block × Split Plot interactions (Section 5.3), with the split plot error coming from pooling B × Blocks and A × B × Blocks.

A.7 Strip Plot

<u>Layout</u>

Block 1 | b ab | (1) a |
         | (1) a |
Block 2 | (1) a |
         | b ab |

Block 3 | ab b | ab b |
         | a (1) |
Block 4 | ab b |
         | a (1) |

<u>Anova</u>

Source	df
Blocks	3
A	1
A × Blocks	3
B	1
B × Blocks	3
A × B	1
A × B × Blocks	3
Total	15

Notes: In each block *A* is randomized in columns and *B* is randomized in rows. Separately, this is an RCB on each of *A* and *B*.

A.8 Confounding in Blocks - No Interaction Test_____

Layout	Anova

Layout

I [a | b] [(1)|ab]

Reps II [(1)|ab] [b | a]

III [a | b] [ab|(1)]

IV [b | a] [ab|(1)]

Source	df
Reps	3
Blocks (in Reps)	4
A	1
B	1
Residual	6
Total	15

Notes: This is not a great design unless there is no chance of the A × B interaction being significant. In each rep the interaction is confounded with blocks, so there is no test on interaction.

A.9 Confounding in Blocks - With Interaction Test_____

Layout

I [a | b] [(1)| ab]

Reps II [(1)| ab] [a | b]

III [(1)| a] [b | ab]

IV [(1)| b] [a | ab]

Source	df
Reps	3
Blocks (in Reps)	4
A	1
B	1
A × B	1
Residual	5
Total	15

Notes: In Reps I and II the interaction is confounded with Blocks. In Rep III, B is confounded with blocks, and in Rep IV, A is confounded with blocks. The interaction can be estimated with information from two Reps, while the main effects use information from three Reps. Note that Reps II, III, and IV (or I, III and IV) are a BIBD with $t = 4$, $b = 6$, $\lambda = 1$, and $k = 2$.

References

Albert, A. (1976). When is Sum of Squares an Analysis of Variance? *Ann. Statist.* **4** 775-778.

Belko, A. Z. *et al.* (1984). The effects of aerobic exercise and weight loss on riboflavin requirements of moderately obese marginally deficient young women. *Am. J. Clin. Nutr.* **40** 553-561.

Benjamini, Y. and Hochberg, Y. (1995). Controlling the False Discovery Rate: A Practical and Powerful Approach to Multiple Testing. *J. Roy. Statist. Soc. Ser. B* **57** 289-300.

Berger, R. L. and Casella, G. (1992). Deriving Generalized Means as Least Squares and Maximum Likelihood Estimates. *Am. Statist.* **46** 279-282.

Berger, R. L. and Hsu, J. C. (1996). Bioequivalence Trials, Intersection – Union Tests and Equivalence Confidence Sets. *Statist. Sci.* **11** 283-302.

Berkowitz, R., Njus, G., and Vrabec, G. (2005). Pullout Strength of Self-Tapping Screws Inserted to Different Depths. *J. Orthop. Trauma* **19** 462-465.

Bhattacharya, C. G. (1980). Estimation of a Common Mean and Recovery of Interblock Information. *Ann. Statist.* **8** 205-211.

Bradley, E. L. (1973). The Equivalence of Maximum likelihood and Weighted Least Squares Estimates in the the Exponential Family. *J. Am. Statist. Assoc.* **69** 199-200.

Brogan, D. R. and Kutner, M. H. (1980). Comparative Analysis of Pretest-Postest Research Designs. *Am. Statist.* **34** 229-232.

Brown, L. D. and Cohen, A. (1074) Point and Confidence Estimation of a Common Mean and Recovery of Interblock Information. *Ann. Statist.* **2** 963-976.

Casella, G., and Berger, R. L. (2001). *Statistical Inference, Second Edition.* Monterey, CA: Duxbury Press.

Chang, J. J., Rabinowitz, D. and Shea, S. (2003). Sources of Variability in Blood Pressure Measurement using the Dinamap PRO 100 Automated Oscillometric Device. *Am. J. Epidemiol.* **158** 1218-1226.

294   References

Churchill, G. A. and Doerge, R. W. (1994). Empirical Threshold Values for Quantitative Trait Mapping. *Genetics* **138** 963-971.

Cochran, W.G. (1934). The Distribution of Quadratic Forms in a Normal System with Applications to the Analysis of Covariance. *Proc. Cam. Phil. Soc.* **30** 178-191.

Cochran, W. G. and Cox, G. M. (1957). *Experimental Designs, Second Edition.* New York: John Wiley

Cox, D. R. and Reid, N. (2000). *The Theory of the Design of Experiments.* Chapman & Hall, London.

Cox, D. R. and Solomon, P.J. (2003). *Components of Variance.* Chapman & Hall, London.

Cui, X., Hwang, J.T.G., Qiu, J., Blades, N. J. and Churchill, G. A. (2005). Improved Statistical Tests for Differential Gene Expression by Shrinking Variance Components Estimates. *Biostatistics* **6** 59-75.

Davis, C. S. (2002). *Statistical Methods for the Analysis of Repeated Measurements.* Springer–Verlag, New York.

Dean, A. and Voss, D. (1999). *Design and Analysis of Experiments.* Springer–Verlag, New York.

Edgington, E. and Onghena, P. (2007). *Randomization Tests, Fourth Edition.* Chapman & Hall, London.

Efron, B. and Tibshirani, R. J. (1993). *An Introduction to the Bootstrap.* Chapman & Hall, London.

Federer, W. T. (1955). *Experimental Design: Theory and Applications.* Calcutta: Oxford and IBH Publishing Co.

Federer, W. T. (1998). Recovery of Interblock, Intergradient, and Intervariety Information in Incomplete Block and Lattice Rectangle Designs. *Biometrics* **54** 471-481.

Federer, W. T. and Meredith, M. P. (1992). Covariance Analysis for Split-Plot and Split-Block Designs. *Am. Statist.* **46** 155-162.

Feynman, R. P. (1985). *Surely You're Joking Mr. Feynman.* New York: W.W. Norton.

Finney, D. J. (1946). Standard Errors of Yields Adjusted for Regression on an Independent Measurement. *Biometrics* **2** 53-55.

Fisher, R. A. and Wishart, J. (1930). The Arrangement of Field Experiments and the Statistical Reduction of the Results. Imperial Bureau of Soil Science Technical Communication No.10. (All of Fisher's papers are available at http://digital.library.adelaide.edu.au/coll/special/fisher.)

Fisher, R. A. (1934). Discussion of the paper by J. Wishart. *J. Roy. Statist. Soc. Ser. B* Suppl. **1** 51-53.

Fisher, R. A. (1947). Development of the Theory of Experimental Design. *Proc.of the Int. Statist. Conf., Washington* **3** 434-439.

Fisher, R. A. (1962). The Place of the Design of Experiments in the Logic of Scientific Inference. *Colloques Int. Centre Natl. Recherche Sci. Paris* **110** 13-19.

Fisher, R. A. (1971). *The Design of Experiments, Eighth Edition.* Reprinted by Hafner Publishing Company, New York. Also contained in *Statistical Methods, Experimental Design, and Scientific Inference.* Oxford University Press, London 1990.

Gail, M. and Simon, R. (1985). Testing for Qualitative Interactions Between Treatment Effects and Patients Subsets. *Biometrics* **41** 361-372.

Gates, C. E. (1995). What Really is the Experimental Error in Block Designs? *Am. Statist.* **49** 362-363.

Geisser, S. and Greenhouse, S. W. (1958). An Extension of Box's Result on the Use of the F Distribution in Multivariate Analysis. *Ann. Math. Statist.* **29** 885-891.

Genovese, C. R., Roeder, K. and Wasserman, L. (2006). False Discovery Control with p-Value Weighting. *Biometrika* **93** 509-524.

Genovese, C. R. and Wasserman, L. (2002). Operating Characteristics and Extensions of the False Discovery Rate Procedure. *J. Roy. Statist. Soc. Ser. B* **64** 499-517.

Genovese, C. R. and Wasserman, L. (2004). A Stochastic Process Approach to False Discovery Control *Ann. Statist.* **32** 1035-1061.

Good, P. (2005). *Permutation, Parametric, and Bootstrap tests of Hypotheses, Third Edition.* Springer–Verlag, New York.

Graybill, F. A. and Hultquist, R. A. (1961). Theorems Concerning Eisenhart's Model II. *Ann. Math. Statist.* **32** 261-269.

Graybill, F. A. and Weeks, D. L. (1959). Combining Inter-Block and Intra-Block Information in Balanced Incomplete Block Designs. *Ann. Math. Statist.* **30** 799-805.

Harville, D. A. (1976). Extensions of the Gauss-Markov Theorem to Include Random Effects. *Ann. Statist.* **2** 384-395.

Hicks, C. R. (1993) *Fundamental Concepts in the Design of Experiments, Fourth Edition.* New York: Oxford University Press.

Hinkelmann, K. and Kempthorne, O. (1994). *Design and Analysis of Experiments, Volume I.* New York: John Wiley

Hocking, R. R. (1973). A Discussion of the Two-way Mixed Model. *Am. Statist.* **27** 148-152

Hocking, R. R. (1985). *The Analysis of Linear Models* Monterey CA: Brooks-Cole.

Hsu, J. C. (1996). *Multiple Comparisons: Theory and Methods.* Chapman & Hall, London.

Huynh, H. and Feldt, L. S. (1970). Conditions Under Which Mean Square Ratios in Repeated Measures Designs Have Exact F-Distributions. *J. Am. Statist. Assoc.* **65** 1582-1589.

Kacker, R. N. and Harville, D. A. (1984). Approximations for Standard Errors of Estimators of Fixed and Random Effects in Mixed Linear Models. *J. Am. Statist. Assoc.* **79** 853-861.

Kempthorne, O. (1952). *Design and Analysis of Experiments.* New York: John Wiley

Kerr, M. K. and Churchill, G. A. (2001a). Statistical Design and the Analysis of Gene Expression Microarray Data. *Genet. Res.* **77** 123-128.

Kerr, M. K. and Churchill, G. A. (2001b). Experimental Design for Gene Expression Microarrays. *Biostatistics* **7** 183-201.

Kerr, M. K. Martin, M., and Churchill, G. A. (2000). Analysis of Variance for Gene expression Microarray Data. *J. Comp. Biol.* **7** 819-837.

Khuri, A. I. (1982). Direct Products: A Powerful Tool for the Analysis of Balanced Data. *Commun. Statist. Theor. Methods* **11** 2903-2920.

Khuri, A. I. and Cornell, J. A. (1996). *Response Surfaces: Design and Analyses.* New York: Marcel Dekker

Kuehl, R. O. (1994). *Statistical Principles of Research Design and Analysis.* New York: Wadsworth.

Lee, Y. J., Lee, M. G., Chung, S. J., Lee, M. H., and Shim, C. K. (1998). Statistical Analysis of Three Sequence Three Periods Bioequivalence Model: Example of Bioequivalence Test of Ondansetron Formulations. *J. Korean Pharma. Sci.* **28** 3542.

Levey, W. A., Manore, M. M., Vaughan, L. A., Carroll, S. S., vanHalderen, L. and Felicetta, J. (1995). Blood Pressure Responses of White Men with Hypertension to Two Low-Sodium Metabolic Diets with Different Levels of Dietary Calcium. *J. Am. Diet. Assoc.* **95** 1280-1288.

Little, R. J. A. and Rubin, D. B. (2002). *Statistical Analysis with Missing Data, Second Edition.* New York: John Wiley

Mead, R. (1988). *The Design of Experiments: Statistical Principles for Practical Application.* New York: Cambridge University Press.

McLean, R. A., Sanders, W. L., and Stroup, W. W. (1991). A Unified Approach to Mixed Linear Models. *Am. Statist.* **45** 54-64.

Miller, R. G. (1981). *Simultaneous Statistical Inference, Second Edition.* Springer–Verlag, New York.

Morgan, J. P. (1998). Orthogonal Collections of Latin Squares. *Technometrics* **40** 327-333.

Morrison, D. F. (2005). *Multivariate Statistical Methods, Fourth Edition.* Monterey, CA: Duxbury Press.

Ott, R. L., and Longnecker, M.T. (2000). *An Introduction to Statistical Methods and Data Analysis.* Monterey: Duxbury Press.

Piantidosi, S. and Gail, M. H. (1993). A Comparison of Power of Two Tests for Qualitatitive Interaction. *Statist. Med.* **12** 1239-1248.

Puntanen, S. and Styan, G. (1989). The Equality of the Ordinary Least Squares Estimator and the Best Linear Unbiased Estimator. *Am. Statist.* **43** 153-161

Ratkowsky, D. A. and Reedy, T. J. (1986). Choosing Near-Linear Parameters in the Four-Parameter Logistic Model for Radioligand and Related Assays. *Biometrics* **42** 575-582.

Rawlings, J. O., Pantula, S., and Dickey, D. (1998). *Applied Regression Analysis: A Research Tool.* Springer–Verlag, New York.

Robert, C. P. and Casella, G. (2004). *Monte Carlo Statistical Methods, Second Edition.* Springer–Verlag, New York.

Ruberg, S. (1989). Contrasts for Identifying Minimum Effective Dose. *J. Am. Statist. Assoc.* **84** 816-822

Samuels, M. L., Casella, G., and McCabe, G.P. (1991). Interpreting Blocks and Random Factors. *J. Am. Statist. Assoc.* **86** 798-821 (with discussion)

Samuels, M. L., Casella, G. and McCabe, G.P. (1993). Evaluating the Efficiency of Blocking Without Assuming Compound Symmetry. *J. Statist. Plan. Inf.* **38** 237-248.

Satterthwaite, F. E. (1946). An Approximate Distribution of Estimates of Variance Components. *Biometrics Bull.* **2** 110–114.

Scheffé, H. (1959). *The Analysis of Variance.* New York: John Wiley

Schulze, A. and Downward, J. (2001) Navigating Gene Expression Using Microarrays A Technology Review. *Nature Cell Biol.* **3** E190-E195.http:// cellbio.nature.com

Searle, S. R., Casella, G. and McCulloch, C. E. (1992). *Variance Components.* New York: John Wiley

Shah, K. R. (1970). On the Loss of Information in Combined Inter- and Intra-Block Estimation. *J. Am. Statist. Assoc.* **65** 1562-1564.

Simon, R. M., Korn, E. L., McShane, L. M., Radmacher, M. D., Wright, G. W. and Zhao, Y. (2003). *Design and Analysis of DNA Microarry Investigations.* New York: Springer-Verlag.

Snedecor, G. W. and Cochran, W. G. (1989). *Statistical Methods, Eighth Edition.* Ames. IA: Iowa State University Press.

Speed, T. P. (1987). What is an Analysis of Variance? *Ann. Statist.* **15** 885-910 (with discussion).

Speed, T. P., Williams, E. R., and Patterson, H. D. (1985). A Note on the Analysis of Resolvable Designs. *J. Roy. Statist. Soc. Ser. B* **47** 357-361.

Stein, C. (1966). An Approach to the Recovery of Inter-block Information in Balanced Incomplete Block Designs. *Research Papers in Statistics* Neymann festschrift, F. N. David, ed. New York: John Wileypp. 351-366.

Stein, C. (1956). Inadmissibility of the Usual Estimator for the Mean of a Multivariate Normal Distribution. *Proc. Third Berkeley Symp. Math. Statist. Prob.* **1**, University of California Press, Berkeley pp.197-206.

Storey, J. D. (2003). The Positive False Discovery Rate: A Bayesian Interpretation and the q-Value. *Ann. Statist.* **6** 2013-2035.

Storey, J. D. (2002). A Direct Approach to False Discovery Rates *J. Roy. Statist. Soc. Ser. B* **64** 479-498.

Storey, J. D. and Tibshirani, R. (2003). Statistical Significance for Genome-wide Studies *Proc. Natl. Acad. Sci.* **100** 9440-9445 www.pnas.org/cgi/ doi/10.1073/pnas.1530509100

Stuart, A. J. and Ord, J. K. (1987). *Kendall's Advanced Theory of Statistics, Volume I: Distribution Theory, Fifth Edition.* New York: Oxford University Press.

298   References

Tukey, J. W. (1949). One Degree of Freedom for Additivity. *Biometrics* **5** 232-244.

Tusher, V. G., Tibshirani, R. and Chu, G. (2001). Significance Analysis of Microarrays Applied to the Ionizing Radiation Response. *Proc. Natl. Acad. Sci.* **98** 51165121 www.pnas.org/cgi/doi/10.1073/pnas.091062498

Winer, B. J. (1971). *Statistical Principles in Experimental Design, Second Edition.* New York: McGraw-Hill

Yates, F. (1936). A New Method of Arranging Variety Trails Involving a Large Number of Varieties. *J. Agric. Sci.* **26** 424-455.

Yates, F. (1940). Lattice Squares. *J. Ag. Sci.* **30** 672-687.

Zyskind, G. (1967). On Canonical Forms, Non-Negative Covariance Matrices, and Best Simple Least Squares Linear Estimators in Linear Models. *Ann. Math. Statist.* **38** 1092-1109.

Author Index

Subject Index

Springer Texts in Statistics

(continued from p. ii)

Robert: The Bayesian Choice: From Decision-Theoretic Foundations to Computational Implementation, Second Edition

Robert/Casella: Monte Carlo Statistical Methods, Second Edition

Rose/Smith: Mathematical Statistics with *Mathematica*

Ruppert: Statistics and Finance: An Introduction

Sen/Srivastava: Regression Analysis: Theory, Methods, and Applications

Shao: Mathematical Statistics, Second Edition

Shorack: Probability for Statisticians

Shumway/Stoffer: Time Series Analysis and Its Applications, Second Edition

Simonoff: Analyzing Categorical Data

Terrell: Mathematical Statistics: A Unified Introduction

Timm: Applied Multivariate Analysis

Toutenberg: Statistical Analysis of Designed Experiments, Second Edition

Wasserman: All of Nonparametric Statistics

Wasserman: All of Statistics: A Concise Course in Statistical Inference

Weiss: Modeling Longitudinal Data

Whittle: Probability via Expectation, Fourth Edition

Monte Carlo Statistical Methods
Second Edition

Christian P. Robert and George Casella

Monte Carlo statistical methods, particularly those based on Markov chains, are now an essential component of the standard set of techniques used by statisticians. This new edition has been revised towards a coherent and flowing coverage of these simulation techniques, with incorporation of the most recent developments in the field. In particular, the introductory coverage of random variable generation has been totally revised, with many concepts being unified through a fundamental theorem of simulation

2004. 645 pp. Hardcover ISBN 978-0-387-21239-5

Statistical Genetics of Quantitative Traits
Linkage, Maps, and UTL

Rongling Wu, Changxing Ma, and George Casella

This book gives, for the first time in book form, a comprehensive and up-to-date account of this modern theory. Many major classes of designs are covered in the book. While maintaining a high level of mathematical rigor, it also provides extensive design tables for research and practical purposes. Apart from being useful to researchers and practitioners, the book can form the core of a graduate level course in experimental design.

2007. 368 pp. (Statistics for Biology and Health) Hardcover
ISBN 978-0-387-20334-8

Theory of Point Estimation
Second Edition

E.L. Lehmann and George Casella

This second, much enlarged edition by Lehmann and Casella of Lehmann's classic text on point estimation maintains the outlook and general style of the first edition. All of the topics are updated. An entirely new chapter on Bayesian and hierarchical Bayesian approaches is provided, and there is much new material on simultaneous estimation. Each chapter concludes with a Notes section which contains suggestions for further study. The book is a companion volume to the second edition of Lehmann's "*Testing Statistical Hypotheses*"

2003. 589 pp. (Springer Texts in Statistics) Hardcover
ISBN 978-0-387-98502-2

Easy Ways to Order▶ Call: Toll-Free 1-800-SPRINGER • E-mail: orders-ny@springer.com • Write: Springer, Dept. S8113, PO Box 2485, Secaucus, NJ 07096-2485 • Visit: Your local scientific bookstore or urge your librarian to order.